OpenIntro Statistics
Third Edition

David M Diez
Quantitative Analyst
david@openintro.org

Christopher D Barr
Graduate Student
Yale School of Management
chris@openintro.org

Mine Çetinkaya-Rundel
Assistant Professor of the Practice
Department of Statistics
Duke University
mine@openintro.org

Contents

Preface

This book may be downloaded as a free PDF at **openintro.org**.

We hope readers will take away three ideas from this book in addition to forming a foundation of statistical thinking and methods.

(1) Statistics is an applied field with a wide range of practical applications.

(2) You don't have to be a math guru to learn from real, interesting data.

(3) Data are messy, and statistical tools are imperfect. But, when you understand the strengths and weaknesses of these tools, you can use them to learn about the real world.

Textbook overview

The chapters of this book are as follows:

1. **Introduction to data.** Data structures, variables, summaries, graphics, and basic data collection techniques.

2. **Probability (special topic).** The basic principles of probability. An understanding of this chapter is not required for the main content in Chapters 3-8.

3. **Distributions of random variables.** Introduction to the normal model and other key distributions.

4. **Foundations for inference.** General ideas for statistical inference in the context of estimating the population mean.

5. **Inference for numerical data.** Inference for one or two sample means using the t-distribution, and also comparisons of many means using ANOVA.

6. **Inference for categorical data.** Inference for proportions using the normal and chi-square distributions, as well as simulation and randomization techniques.

7. **Introduction to linear regression.** An introduction to regression with two variables. Most of this chapter could be covered after Chapter 1.

8. **Multiple and logistic regression.** A light introduction to multiple regression and logistic regression for an accelerated course.

OpenIntro Statistics was written to allow flexibility in choosing and ordering course topics. The material is divided into two pieces: main text and special topics. The main text has been structured to bring statistical inference and modeling closer to the front of a course. Special topics, labeled in the table of contents and in section titles, may be added to a course as they arise naturally in the curriculum.

Videos for sections and calculators

The ▶ icon indicates that a section or topic has a video overview readily available. The icons are hyperlinked in the textbook PDF, and the videos may also be found at

www.openintro.org/stat/videos.php

Examples, exercises, and appendices

Examples and Guided Practice throughout the textbook may be identified by their distinctive bullets:

● **Example 0.1** Large filled bullets signal the start of an example.

Full solutions to examples are provided and may include an accompanying table or figure.

⊙ **Guided Practice 0.2** Large empty bullets signal to readers that an exercise has been inserted into the text for additional practice and guidance. Students may find it useful to fill in the bullet after understanding or successfully completing the exercise. Solutions are provided for all Guided Practice in footnotes.[1]

There are exercises at the end of each chapter for practice or homework assignments. Odd-numbered exercise solutions are in Appendix A. Probability tables for the normal, t, and chi-square distributions are in Appendix B.

OpenIntro, online resources, and getting involved

OpenIntro is an organization focused on developing free and affordable education materials. *OpenIntro Statistics* is intended for introductory statistics courses at the college level. We offer another title, *Advanced High School Statistics*, for high school courses.

We encourage anyone learning or teaching statistics to visit **openintro.org** and get involved. We also provide many free online resources, including free course software. Data sets for this textbook are available on the website and through a companion R package.[2] All of these resources are free and may be used with or without this textbook as a companion.

We value your feedback. If there is a particular component of the project you especially like or think needs improvement, we want to hear from you. You may find our contact information on the title page of this book or on the About section of **openintro.org**.

Acknowledgements

This project would not be possible without the passion and dedication of all those involved. The authors would like to thank the OpenIntro Staff for their involvement and ongoing contributions. We are also very grateful to the hundreds of students and instructors who have provided us with valuable feedback over the last several years.

---

[1] Full solutions are located down here in the footnote!

[2] Diez DM, Barr CD, Çetinkaya-Rundel M. 2015. openintro: OpenIntro data sets and supplement functions. github.com/OpenIntroOrg/openintro-r-package.

Chapter 1

Introduction to data

Scientists seek to answer questions using rigorous methods and careful observations. These observations – collected from the likes of field notes, surveys, and experiments – form the backbone of a statistical investigation and are called **data**. Statistics is the study of how best to collect, analyze, and draw conclusions from data. It is helpful to put statistics in the context of a general process of investigation:

1. Identify a question or problem.

2. Collect relevant data on the topic.

3. Analyze the data.

4. Form a conclusion.

Statistics as a subject focuses on making stages 2-4 objective, rigorous, and efficient. That is, statistics has three primary components: How best can we collect data? How should it be analyzed? And what can we infer from the analysis?

The topics scientists investigate are as diverse as the questions they ask. However, many of these investigations can be addressed with a small number of data collection techniques, analytic tools, and fundamental concepts in statistical inference. This chapter provides a glimpse into these and other themes we will encounter throughout the rest of the book. We introduce the basic principles of each branch and learn some tools along the way. We will encounter applications from other fields, some of which are not typically associated with science but nonetheless can benefit from statistical study.

1.1   Case study: using stents to prevent strokes

Section 1.1 introduces a classic challenge in statistics: evaluating the efficacy of a medical treatment. Terms in this section, and indeed much of this chapter, will all be revisited later in the text. The plan for now is simply to get a sense of the role statistics can play in practice.

In this section we will consider an experiment that studies effectiveness of stents in treating patients at risk of stroke.[1] Stents are devices put inside blood vessels that assist

---

[1]Chimowitz MI, Lynn MJ, Derdeyn CP, et al. 2011. Stenting versus Aggressive Medical Therapy for Intracranial Arterial Stenosis. New England Journal of Medicine 365:993-1003. www.nejm.org/doi/full/10.1056/NEJMoa1105335. NY Times article reporting on the study: www.nytimes.com/2011/09/08/health/research/08stent.html.

in patient recovery after cardiac events and reduce the risk of an additional heart attack or death. Many doctors have hoped that there would be similar benefits for patients at risk of stroke. We start by writing the principal question the researchers hope to answer:

Does the use of stents reduce the risk of stroke?

The researchers who asked this question collected data on 451 at-risk patients. Each volunteer patient was randomly assigned to one of two groups:

Treatment group. Patients in the treatment group received a stent and medical management. The medical management included medications, management of risk factors, and help in lifestyle modification.

Control group. Patients in the control group received the same medical management as the treatment group, but they did not receive stents.

Researchers randomly assigned 224 patients to the treatment group and 227 to the control group. In this study, the control group provides a reference point against which we can measure the medical impact of stents in the treatment group.

Researchers studied the effect of stents at two time points: 30 days after enrollment and 365 days after enrollment. The results of 5 patients are summarized in Table 1.1. Patient outcomes are recorded as "stroke" or "no event", representing whether or not the patient had a stroke at the end of a time period.

Patient	group	0-30 days	0-365 days
1	treatment	no event	no event
2	treatment	stroke	stroke
3	treatment	no event	no event
⋮	⋮	⋮	
450	control	no event	no event
451	control	no event	no event

Table 1.1: Results for five patients from the stent study.

Considering data from each patient individually would be a long, cumbersome path towards answering the original research question. Instead, performing a statistical data analysis allows us to consider all of the data at once. Table 1.2 summarizes the raw data in a more helpful way. In this table, we can quickly see what happened over the entire study. For instance, to identify the number of patients in the treatment group who had a stroke within 30 days, we look on the left-side of the table at the intersection of the treatment and stroke: 33.

	0-30 days		0-365 days	
	stroke	no event	stroke	no event
treatment	33	191	45	179
control	13	214	28	199
Total	46	405	73	378

Table 1.2: Descriptive statistics for the stent study.

⊙ **Guided Practice 1.1** Of the 224 patients in the treatment group, 45 had a stroke by the end of the first year. Using these two numbers, compute the proportion of patients in the treatment group who had a stroke by the end of their first year. (Please note: answers to all in-text exercises are provided using footnotes.)[2]

We can compute summary statistics from the table. A **summary statistic** is a single number summarizing a large amount of data.[3] For instance, the primary results of the study after 1 year could be described by two summary statistics: the proportion of people who had a stroke in the treatment and control groups.

Proportion who had a stroke in the treatment (stent) group: $45/224 = 0.20 = 20\%$.

Proportion who had a stroke in the control group: $28/227 = 0.12 = 12\%$.

These two summary statistics are useful in looking for differences in the groups, and we are in for a surprise: an additional 8% of patients in the treatment group had a stroke! This is important for two reasons. First, it is contrary to what doctors expected, which was that stents would *reduce* the rate of strokes. Second, it leads to a statistical question: do the data show a "real" difference between the groups?

This second question is subtle. Suppose you flip a coin 100 times. While the chance a coin lands heads in any given coin flip is 50%, we probably won't observe exactly 50 heads. This type of fluctuation is part of almost any type of data generating process. It is possible that the 8% difference in the stent study is due to this natural variation. However, the larger the difference we observe (for a particular sample size), the less believable it is that the difference is due to chance. So what we are really asking is the following: is the difference so large that we should reject the notion that it was due to chance?

While we don't yet have our statistical tools to fully address this question on our own, we can comprehend the conclusions of the published analysis: there was compelling evidence of harm by stents in this study of stroke patients.

Be careful: do not generalize the results of this study to all patients and all stents. This study looked at patients with very specific characteristics who volunteered to be a part of this study and who may not be representative of all stroke patients. In addition, there are many types of stents and this study only considered the self-expanding Wingspan stent (Boston Scientific). However, this study does leave us with an important lesson: we should keep our eyes open for surprises.

1.2 Data basics 🎥

Effective presentation and description of data is a first step in most analyses. This section introduces one structure for organizing data as well as some terminology that will be used throughout this book.

1.2.1 Observations, variables, and data matrices

Table 1.3 displays rows 1, 2, 3, and 50 of a data set concerning 50 emails received during early 2012. These observations will be referred to as the `email50` data set, and they are a random sample from a larger data set that we will see in Section 1.7.

---

[2]The proportion of the 224 patients who had a stroke within 365 days: $45/224 = 0.20$.

[3]Formally, a summary statistic is a value computed from the data. Some summary statistics are more useful than others.

	spam	num_char	line_breaks	format	number
1	no	21,705	551	html	small
2	no	7,011	183	html	big
3	yes	631	28	text	none
⋮	⋮	⋮	⋮	⋮	⋮
50	no	15,829	242	html	small

Table 1.3: Four rows from the email50 data matrix.

variable	description
spam	Specifies whether the message was spam
num_char	The number of characters in the email
line_breaks	The number of line breaks in the email (not including text wrapping)
format	Indicates if the email contained special formatting, such as bolding, tables, or links, which would indicate the message is in HTML format
number	Indicates whether the email contained no number, a small number (under 1 million), or a large number

Table 1.4: Variables and their descriptions for the email50 data set.

Each row in the table represents a single email or **case**.[4] The columns represent characteristics, called **variables**, for each of the emails. For example, the first row represents email 1, which is a not spam, contains 21,705 characters, 551 line breaks, is written in HTML format, and contains only small numbers.

In practice, it is especially important to ask clarifying questions to ensure important aspects of the data are understood. For instance, it is always important to be sure we know what each variable means and the units of measurement. Descriptions of all five email variables are given in Table 1.4.

The data in Table 1.3 represent a **data matrix**, which is a common way to organize data. Each row of a data matrix corresponds to a unique case, and each column corresponds to a variable. A data matrix for the stroke study introduced in Section 1.1 is shown in Table 1.1 on page 8, where the cases were patients and there were three variables recorded for each patient.

Data matrices are a convenient way to record and store data. If another individual or case is added to the data set, an additional row can be easily added. Similarly, another column can be added for a new variable.

⊙ **Guided Practice 1.2**   We consider a publicly available data set that summarizes information about the 3,143 counties in the United States, and we call this the county data set. This data set includes information about each county: its name, the state where it resides, its population in 2000 and 2010, per capita federal spending, poverty rate, and five additional characteristics. How might these data be organized in a data matrix? Reminder: look in the footnotes for answers to in-text exercises.[5]

Seven rows of the county data set are shown in Table 1.5, and the variables are summarized in Table 1.6. These data were collected from the US Census website.[6]

---

[4]A case is also sometimes called a **unit of observation** or an **observational unit**.

[5]Each county may be viewed as a case, and there are eleven pieces of information recorded for each case. A table with 3,143 rows and 11 columns could hold these data, where each row represents a county and each column represents a particular piece of information.

[6]quickfacts.census.gov/qfd/index.html

	name	state	pop2000	pop2010	fed_spend	poverty	homeownership	multiunit	income	med_income	smoking_ban
1	Autauga	AL	43671	54571	6.068	10.6	77.5	7.2	24568	53255	none
2	Baldwin	AL	140415	182265	6.140	12.2	76.7	22.6	26469	50147	none
3	Barbour	AL	29038	27457	8.752	25.0	68.0	11.1	15875	33219	none
4	Bibb	AL	20826	22915	7.122	12.6	82.9	6.6	19918	41770	none
5	Blount	AL	51024	57322	5.131	13.4	82.0	3.7	21070	45549	none
⋮	⋮	⋮					⋮	⋮	⋮	⋮	⋮
3142	Washakie	WY	8289	8533	8.714	5.6	70.9	10.0	28557	48379	none
3143	Weston	WY	6644	7208	6.695	7.9	77.9	6.5	28463	53853	none

Table 1.5: Seven rows from the county data set.

variable	description
name	County name
state	State where the county resides (also including the District of Columbia)
pop2000	Population in 2000
pop2010	Population in 2010
fed_spend	Federal spending per capita
poverty	Percent of the population in poverty
homeownership	Percent of the population that lives in their own home or lives with the owner (e.g. children living with parents who own the home)
multiunit	Percent of living units that are in multi-unit structures (e.g. apartments)
income	Income per capita
med_income	Median household income for the county, where a household's income equals the total income of its occupants who are 15 years or older
smoking_ban	Type of county-wide smoking ban in place at the end of 2011, which takes one of three values: none, partial, or comprehensive, where a comprehensive ban means smoking was not permitted in restaurants, bars, or workplaces, and partial means smoking was banned in at least one of those three locations

Table 1.6: Variables and their descriptions for the county data set.

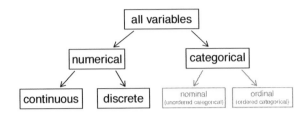

Figure 1.7: Breakdown of variables into their respective types.

1.2.2 Types of variables

Examine the `fed_spend`, `pop2010`, `state`, and `smoking_ban` variables in the `county` data set. Each of these variables is inherently different from the other three yet many of them share certain characteristics.

First consider `fed_spend`, which is said to be a **numerical** variable since it can take a wide range of numerical values, and it is sensible to add, subtract, or take averages with those values. On the other hand, we would not classify a variable reporting telephone area codes as numerical since their average, sum, and difference have no clear meaning.

The `pop2010` variable is also numerical, although it seems to be a little different than `fed_spend`. This variable of the population count can only take whole non-negative numbers (0, 1, 2, ...). For this reason, the population variable is said to be **discrete** since it can only take numerical values with jumps. On the other hand, the federal spending variable is said to be **continuous**.

The variable `state` can take up to 51 values after accounting for Washington, DC: AL, ..., and WY. Because the responses themselves are categories, `state` is called a **categorical** variable, and the possible values are called the variable's **levels**.

Finally, consider the `smoking_ban` variable, which describes the type of county-wide smoking ban and takes values `none`, `partial`, or `comprehensive` in each county. This variable seems to be a hybrid: it is a categorical variable but the levels have a natural ordering. A variable with these properties is called an **ordinal** variable, while a regular categorical variable without this type of special ordering is called a **nominal** variable. To simplify analyses, any ordinal variables in this book will be treated as categorical variables.

● **Example 1.3** Data were collected about students in a statistics course. Three variables were recorded for each student: number of siblings, student height, and whether the student had previously taken a statistics course. Classify each of the variables as continuous numerical, discrete numerical, or categorical.

The number of siblings and student height represent numerical variables. Because the number of siblings is a count, it is discrete. Height varies continuously, so it is a continuous numerical variable. The last variable classifies students into two categories – those who have and those who have not taken a statistics course – which makes this variable categorical.

⊙ **Guided Practice 1.4** Consider the variables `group` and `outcome` (at 30 days) from the stent study in Section 1.1. Are these numerical or categorical variables?[7]

---

[7]There are only two possible values for each variable, and in both cases they describe categories. Thus, each is a categorical variable.

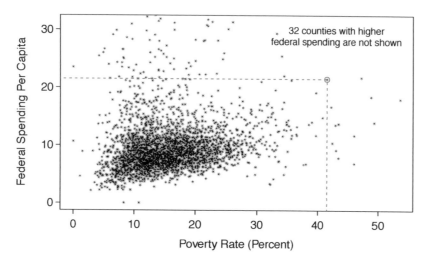

Figure 1.8: A scatterplot showing `fed_spend` against `poverty`. Owsley County of Kentucky, with a poverty rate of 41.5% and federal spending of $21.50 per capita, is highlighted.

1.2.3 Relationships between variables

Many analyses are motivated by a researcher looking for a relationship between two or more variables. A social scientist may like to answer some of the following questions:

(1) Is federal spending, on average, higher or lower in counties with high rates of poverty?

(2) If homeownership is lower than the national average in one county, will the percent of multi-unit structures in that county likely be above or below the national average?

(3) Which counties have a higher average income: those that enact one or more smoking bans or those that do not?

To answer these questions, data must be collected, such as the `county` data set shown in Table 1.5. Examining summary statistics could provide insights for each of the three questions about counties. Additionally, graphs can be used to visually summarize data and are useful for answering such questions as well.

Scatterplots are one type of graph used to study the relationship between two numerical variables. Figure 1.8 compares the variables `fed_spend` and `poverty`. Each point on the plot represents a single county. For instance, the highlighted dot corresponds to County 1088 in the `county` data set: Owsley County, Kentucky, which had a poverty rate of 41.5% and federal spending of $21.50 per capita. The scatterplot suggests a relationship between the two variables: counties with a high poverty rate also tend to have slightly more federal spending. We might brainstorm as to why this relationship exists and investigate each idea to determine which is the most reasonable explanation.

⊙ **Guided Practice 1.5** Examine the variables in the `email50` data set, which are described in Table 1.4 on page 10. Create two questions about the relationships between these variables that are of interest to you.[8]

---

[8]Two sample questions: (1) Intuition suggests that if there are many line breaks in an email then there also would tend to be many characters: does this hold true? (2) Is there a connection between whether an email format is plain text (versus HTML) and whether it is a spam message?

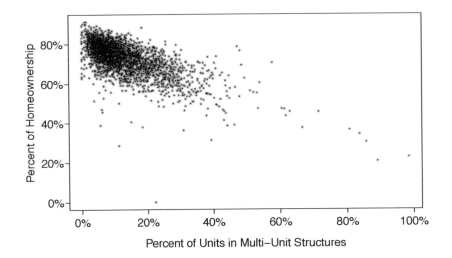

Figure 1.9: A scatterplot of homeownership versus the percent of units that are in multi-unit structures for all 3,143 counties. Interested readers may find an image of this plot with an additional third variable, county population, presented at www.openintro.org/stat/down/MHP.png.

The fed_spend and poverty variables are said to be associated because the plot shows a discernible pattern. When two variables show some connection with one another, they are called **associated** variables. Associated variables can also be called **dependent** variables and vice-versa.

● **Example 1.6** This example examines the relationship between homeownership and the percent of units in multi-unit structures (e.g. apartments, condos), which is visualized using a scatterplot in Figure 1.9. Are these variables associated?

It appears that the larger the fraction of units in multi-unit structures, the lower the homeownership rate. Since there is some relationship between the variables, they are associated.

Because there is a downward trend in Figure 1.9 – counties with more units in multi-unit structures are associated with lower homeownership – these variables are said to be **negatively associated**. A **positive association** is shown in the relationship between the poverty and fed_spend variables represented in Figure 1.8, where counties with higher poverty rates tend to receive more federal spending per capita.

If two variables are not associated, then they are said to be **independent**. That is, two variables are independent if there is no evident relationship between the two.

Associated or independent, not both

A pair of variables are either related in some way (associated) or not (independent). No pair of variables is both associated and independent.

1.3 Overview of data collection principles 📹

The first step in conducting research is to identify topics or questions that are to be investigated. A clearly laid out research question is helpful in identifying what subjects or cases should be studied and what variables are important. It is also important to consider *how* data are collected so that they are reliable and help achieve the research goals.

1.3.1 Populations and samples

Consider the following three research questions:

1. What is the average mercury content in swordfish in the Atlantic Ocean?

2. Over the last 5 years, what is the average time to complete a degree for Duke undergraduate students?

3. Does a new drug reduce the number of deaths in patients with severe heart disease?

Each research question refers to a target **population**. In the first question, the target population is all swordfish in the Atlantic ocean, and each fish represents a case. Often times, it is too expensive to collect data for every case in a population. Instead, a sample is taken. A **sample** represents a subset of the cases and is often a small fraction of the population. For instance, 60 swordfish (or some other number) in the population might be selected, and this sample data may be used to provide an estimate of the population average and answer the research question.

⊙ **Guided Practice 1.7** For the second and third questions above, identify the target population and what represents an individual case.[9]

1.3.2 Anecdotal evidence

Consider the following possible responses to the three research questions:

1. A man on the news got mercury poisoning from eating swordfish, so the average mercury concentration in swordfish must be dangerously high.

2. I met two students who took more than 7 years to graduate from Duke, so it must take longer to graduate at Duke than at many other colleges.

3. My friend's dad had a heart attack and died after they gave him a new heart disease drug, so the drug must not work.

Each conclusion is based on data. However, there are two problems. First, the data only represent one or two cases. Second, and more importantly, it is unclear whether these cases are actually representative of the population. Data collected in this haphazard fashion are called **anecdotal evidence**.

---

[9](2) Notice that the first question is only relevant to students who complete their degree; the average cannot be computed using a student who never finished her degree. Thus, only Duke undergraduate students who have graduated in the last five years represent cases in the population under consideration. Each such student would represent an individual case. (3) A person with severe heart disease represents a case. The population includes all people with severe heart disease.

Figure 1.10: In February 2010, some media pundits cited one large snow storm as valid evidence against global warming. As comedian Jon Stewart pointed out, "It's one storm, in one region, of one country."

Anecdotal evidence

Be careful of data collected in a haphazard fashion. Such evidence may be true and verifiable, but it may only represent extraordinary cases.

Anecdotal evidence typically is composed of unusual cases that we recall based on their striking characteristics. For instance, we are more likely to remember the two people we met who took 7 years to graduate than the six others who graduated in four years. Instead of looking at the most unusual cases, we should examine a sample of many cases that represent the population.

1.3.3 Sampling from a population

We might try to estimate the time to graduation for Duke undergraduates in the last 5 years by collecting a sample of students. All graduates in the last 5 years represent the *population*, and graduates who are selected for review are collectively called the *sample*. In general, we always seek to *randomly* select a sample from a population. The most basic type of random selection is equivalent to how raffles are conducted. For example, in selecting graduates, we could write each graduate's name on a raffle ticket and draw 100 tickets. The selected names would represent a random sample of 100 graduates.

Why pick a sample randomly? Why not just pick a sample by hand? Consider the following scenario.

● **Example 1.8** Suppose we ask a student who happens to be majoring in nutrition to select several graduates for the study. What kind of students do you think she might collect? Do you think her sample would be representative of all graduates?

Perhaps she would pick a disproportionate number of graduates from health-related fields. Or perhaps her selection would be well-representative of the population. When selecting samples by hand, we run the risk of picking a *biased* sample, even if that bias is unintentional or difficult to discern.

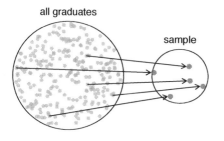

Figure 1.11: In this graphic, five graduates are randomly selected from the population to be included in the sample.

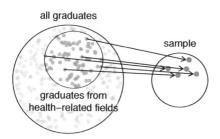

Figure 1.12: Instead of sampling from all graduates equally, a nutrition major might inadvertently pick graduates with health-related majors disproportionately often.

If someone was permitted to pick and choose exactly which graduates were included in the sample, it is entirely possible that the sample could be skewed to that person's interests, which may be entirely unintentional. This introduces **bias** into a sample. Sampling randomly helps resolve this problem. The most basic random sample is called a **simple random sample**, and which is equivalent to using a raffle to select cases. This means that each case in the population has an equal chance of being included and there is no implied connection between the cases in the sample.

The act of taking a simple random sample helps minimize bias, however, bias can crop up in other ways. Even when people are picked at random, e.g. for surveys, caution must be exercised if the **non-response** is high. For instance, if only 30% of the people randomly sampled for a survey actually respond, then it is unclear whether the results are **representative** of the entire population. This **non-response bias** can skew results.

Another common downfall is a **convenience sample**, where individuals who are easily accessible are more likely to be included in the sample. For instance, if a political survey is done by stopping people walking in the Bronx, this will not represent all of New York City. It is often difficult to discern what sub-population a convenience sample represents.

⊙ **Guided Practice 1.9** We can easily access ratings for products, sellers, and companies through websites. These ratings are based only on those people who go out of their way to provide a rating. If 50% of online reviews for a product are negative, do you think this means that 50% of buyers are dissatisfied with the product?[10]

---

[10]Answers will vary. From our own anecdotal experiences, we believe people tend to rant more about products that fell below expectations than rave about those that perform as expected. For this reason, we suspect there is a negative bias in product ratings on sites like Amazon. However, since our experiences may not be representative, we also keep an open mind.

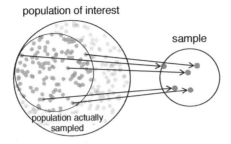

Figure 1.13: Due to the possibility of non-response, surveys studies may only reach a certain group within the population. It is difficult, and often times impossible, to completely fix this problem.

1.3.4   Explanatory and response variables

Consider the following question from page 13 for the `county` data set:

(1) Is federal spending, on average, higher or lower in counties with high rates of poverty?

If we suspect poverty might affect spending in a county, then poverty is the **explanatory** variable and federal spending is the **response** variable in the relationship.[11] If there are many variables, it may be possible to consider a number of them as explanatory variables.

TIP: Explanatory and response variables

To identify the explanatory variable in a pair of variables, identify which of the two is suspected of affecting the other and plan an appropriate analysis.

explanatory   *might affect*   response
variable $\longrightarrow$ variable

Caution: association does not imply causation

Labeling variables as *explanatory* and *response* does not guarantee the relationship between the two is actually causal, even if there is an association identified between the two variables. We use these labels only to keep track of which variable we suspect affects the other.

In some cases, there is no explanatory or response variable. Consider the following question from page 13:

(2) If homeownership is lower than the national average in one county, will the percent of multi-unit structures in that county likely be above or below the national average?

It is difficult to decide which of these variables should be considered the explanatory and response variable, i.e. the direction is ambiguous, so no explanatory or response labels are suggested here.

---

[11]Sometimes the explanatory variable is called the **independent** variable and the response variable is called the **dependent** variable. However, this becomes confusing since a *pair* of variables might be independent or dependent, so we avoid this language.

1.3.5 Introducing observational studies and experiments

There are two primary types of data collection: observational studies and experiments.

Researchers perform an **observational study** when they collect data in a way that does not directly interfere with how the data arise. For instance, researchers may collect information via surveys, review medical or company records, or follow a **cohort** of many similar individuals to study why certain diseases might develop. In each of these situations, researchers merely observe the data that arise. In general, observational studies can provide evidence of a naturally occurring association between variables, but they cannot by themselves show a causal connection.

When researchers want to investigate the possibility of a causal connection, they conduct an **experiment**. Usually there will be both an explanatory and a response variable. For instance, we may suspect administering a drug will reduce mortality in heart attack patients over the following year. To check if there really is a causal connection between the explanatory variable and the response, researchers will collect a sample of individuals and split them into groups. The individuals in each group are *assigned* a treatment. When individuals are randomly assigned to a group, the experiment is called a **randomized experiment**. For example, each heart attack patient in the drug trial could be randomly assigned, perhaps by flipping a coin, into one of two groups: the first group receives a **placebo** (fake treatment) and the second group receives the drug. See the case study in Section 1.1 for another example of an experiment, though that study did not employ a placebo.

TIP: association ≠ causation

In general, association does not imply causation, and causation can only be inferred from a randomized experiment.

1.4 Observational studies and sampling strategies

1.4.1 Observational studies

Generally, data in observational studies are collected only by monitoring what occurs, while experiments require the primary explanatory variable in a study be assigned for each subject by the researchers.

Making causal conclusions based on experiments is often reasonable. However, making the same causal conclusions based on observational data can be treacherous and is not recommended. Thus, observational studies are generally only sufficient to show associations.

⊙ **Guided Practice 1.10** Suppose an observational study tracked sunscreen use and skin cancer, and it was found that the more sunscreen someone used, the more likely the person was to have skin cancer. Does this mean sunscreen *causes* skin cancer?[12]

Some previous research tells us that using sunscreen actually reduces skin cancer risk, so maybe there is another variable that can explain this hypothetical association between sunscreen usage and skin cancer. One important piece of information that is absent is sun exposure. If someone is out in the sun all day, she is more likely to use sunscreen *and* more likely to get skin cancer. Exposure to the sun is unaccounted for in the simple investigation.

---

[12]No. See the paragraph following the exercise for an explanation.

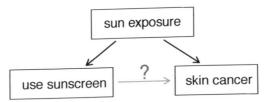

Sun exposure is what is called a **confounding variable**,[13] which is a variable that is correlated with both the explanatory and response variables. While one method to justify making causal conclusions from observational studies is to exhaust the search for confounding variables, there is no guarantee that all confounding variables can be examined or measured.

In the same way, the `county` data set is an observational study with confounding variables, and its data cannot easily be used to make causal conclusions.

⊙ **Guided Practice 1.11** Figure 1.9 shows a negative association between the home-ownership rate and the percentage of multi-unit structures in a county. However, it is unreasonable to conclude that there is a causal relationship between the two variables. Suggest one or more other variables that might explain the relationship visible in Figure 1.9.[14]

Observational studies come in two forms: prospective and retrospective studies. A **prospective study** identifies individuals and collects information as events unfold. For instance, medical researchers may identify and follow a group of similar individuals over many years to assess the possible influences of behavior on cancer risk. One example of such a study is The Nurses' Health Study, started in 1976 and expanded in 1989.[15] This prospective study recruits registered nurses and then collects data from them using questionnaires. **Retrospective studies** collect data after events have taken place, e.g. researchers may review past events in medical records. Some data sets, such as `county`, may contain both prospectively- and retrospectively-collected variables. Local governments prospectively collect some variables as events unfolded (e.g. retails sales) while the federal government retrospectively collected others during the 2010 census (e.g. county population counts).

1.4.2   Four sampling methods (special topic)

Almost all statistical methods are based on the notion of implied randomness. If observational data are not collected in a random framework from a population, these statistical methods – the estimates and errors associated with the estimates – are not reliable. Here we consider four random sampling techniques: simple, stratified, cluster, and multistage sampling. Figures 1.14 and 1.15 provide graphical representations of these techniques.

Simple random sampling is probably the most intuitive form of random sampling. Consider the salaries of Major League Baseball (MLB) players, where each player is a member of one of the league's 30 teams. To take a simple random sample of 120 baseball players and their salaries from the 2010 season, we could write the names of that season's

---

[13]Also called a **lurking variable**, **confounding factor**, or a **confounder**.

[14]Answers will vary. Population density may be important. If a county is very dense, then this may require a larger fraction of residents to live in multi-unit structures. Additionally, the high density may contribute to increases in property value, making homeownership infeasible for many residents.

[15]www.channing.harvard.edu/nhs

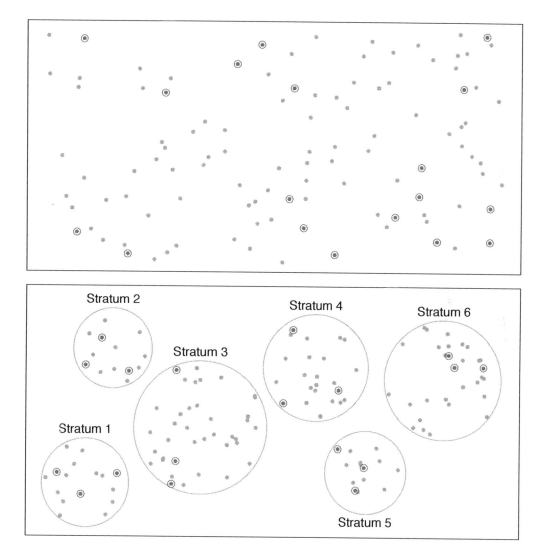

Figure 1.14: Examples of simple random and stratified sampling. In the top panel, simple random sampling was used to randomly select the 18 cases. In the bottom panel, stratified sampling was used: cases were grouped into strata, then simple random sampling was employed within each stratum.

828 players onto slips of paper, drop the slips into a bucket, shake the bucket around until we are sure the names are all mixed up, then draw out slips until we have the sample of 120 players. In general, a sample is referred to as "simple random" if each case in the population has an equal chance of being included in the final sample *and* knowing that a case is included in a sample does not provide useful information about which other cases are included.

Stratified sampling is a divide-and-conquer sampling strategy. The population is divided into groups called **strata**. The strata are chosen so that similar cases are grouped together, then a second sampling method, usually simple random sampling, is employed within each stratum. In the baseball salary example, the teams could represent the strata, since some teams have a lot more money (up to 4 times as much!). Then we might randomly sample 4 players from each team for a total of 120 players.

Stratified sampling is especially useful when the cases in each stratum are very similar with respect to the outcome of interest. The downside is that analyzing data from a stratified sample is a more complex task than analyzing data from a simple random sample. The analysis methods introduced in this book would need to be extended to analyze data collected using stratified sampling.

● **Example 1.12** Why would it be good for cases within each stratum to be very similar?

We might get a more stable estimate for the subpopulation in a stratum if the cases are very similar. These improved estimates for each subpopulation will help us build a reliable estimate for the full population.

In a **cluster sample**, we break up the population into many groups, called **clusters**. Then we sample a fixed number of clusters and include all observations from each of those clusters in the sample. A **multistage sample** is like a cluster sample, but rather than keeping all observations in each cluster, we collect a random sample within each selected cluster.

Sometimes cluster or multistage sampling can be more economical than the alternative sampling techniques. Also, unlike stratified sampling, these approaches are most helpful when there is a lot of case-to-case variability within a cluster but the clusters themselves don't look very different from one another. For example, if neighborhoods represented clusters, then cluster or multistage sampling work best when the neighborhoods are very diverse. A downside of these methods is that more advanced analysis techniques are typically required, though the methods in this book can be extended to handle such data.

● **Example 1.13** Suppose we are interested in estimating the malaria rate in a densely tropical portion of rural Indonesia. We learn that there are 30 villages in that part of the Indonesian jungle, each more or less similar to the next. Our goal is to test 150 individuals for malaria. What sampling method should be employed?

A simple random sample would likely draw individuals from all 30 villages, which could make data collection extremely expensive. Stratified sampling would be a challenge since it is unclear how we would build strata of similar individuals. However, cluster sampling or multistage sampling seem like very good ideas. If we decided to use multistage sampling, we might randomly select half of the villages, then randomly select 10 people from each. This would probably reduce our data collection costs substantially in comparison to a simple random sample, and this approach would still give us reliable information.

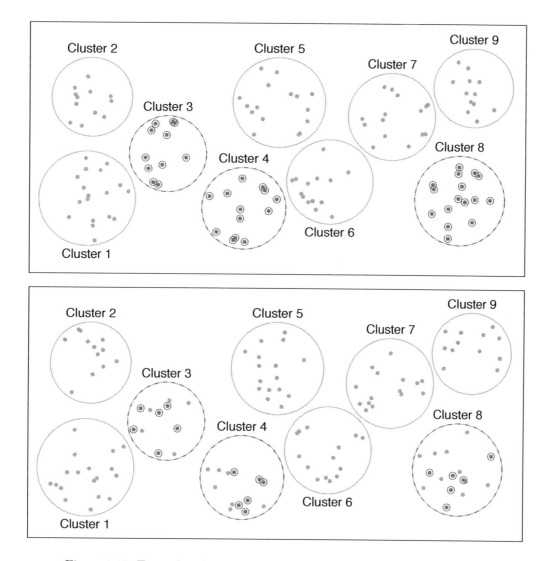

Figure 1.15: Examples of cluster and multistage sampling. In the top panel, cluster sampling was used. Here, data were binned into nine clusters, three of these clusters were sampled, and all observations within these three cluster were included in the sample. In the bottom panel, multistage sampling was used. It differs from cluster sampling in that of the clusters selected, we randomly select a subset of each cluster to be included in the sample.

1.5   Experiments 📹

Studies where the researchers assign treatments to cases are called **experiments**. When this assignment includes randomization, e.g. using a coin flip to decide which treatment a patient receives, it is called a **randomized experiment**. Randomized experiments are fundamentally important when trying to show a causal connection between two variables.

1.5.1   Principles of experimental design

Randomized experiments are generally built on four principles.

Controlling. Researchers assign treatments to cases, and they do their best to **control** any other differences in the groups. For example, when patients take a drug in pill form, some patients take the pill with only a sip of water while others may have it with an entire glass of water. To control for the effect of water consumption, a doctor may ask all patients to drink a 12 ounce glass of water with the pill.

Randomization. Researchers randomize patients into treatment groups to account for variables that cannot be controlled. For example, some patients may be more susceptible to a disease than others due to their dietary habits. Randomizing patients into the treatment or control group helps even out such differences, and it also prevents accidental bias from entering the study.

Replication. The more cases researchers observe, the more accurately they can estimate the effect of the explanatory variable on the response. In a single study, we **replicate** by collecting a sufficiently large sample. Additionally, a group of scientists may replicate an entire study to verify an earlier finding.

Blocking. Researchers sometimes know or suspect that variables, other than the treatment, influence the response. Under these circumstances, they may first group individuals based on this variable into **blocks** and then randomize cases within each block to the treatment groups. This strategy is often referred to as **blocking**. For instance, if we are looking at the effect of a drug on heart attacks, we might first split patients in the study into low-risk and high-risk blocks, then randomly assign half the patients from each block to the control group and the other half to the treatment group, as shown in Figure 1.16. This strategy ensures each treatment group has an equal number of low-risk and high-risk patients.

It is important to incorporate the first three experimental design principles into any study, and this book describes applicable methods for analyzing data from such experiments. Blocking is a slightly more advanced technique, and statistical methods in this book may be extended to analyze data collected using blocking.

1.5.2   Reducing bias in human experiments

Randomized experiments are the gold standard for data collection, but they do not ensure an unbiased perspective into the cause and effect relationships in all cases. Human studies are perfect examples where bias can unintentionally arise. Here we reconsider a study where a new drug was used to treat heart attack patients.[16] In particular, researchers wanted to know if the drug reduced deaths in patients.

---

[16]Anturane Reinfarction Trial Research Group. 1980. Sulfinpyrazone in the prevention of sudden death after myocardial infarction. New England Journal of Medicine 302(5):250-256.

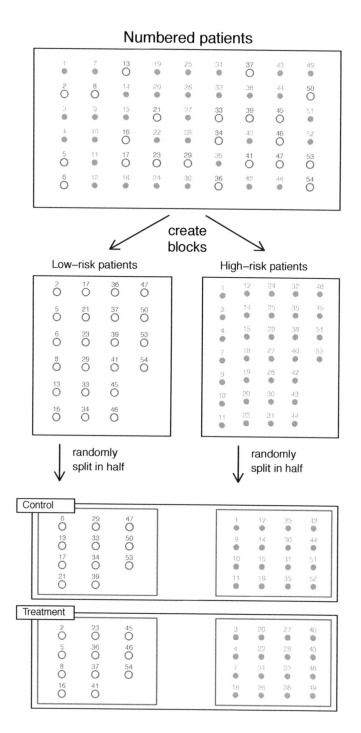

Figure 1.16: Blocking using a variable depicting patient risk. Patients are first divided into low-risk and high-risk blocks, then each block is evenly separated into the treatment groups using randomization. This strategy ensures an equal representation of patients in each treatment group from both the low-risk and high-risk categories.

These researchers designed a randomized experiment because they wanted to draw causal conclusions about the drug's effect. Study volunteers[17] were randomly placed into two study groups. One group, the **treatment group**, received the drug. The other group, called the **control group**, did not receive any drug treatment.

Put yourself in the place of a person in the study. If you are in the treatment group, you are given a fancy new drug that you anticipate will help you. On the other hand, a person in the other group doesn't receive the drug and sits idly, hoping her participation doesn't increase her risk of death. These perspectives suggest there are actually two effects: the one of interest is the effectiveness of the drug, and the second is an emotional effect that is difficult to quantify.

Researchers aren't usually interested in the emotional effect, which might bias the study. To circumvent this problem, researchers do not want patients to know which group they are in. When researchers keep the patients uninformed about their treatment, the study is said to be **blind**. But there is one problem: if a patient doesn't receive a treatment, she will know she is in the control group. The solution to this problem is to give fake treatments to patients in the control group. A fake treatment is called a **placebo**, and an effective placebo is the key to making a study truly blind. A classic example of a placebo is a sugar pill that is made to look like the actual treatment pill. Often times, a placebo results in a slight but real improvement in patients. This effect has been dubbed the **placebo effect**.

The patients are not the only ones who should be blinded: doctors and researchers can accidentally bias a study. When a doctor knows a patient has been given the real treatment, she might inadvertently give that patient more attention or care than a patient that she knows is on the placebo. To guard against this bias, which again has been found to have a measurable effect in some instances, most modern studies employ a **double-blind** setup where doctors or researchers who interact with patients are, just like the patients, unaware of who is or is not receiving the treatment.[18]

⊙ **Guided Practice 1.14**   Look back to the study in Section 1.1 where researchers were testing whether stents were effective at reducing strokes in at-risk patients. Is this an experiment? Was the study blinded? Was it double-blinded?[19]

1.6   Examining numerical data 📹

In this section we will be introduced to techniques for exploring and summarizing numerical variables. The email50 and county data sets from Section 1.2 provide rich opportunities for examples. Recall that outcomes of numerical variables are numbers on which it is reasonable to perform basic arithmetic operations. For example, the pop2010 variable, which represents the populations of counties in 2010, is numerical since we can sensibly discuss the difference or ratio of the populations in two counties. On the other hand, area codes and zip codes are not numerical, but rather they are categorical variables.

---

[17]Human subjects are often called **patients**, **volunteers**, or **study participants**.

[18]There are always some researchers involved in the study who do know which patients are receiving which treatment. However, they do not interact with the study's patients and do not tell the blinded health care professionals who is receiving which treatment.

[19]The researchers assigned the patients into their treatment groups, so this study was an experiment. However, the patients could distinguish what treatment they received, so this study was not blind. The study could not be double-blind since it was not blind.

1.6.1 Scatterplots for paired data

A **scatterplot** provides a case-by-case view of data for two numerical variables. In Figure 1.8 on page 13, a scatterplot was used to examine how federal spending and poverty were related in the `county` data set. Another scatterplot is shown in Figure 1.17, comparing the number of line breaks (`line_breaks`) and number of characters (`num_char`) in emails for the `email50` data set. In any scatterplot, each point represents a single case. Since there are 50 cases in `email50`, there are 50 points in Figure 1.17.

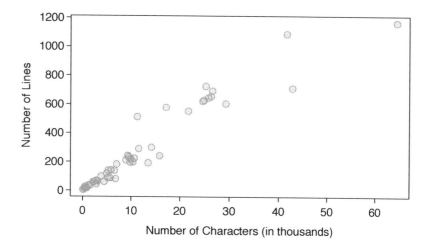

Figure 1.17: A scatterplot of `line_breaks` versus `num_char` for the `email50` data.

To put the number of characters in perspective, this paragraph has 363 characters. Looking at Figure 1.17, it seems that some emails are incredibly verbose! Upon further investigation, we would actually find that most of the long emails use the HTML format, which means most of the characters in those emails are used to format the email rather than provide text.

⊙ **Guided Practice 1.15** What do scatterplots reveal about the data, and how might they be useful?[20]

● **Example 1.16** Consider a new data set of 54 cars with two variables: vehicle price and weight.[21] A scatterplot of vehicle price versus weight is shown in Figure 1.18. What can be said about the relationship between these variables?

The relationship is evidently nonlinear, as highlighted by the dashed line. This is different from previous scatterplots we've seen, such as Figure 1.8 on page 13 and Figure 1.17, which show relationships that are very linear.

⊙ **Guided Practice 1.17** Describe two variables that would have a horseshoe shaped association in a scatterplot.[22]

---

[20]Answers may vary. Scatterplots are helpful in quickly spotting associations relating variables, whether those associations come in the form of simple trends or whether those relationships are more complex.

[21]Subset of data from www.amstat.org/publications/jse/v1n1/datasets.lock.html

[22]Consider the case where your vertical axis represents something "good" and your horizontal axis represents something that is only good in moderation. Health and water consumption fit this description since water becomes toxic when consumed in excessive quantities.

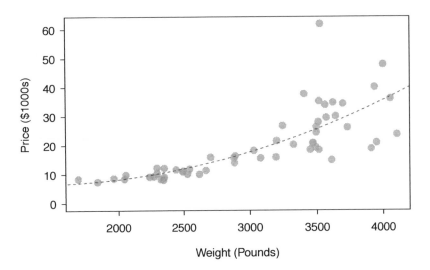

Figure 1.18: A scatterplot of `price` versus `weight` for 54 cars.

1.6.2   Dot plots and the mean

Sometimes two variables are one too many: only one variable may be of interest. In these cases, a dot plot provides the most basic of displays. A **dot plot** is a one-variable scatterplot; an example using the number of characters from 50 emails is shown in Figure 1.19. A stacked version of this dot plot is shown in Figure 1.20.

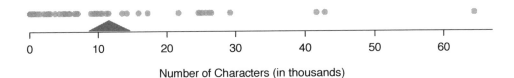

Figure 1.19: A dot plot of `num_char` for the `email50` data set.

The **mean**, sometimes called the average, is a common way to measure the center of a **distribution** of data. To find the mean number of characters in the 50 emails, we add up all the character counts and divide by the number of emails. For computational convenience, the number of characters is listed in the thousands and rounded to the first decimal.

$$\bar{x} = \frac{21.7 + 7.0 + \cdots + 15.8}{50} = 11.6 \tag{1.18}$$

$\bar{x}$
sample
mean

The sample mean is often labeled $\bar{x}$. The letter x is being used as a generic placeholder for the variable of interest, `num_char`, and the bar over on the x communicates that the average number of characters in the 50 emails was 11,600. It is useful to think of the mean as the balancing point of the distribution. The sample mean is shown as a triangle in Figures 1.19 and 1.20.

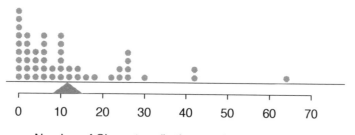

Number of Characters (in thousands, with rounding)

Figure 1.20: A stacked dot plot of `num_char` for the `email50` data set. The values have been rounded to the nearest 2,000 in this plot.

Mean

The sample mean of a numerical variable is computed as the sum of all of the observations divided by the number of observations:

$$\bar{x} = \frac{x_1 + x_2 + \cdots + x_n}{n} \tag{1.19}$$

where $x_1, x_2, \ldots, x_n$ represent the n observed values.

---

n
sample size

⊙ **Guided Practice 1.20**   Examine Equations (1.18) and (1.19) above. What does x_1 correspond to? And x_2? Can you infer a general meaning to what x_i might represent?[23]

⊙ **Guided Practice 1.21**   What was n in this sample of emails?[24]

The `email50` data set represents a sample from a larger population of emails that were received in January and March. We could compute a mean for this population in the same way as the sample mean, however, the population mean has a special label: μ. The symbol μ is the Greek letter *mu* and represents the average of all observations in the population. Sometimes a subscript, such as $_x$, is used to represent which variable the population mean refers to, e.g. μ_x.

μ
population
mean

● **Example 1.22**   The average number of characters across all emails can be estimated using the sample data. Based on the sample of 50 emails, what would be a reasonable estimate of μ_x, the mean number of characters in all emails in the `email` data set? (Recall that `email50` is a sample from `email`.)

The sample mean, 11,600, may provide a reasonable estimate of μ_x. While this number will not be perfect, it provides a *point estimate* of the population mean. In Chapter 4 and beyond, we will develop tools to characterize the accuracy of point estimates, and we will find that point estimates based on larger samples tend to be more accurate than those based on smaller samples.

---

[23]x_1 corresponds to the number of characters in the first email in the sample (21.7, in thousands), x_2 to the number of characters in the second email (7.0, in thousands), and x_i corresponds to the number of characters in the ith email in the data set.

[24]The sample size was $n = 50$.

● **Example 1.23** We might like to compute the average income per person in the US. To do so, we might first think to take the mean of the per capita incomes across the 3,143 counties in the county data set. What would be a better approach?

The county data set is special in that each county actually represents many individual people. If we were to simply average across the income variable, we would be treating counties with 5,000 and 5,000,000 residents equally in the calculations. Instead, we should compute the total income for each county, add up all the counties' totals, and then divide by the number of people in all the counties. If we completed these steps with the county data, we would find that the per capita income for the US is $27,348.43. Had we computed the *simple* mean of per capita income across counties, the result would have been just $22,504.70!

Example 1.23 used what is called a **weighted mean**, which will not be a key topic in this textbook. However, we have provided an online supplement on weighted means for interested readers:

www.openintro.org/stat/down/supp/wtdmean.pdf

1.6.3   Histograms and shape

Dot plots show the exact value for each observation. This is useful for small data sets, but they can become hard to read with larger samples. Rather than showing the value of each observation, we prefer to think of the value as belonging to a *bin*. For example, in the email50 data set, we create a table of counts for the number of cases with character counts between 0 and 5,000, then the number of cases between 5,000 and 10,000, and so on. Observations that fall on the boundary of a bin (e.g. 5,000) are allocated to the lower bin. This tabulation is shown in Table 1.21. These binned counts are plotted as bars in Figure 1.22 into what is called a **histogram**, which resembles the stacked dot plot shown in Figure 1.20.

Characters (in thousands)	0-5	5-10	10-15	15-20	20-25	25-30	⋯	55-60	60-65
Count	19	12	6	2	3	5	⋯	0	1

Table 1.21: The counts for the binned num_char data.

Histograms provide a view of the **data density**. Higher bars represent where the data are relatively more common. For instance, there are many more emails with fewer than 20,000 characters than emails with at least 20,000 in the data set. The bars make it easy to see how the density of the data changes relative to the number of characters.

Histograms are especially convenient for describing the shape of the data distribution. Figure 1.22 shows that most emails have a relatively small number of characters, while fewer emails have a very large number of characters. When data trail off to the right in this way and have a longer right tail, the shape is said to be **right skewed**.[25]

Data sets with the reverse characteristic – a long, thin tail to the left – are said to be **left skewed**. We also say that such a distribution has a long left tail. Data sets that show roughly equal trailing off in both directions are called **symmetric**.

---

[25]Other ways to describe data that are skewed to the right: **skewed to the right**, **skewed to the high end**, or **skewed to the positive end**.

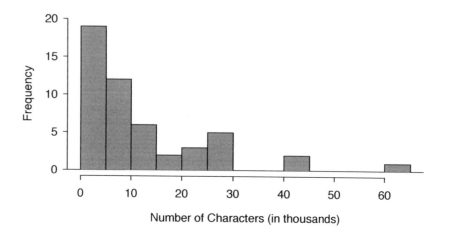

Figure 1.22: A histogram of `num_char`. This distribution is very strongly skewed to the right.

Long tails to identify skew

When data trail off in one direction, the distribution has a **long tail**. If a distribution has a long left tail, it is left skewed. If a distribution has a long right tail, it is right skewed.

⊙ **Guided Practice 1.24** Take a look at the dot plots in Figures 1.19 and 1.20. Can you see the skew in the data? Is it easier to see the skew in this histogram or the dot plots?[26]

⊙ **Guided Practice 1.25** Besides the mean (since it was labeled), what can you see in the dot plots that you cannot see in the histogram?[27]

In addition to looking at whether a distribution is skewed or symmetric, histograms can be used to identify modes. A **mode** is represented by a prominent peak in the distribution.[28] There is only one prominent peak in the histogram of `num_char`.

Figure 1.23 shows histograms that have one, two, or three prominent peaks. Such distributions are called **unimodal**, **bimodal**, and **multimodal**, respectively. Any distribution with more than 2 prominent peaks is called multimodal. Notice that there was one prominent peak in the unimodal distribution with a second less prominent peak that was not counted since it only differs from its neighboring bins by a few observations.

⊙ **Guided Practice 1.26** Figure 1.22 reveals only one prominent mode in the number of characters. Is the distribution unimodal, bimodal, or multimodal?[29]

---

[26]The skew is visible in all three plots, though the flat dot plot is the least useful. The stacked dot plot and histogram are helpful visualizations for identifying skew.

[27]Character counts for individual emails.

[28]Another definition of mode, which is not typically used in statistics, is the value with the most occurrences. It is common to have *no* observations with the same value in a data set, which makes this other definition useless for many real data sets.

[29]Unimodal. Remember that *uni* stands for 1 (think *unicycles*). Similarly, *bi* stands for 2 (think *bicycles*). (We're hoping a *multicycle* will be invented to complete this analogy.)

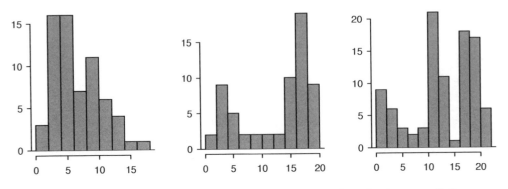

Figure 1.23: Counting only prominent peaks, the distributions are (left to right) unimodal, bimodal, and multimodal.

⊙ **Guided Practice 1.27**   Height measurements of young students and adult teachers at a K-3 elementary school were taken. How many modes would you anticipate in this height data set?[30]

TIP: Looking for modes

Looking for modes isn't about finding a clear and correct answer about the number of modes in a distribution, which is why *prominent* is not rigorously defined in this book. The important part of this examination is to better understand your data and how it might be structured.

1.6.4   Variance and standard deviation

The mean was introduced as a method to describe the center of a data set, but the variability in the data is also important. Here, we introduce two measures of variability: the variance and the standard deviation. Both of these are very useful in data analysis, even though their formulas are a bit tedious to calculate by hand. The standard deviation is the easier of the two to understand, and it roughly describes how far away the typical observation is from the mean.

We call the distance of an observation from its mean its **deviation**. Below are the deviations for the 1st, 2nd, 3rd, and 50th observations in the num_char variable. For computational convenience, the number of characters is listed in the thousands and rounded to the first decimal.

$$x_1 - \bar{x} = 21.7 - 11.6 = 10.1$$
$$x_2 - \bar{x} = 7.0 - 11.6 = -4.6$$
$$x_3 - \bar{x} = 0.6 - 11.6 = -11.0$$
$$\vdots$$
$$x_{50} - \bar{x} = 15.8 - 11.6 = 4.2$$

---

[30]There might be two height groups visible in the data set: one of the students and one of the adults. That is, the data are probably bimodal.

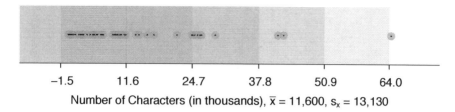

Number of Characters (in thousands), $\bar{x} = 11{,}600$, $s_x = 13{,}130$

Figure 1.24: In the num_char data, 41 of the 50 emails (82%) are within 1 standard deviation of the mean, and 47 of the 50 emails (94%) are within 2 standard deviations. Usually about 70% of the data are within 1 standard deviation of the mean and 95% are within 2 standard deviations, though this rule of thumb is less accurate for skewed data, as shown in this example.

If we square these deviations and then take an average, the result is about equal to the sample **variance**, denoted by s^2:

$$
\begin{aligned}
s^2 &= \frac{10.1^2 + (-4.6)^2 + (-11.0)^2 + \cdots + 4.2^2}{50 - 1} \\
&= \frac{102.01 + 21.16 + 121.00 + \cdots + 17.64}{49} \\
&= 172.44
\end{aligned}
$$

s^2
sample
variance

We divide by $n - 1$, rather than dividing by n, when computing the variance; you need not worry about this mathematical nuance for the material in this textbook. Notice that squaring the deviations does two things. First, it makes large values much larger, seen by comparing 10.1^2, $(-4.6)^2$, $(-11.0)^2$, and 4.2^2. Second, it gets rid of any negative signs.

The **standard deviation** is defined as the square root of the variance:

$$
s = \sqrt{172.44} = 13.13
$$

s
sample
standard
deviation

The standard deviation of the number of characters in an email is about 13.13 thousand. A subscript of x may be added to the variance and standard deviation, i.e. s_x^2 and s_x, as a reminder that these are the variance and standard deviation of the observations represented by $x_1, x_2, ..., x_n$. The x subscript is usually omitted when it is clear which data the variance or standard deviation is referencing.

Variance and standard deviation

The variance is roughly the average squared distance from the mean. The standard deviation is the square root of the variance. The standard deviation is useful when considering how close the data are to the mean.

Formulas and methods used to compute the variance and standard deviation for a population are similar to those used for a sample.[31] However, like the mean, the population values have special symbols: σ^2 for the variance and σ for the standard deviation. The symbol σ is the Greek letter *sigma*.

σ^2
population
variance

σ
population
standard
deviation

---

[31]The only difference is that the population variance has a division by n instead of $n - 1$.

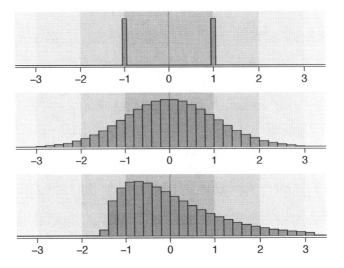

Figure 1.25: Three very different population distributions with the same mean $\mu = 0$ and standard deviation $\sigma = 1$.

TIP: standard deviation describes variability

Focus on the conceptual meaning of the standard deviation as a descriptor of variability rather than the formulas. Usually 70% of the data will be within one standard deviation of the mean and about 95% will be within two standard deviations. However, as seen in Figures 1.24 and 1.25, these percentages are not strict rules.

⊙ **Guided Practice 1.28**   On page 30, the concept of shape of a distribution was introduced. A good description of the shape of a distribution should include modality and whether the distribution is symmetric or skewed to one side. Using Figure 1.25 as an example, explain why such a description is important.[32]

● **Example 1.29**   Describe the distribution of the num_char variable using the histogram in Figure 1.22 on page 31. The description should incorporate the center, variability, and shape of the distribution, and it should also be placed in context: the number of characters in emails. Also note any especially unusual cases.

The distribution of email character counts is unimodal and very strongly skewed to the high end. Many of the counts fall near the mean at 11,600, and most fall within one standard deviation (13,130) of the mean. There is one exceptionally long email with about 65,000 characters.

In practice, the variance and standard deviation are sometimes used as a means to an end, where the "end" is being able to accurately estimate the uncertainty associated with a sample statistic. For example, in Chapter 4 we will use the variance and standard deviation to assess how close the sample mean is to the population mean.

---

[32]Figure 1.25 shows three distributions that look quite different, but all have the same mean, variance, and standard deviation. Using modality, we can distinguish between the first plot (bimodal) and the last two (unimodal). Using skewness, we can distinguish between the last plot (right skewed) and the first two. While a picture, like a histogram, tells a more complete story, we can use modality and shape (symmetry/skew) to characterize basic information about a distribution.

1.6.5   Box plots, quartiles, and the median

A **box plot** summarizes a data set using five statistics while also plotting unusual observations. Figure 1.26 provides a vertical dot plot alongside a box plot of the `num_char` variable from the `email50` data set.

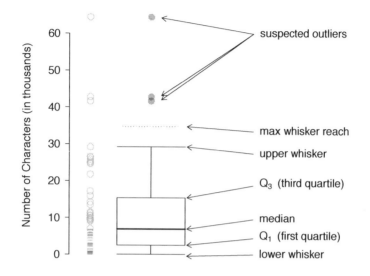

Figure 1.26: A vertical dot plot next to a labeled box plot for the number of characters in 50 emails. The median (6,890), splits the data into the bottom 50% and the top 50%, marked in the dot plot by horizontal dashes and open circles, respectively.

The first step in building a box plot is drawing a dark line denoting the **median**, which splits the data in half. Figure 1.26 shows 50% of the data falling below the median (dashes) and other 50% falling above the median (open circles). There are 50 character counts in the data set (an even number) so the data are perfectly split into two groups of 25. We take the median in this case to be the average of the two observations closest to the 50th percentile: $(6,768 + 7,012)/2 = 6,890$. When there are an odd number of observations, there will be exactly one observation that splits the data into two halves, and in this case that observation is the median (no average needed).

Median: the number in the middle

If the data are ordered from smallest to largest, the **median** is the observation right in the middle. If there are an even number of observations, there will be two values in the middle, and the median is taken as their average.

The second step in building a box plot is drawing a rectangle to represent the middle 50% of the data. The total length of the box, shown vertically in Figure 1.26, is called the **interquartile range** (IQR, for short). It, like the standard deviation, is a measure of variability in data. The more variable the data, the larger the standard deviation and IQR. The two boundaries of the box are called the **first quartile** (the 25th percentile, i.e. 25% of the data fall below this value) and the **third quartile** (the 75th percentile), and these are often labeled Q_1 and Q_3, respectively.

Interquartile range (IQR)

The IQR is the length of the box in a box plot. It is computed as

$$IQR = Q_3 - Q_1$$

where Q_1 and Q_3 are the 25th and 75th percentiles.

⊙ **Guided Practice 1.30** What percent of the data fall between Q_1 and the median? What percent is between the median and Q_3?[33]

Extending out from the box, the **whiskers** attempt to capture the data outside of the box, however, their reach is never allowed to be more than $1.5 \times IQR$.[34] They capture everything within this reach. In Figure 1.26, the upper whisker does not extend to the last three points, which is beyond $Q_3 + 1.5 \times IQR$, and so it extends only to the last point below this limit. The lower whisker stops at the lowest value, 33, since there is no additional data to reach; the lower whisker's limit is not shown in the figure because the plot does not extend down to $Q_1 - 1.5 \times IQR$. In a sense, the box is like the body of the box plot and the whiskers are like its arms trying to reach the rest of the data.

Any observation that lies beyond the whiskers is labeled with a dot. The purpose of labeling these points – instead of just extending the whiskers to the minimum and maximum observed values – is to help identify any observations that appear to be unusually distant from the rest of the data. Unusually distant observations are called **outliers**. In this case, it would be reasonable to classify the emails with character counts of 41,623, 42,793, and 64,401 as outliers since they are numerically distant from most of the data.

Outliers are extreme

An **outlier** is an observation that appears extreme relative to the rest of the data.

TIP: Why it is important to look for outliers

Examination of data for possible outliers serves many useful purposes, including

1. Identifying strong skew in the distribution.

2. Identifying data collection or entry errors. For instance, we re-examined the email purported to have 64,401 characters to ensure this value was accurate.

3. Providing insight into interesting properties of the data.

⊙ **Guided Practice 1.31** The observation 64,401, a suspected outlier, was found to be an accurate observation. What would such an observation suggest about the nature of character counts in emails?[35]

---

[33]Since Q_1 and Q_3 capture the middle 50% of the data and the median splits the data in the middle, 25% of the data fall between Q_1 and the median, and another 25% falls between the median and Q_3.

[34]While the choice of exactly 1.5 is arbitrary, it is the most commonly used value for box plots.

[35]That occasionally there may be very long emails.

⊙ **Guided Practice 1.32** Using Figure 1.26, estimate the following values for num_char in the email50 data set: (a) Q_1, (b) Q_3, and (c) IQR.[36]

> ┌───┐
> **▶ Calculator videos**
> Videos covering how to create statistical summaries and box plots using TI and Casio graphing calculators are available at openintro.org/videos.

1.6.6 Robust statistics

How are the sample statistics of the num_char data set affected by the observation, 64,401? What would have happened if this email wasn't observed? What would happen to these summary statistics if the observation at 64,401 had been even larger, say 150,000? These scenarios are plotted alongside the original data in Figure 1.27, and sample statistics are computed under each scenario in Table 1.28.

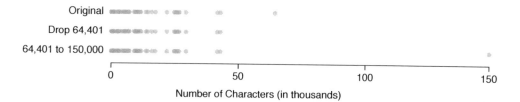

Figure 1.27: Dot plots of the original character count data and two modified data sets.

scenario	robust		not robust	
	median	IQR	$\bar{x}$	s
original num_char data	6,890	12,875	11,600	13,130
drop 64,401 observation	6,768	11,702	10,521	10,798
move 64,401 to 150,000	6,890	12,875	13,310	22,434

Table 1.28: A comparison of how the median, IQR, mean ($\bar{x}$), and standard deviation (s) change when extreme observations are present.

⊙ **Guided Practice 1.33** (a) Which is more affected by extreme observations, the mean or median? Table 1.28 may be helpful. (b) Is the standard deviation or IQR more affected by extreme observations?[37]

The median and IQR are called **robust estimates** because extreme observations have little effect on their values. The mean and standard deviation are much more affected by changes in extreme observations.

---

[36]These visual estimates will vary a little from one person to the next: $Q_1 = 3,000$, $Q_3 = 15,000$, IQR $= Q_3 - Q_1 = 12,000$. (The true values: $Q_1 = 2,536$, $Q_3 = 15,411$, IQR $= 12,875$.)

[37](a) Mean is affected more. (b) Standard deviation is affected more. Complete explanations are provided in the material following Guided Practice 1.33.

● **Example 1.34** The median and IQR do not change much under the three scenarios in Table 1.28. Why might this be the case?

The median and IQR are only sensitive to numbers near Q_1, the median, and Q_3. Since values in these regions are relatively stable – there aren't large jumps between observations – the median and IQR estimates are also quite stable.

⊙ **Guided Practice 1.35** The distribution of vehicle prices tends to be right skewed, with a few luxury and sports cars lingering out into the right tail. If you were searching for a new car and cared about price, should you be more interested in the mean or median price of vehicles sold, assuming you are in the market for a regular car?[38]

1.6.7 Transforming data (special topic)

When data are very strongly skewed, we sometimes transform them so they are easier to model. Consider the histogram of salaries for Major League Baseball players' salaries from 2010, which is shown in Figure 1.29(a).

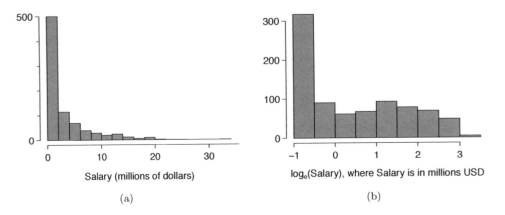

Figure 1.29: (a) Histogram of MLB player salaries for 2010, in millions of dollars. (b) Histogram of the log-transformed MLB player salaries for 2010.

● **Example 1.36** The histogram of MLB player salaries is useful in that we can see the data are extremely skewed and centered (as gauged by the median) at about \$1 million. What isn't useful about this plot?

Most of the data are collected into one bin in the histogram and the data are so strongly skewed that many details in the data are obscured.

There are some standard transformations that are often applied when much of the data cluster near zero (relative to the larger values in the data set) and all observations are positive. A **transformation** is a rescaling of the data using a function. For instance, a plot of the natural logarithm[39] of player salaries results in a new histogram in Figure 1.29(b).

---

[38]Buyers of a "regular car" should be concerned about the median price. High-end car sales can drastically inflate the mean price while the median will be more robust to the influence of those sales.

[39]Statisticians often write the natural logarithm as log. You might be more familiar with it being written as ln.

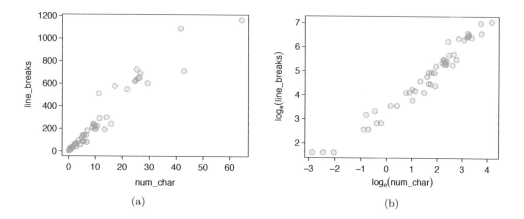

Figure 1.30: (a) Scatterplot of `line_breaks` against `num_char` for 50 emails. (b) A scatterplot of the same data but where each variable has been log-transformed.

Transformed data are sometimes easier to work with when applying statistical models because the transformed data are much less skewed and outliers are usually less extreme.

Transformations can also be applied to one or both variables in a scatterplot. A scatterplot of the `line_breaks` and `num_char` variables is shown in Figure 1.30(a), which was earlier shown in Figure 1.17. We can see a positive association between the variables and that many observations are clustered near zero. In Chapter 7, we might want to use a straight line to model the data. However, we'll find that the data in their current state cannot be modeled very well. Figure 1.30(b) shows a scatterplot where both the `line_breaks` and `num_char` variables have been transformed using a log (base e) transformation. While there is a positive association in each plot, the transformed data show a steadier trend, which is easier to model than the untransformed data.

Transformations other than the logarithm can be useful, too. For instance, the square root ($\sqrt{\text{original observation}}$) and inverse ($\frac{1}{\text{original observation}}$) are used by statisticians. Common goals in transforming data are to see the data structure differently, reduce skew, assist in modeling, or straighten a nonlinear relationship in a scatterplot.

1.6.8 Mapping data (special topic)

The `county` data set offers many numerical variables that we could plot using dot plots, scatterplots, or box plots, but these miss the true nature of the data. Rather, when we encounter geographic data, we should map it using an **intensity map**, where colors are used to show higher and lower values of a variable. Figures 1.31 and 1.32 shows intensity maps for federal spending per capita (`fed_spend`), poverty rate in percent (`poverty`), homeownership rate in percent (`homeownership`), and median household income (`med_income`). The color key indicates which colors correspond to which values. Note that the intensity maps are not generally very helpful for getting precise values in any given county, but they are very helpful for seeing geographic trends and generating interesting research questions.

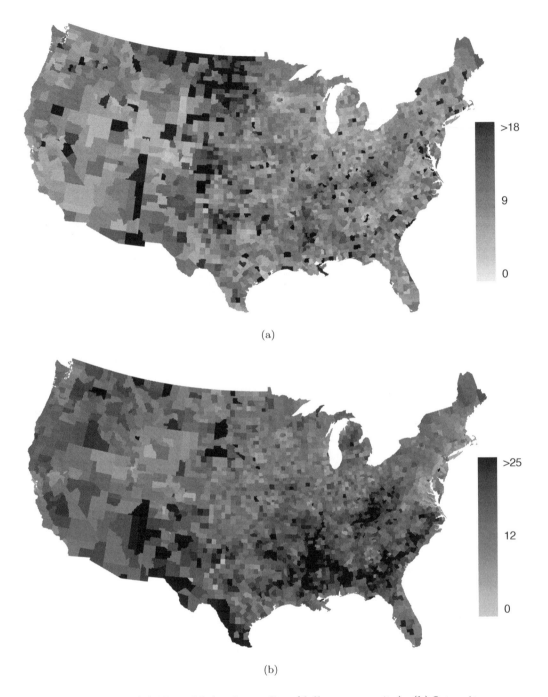

Figure 1.31: (a) Map of federal spending (dollars per capita). (b) Intensity map of poverty rate (percent).

(a)

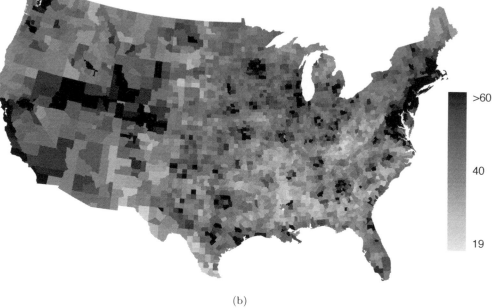

(b)

Figure 1.32: (a) Intensity map of homeownership rate (percent). (b) Intensity map of median household income ($1000s).

● **Example 1.37** What interesting features are evident in the fed_spend and poverty intensity maps?

The federal spending intensity map shows substantial spending in the Dakotas and along the central-to-western part of the Canadian border, which may be related to the oil boom in this region. There are several other patches of federal spending, such as a vertical strip in eastern Utah and Arizona and the area where Colorado, Nebraska, and Kansas meet. There are also seemingly random counties with very high federal spending relative to their neighbors. If we did not cap the federal spending range at $18 per capita, we would actually find that some counties have extremely high federal spending while there is almost no federal spending in the neighboring counties. These high-spending counties might contain military bases, companies with large government contracts, or other government facilities with many employees.

Poverty rates are evidently higher in a few locations. Notably, the deep south shows higher poverty rates, as does the southwest border of Texas. The vertical strip of eastern Utah and Arizona, noted above for its higher federal spending, also appears to have higher rates of poverty (though generally little correspondence is seen between the two variables). High poverty rates are evident in the Mississippi flood plains a little north of New Orleans and also in a large section of Kentucky and West Virginia.

⊙ **Guided Practice 1.38** What interesting features are evident in the med_income intensity map in Figure 1.32(b)?[40]

---

[40]Note: answers will vary. There is a very strong correspondence between high earning and metropolitan areas. You might look for large cities you are familiar with and try to spot them on the map as dark spots.

1.7 Considering categorical data 📹

Like numerical data, categorical data can also be organized and analyzed. In this section, we will introduce tables and other basic tools for categorical data that are used throughout this book. The email50 data set represents a sample from a larger email data set called email. This larger data set contains information on 3,921 emails. In this section we will examine whether the presence of numbers, small or large, in an email provides any useful value in classifying email as spam or not spam.

1.7.1 Contingency tables and bar plots

Table 1.33 summarizes two variables: spam and number. Recall that number is a categorical variable that describes whether an email contains no numbers, only small numbers (values under 1 million), or at least one big number (a value of 1 million or more). A table that summarizes data for two categorical variables in this way is called a **contingency table**. Each value in the table represents the number of times a particular combination of variable outcomes occurred. For example, the value 149 corresponds to the number of emails in the data set that are spam *and* had no number listed in the email. Row and column totals are also included. The **row totals** provide the total counts across each row (e.g. $149 + 168 + 50 = 367$), and **column totals** are total counts down each column.

A table for a single variable is called a **frequency table**. Table 1.34 is a frequency table for the number variable. If we replaced the counts with percentages or proportions, the table would be called a **relative frequency table**.

		number			
		none	small	big	Total
spam	spam	149	168	50	367
	not spam	400	2659	495	3554
	Total	549	2827	545	3921

Table 1.33: A contingency table for spam and number.

none	small	big	Total
549	2827	545	3921

Table 1.34: A frequency table for the number variable.

A bar plot is a common way to display a single categorical variable. The left panel of Figure 1.35 shows a **bar plot** for the number variable. In the right panel, the counts are converted into proportions (e.g. $549/3921 = 0.140$ for none), showing the proportion of observations that are in each level (i.e. in each category).

1.7.2 Row and column proportions

Table 1.36 shows the row proportions for Table 1.33. The **row proportions** are computed as the counts divided by their row totals. The value 149 at the intersection of spam and none is replaced by $149/367 = 0.406$, i.e. 149 divided by its row total, 367. So what does 0.406 represent? It corresponds to the proportion of spam emails in the sample that do not have any numbers.

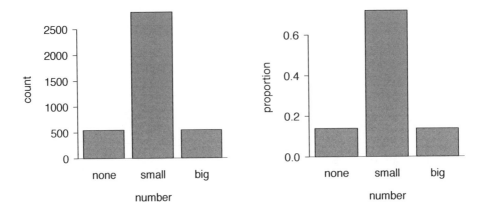

Figure 1.35: Two bar plots of number. The left panel shows the counts, and the right panel shows the proportions in each group.

	none	small	big	Total
spam	$149/367 = 0.406$	$168/367 = 0.458$	$50/367 = 0.136$	1.000
not spam	$400/3554 = 0.113$	$2657/3554 = 0.748$	$495/3554 = 0.139$	1.000
Total	$549/3921 = 0.140$	$2827/3921 = 0.721$	$545/3921 = 0.139$	1.000

Table 1.36: A contingency table with row proportions for the spam and number variables.

A contingency table of the column proportions is computed in a similar way, where each **column proportion** is computed as the count divided by the corresponding column total. Table 1.37 shows such a table, and here the value 0.271 indicates that 27.1% of emails with no numbers were spam. This rate of spam is much higher compared to emails with only small numbers (5.9%) or big numbers (9.2%). Because these spam rates vary between the three levels of number (none, small, big), this provides evidence that the spam and number variables are associated.

	none	small	big	Total
spam	$149/549 = 0.271$	$168/2827 = 0.059$	$50/545 = 0.092$	$367/3921 = 0.094$
not spam	$400/549 = 0.729$	$2659/2827 = 0.941$	$495/545 = 0.908$	$3684/3921 = 0.906$
Total	1.000	1.000	1.000	1.000

Table 1.37: A contingency table with column proportions for the spam and number variables.

We could also have checked for an association between spam and number in Table 1.36 using row proportions. When comparing these row proportions, we would look down columns to see if the fraction of emails with no numbers, small numbers, and big numbers varied from spam to not spam.

⊙ **Guided Practice 1.39**   What does 0.458 represent in Table 1.36? What does 0.059 represent in Table 1.37?[41]

---

[41] 0.458 represents the proportion of spam emails that had a small number. 0.059 represents the fraction of emails with small numbers that are spam.

⊙ **Guided Practice 1.40** What does 0.139 at the intersection of `not spam` and `big` represent in Table 1.36? What does 0.908 represent in the Table 1.37?[42]

● **Example 1.41** Data scientists use statistics to filter spam from incoming email messages. By noting specific characteristics of an email, a data scientist may be able to classify some emails as spam or not spam with high accuracy. One of those characteristics is whether the email contains no numbers, small numbers, or big numbers. Another characteristic is whether or not an email has any HTML content. A contingency table for the `spam` and `format` variables from the `email` data set are shown in Table 1.38. Recall that an HTML email is an email with the capacity for special formatting, e.g. bold text. In Table 1.38, which would be more helpful to someone hoping to classify email as spam or regular email: row or column proportions?

Such a person would be interested in how the proportion of spam changes within each email format. This corresponds to column proportions: the proportion of spam in plain text emails and the proportion of spam in HTML emails.

If we generate the column proportions, we can see that a higher fraction of plain text emails are spam ($209/1195 = 17.5\%$) than compared to HTML emails ($158/2726 = 5.8\%$). This information on its own is insufficient to classify an email as spam or not spam, as over 80% of plain text emails are not spam. Yet, when we carefully combine this information with many other characteristics, such as `number` and other variables, we stand a reasonable chance of being able to classify some email as spam or not spam. This is a topic we will return to in Chapter 8.

	text	HTML	Total
spam	209	158	367
not spam	986	2568	3554
Total	1195	2726	3921

Table 1.38: A contingency table for `spam` and `format`.

Example 1.41 points out that row and column proportions are not equivalent. Before settling on one form for a table, it is important to consider each to ensure that the most useful table is constructed.

⊙ **Guided Practice 1.42** Look back to Tables 1.36 and 1.37. Which would be more useful to someone hoping to identify spam emails using the `number` variable?[43]

---

[42]0.139 represents the fraction of non-spam email that had a big number. 0.908 represents the fraction of emails with big numbers that are non-spam emails.

[43]The column proportions in Table 1.37 will probably be most useful, which makes it easier to see that emails with small numbers are spam about 5.9% of the time (relatively rare). We would also see that about 27.1% of emails with no numbers are spam, and 9.2% of emails with big numbers are spam.

1.7.3 Segmented bar and mosaic plots

Contingency tables using row or column proportions are especially useful for examining how two categorical variables are related. Segmented bar and mosaic plots provide a way to visualize the information in these tables.

A **segmented bar plot** is a graphical display of contingency table information. For example, a segmented bar plot representing Table 1.37 is shown in Figure 1.39(a), where we have first created a bar plot using the `number` variable and then divided each group by the levels of `spam`. The column proportions of Table 1.37 have been translated into a standardized segmented bar plot in Figure 1.39(b), which is a helpful visualization of the fraction of spam emails in each level of `number`.

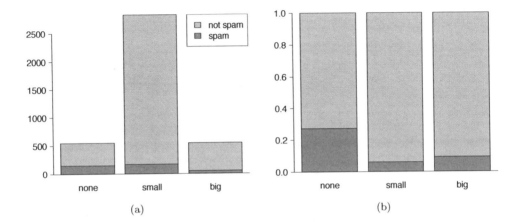

Figure 1.39: (a) Segmented bar plot for numbers found in emails, where the counts have been further broken down by `spam`. (b) Standardized version of Figure (a).

● **Example 1.43** Examine both of the segmented bar plots. Which is more useful?

Figure 1.39(a) contains more information, but Figure 1.39(b) presents the information more clearly. This second plot makes it clear that emails with no number have a relatively high rate of spam email – about 27%! On the other hand, less than 10% of email with small or big numbers are spam.

Since the proportion of spam changes across the groups in Figure 1.39(b), we can conclude the variables are dependent, which is something we were also able to discern using table proportions. Because both the `none` and `big` groups have relatively few observations compared to the `small` group, the association is more difficult to see in Figure 1.39(a).

In some other cases, a segmented bar plot that is not standardized will be more useful in communicating important information. Before settling on a particular segmented bar plot, create standardized and non-standardized forms and decide which is more effective at communicating features of the data.

A **mosaic plot** is a graphical display of contingency table information that is similar to a bar plot for one variable or a segmented bar plot when using two variables. Figure 1.40(a) shows a mosaic plot for the `number` variable. Each column represents a level of `number`, and the column widths correspond to the proportion of emails for each number type.

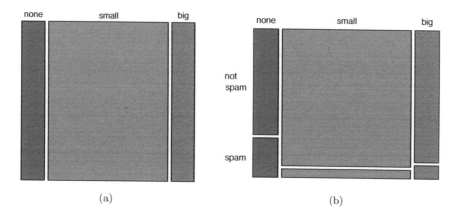

Figure 1.40: The one-variable mosaic plot for number and the two-variable mosaic plot for both number and spam.

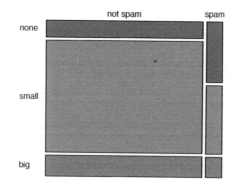

Figure 1.41: Mosaic plot where emails are grouped by the number variable after they've been divided into spam and not spam.

For instance, there are fewer emails with no numbers than emails with only small numbers, so the no number email column is slimmer. In general, mosaic plots use box *areas* to represent the number of observations that box represents.

This one-variable mosaic plot is further divided into pieces in Figure 1.40(b) using the spam variable. Each column is split proportionally according to the fraction of emails that were spam in each number category. For example, the second column, representing emails with only small numbers, was divided into emails that were spam (lower) and not spam (upper). As another example, the bottom of the third column represents spam emails that had big numbers, and the upper part of the third column represents regular emails that had big numbers. We can again use this plot to see that the spam and number variables are associated since some columns are divided in different vertical locations than others, which was the same technique used for checking an association in the standardized version of the segmented bar plot.

In a similar way, a mosaic plot representing row proportions of Table 1.33 could be constructed, as shown in Figure 1.41. However, because it is more insightful for this application to consider the fraction of spam in each category of the number variable, we prefer Figure 1.40(b).

1.7.4 The only pie chart you will see in this book

While pie charts are well known, they are not typically as useful as other charts in a data analysis. A **pie chart** is shown in Figure 1.42 alongside a bar plot. It is generally more difficult to compare group sizes in a pie chart than in a bar plot, especially when categories have nearly identical counts or proportions. In the case of the `none` and `big` categories, the difference is so slight you may be unable to distinguish any difference in group sizes for either plot!

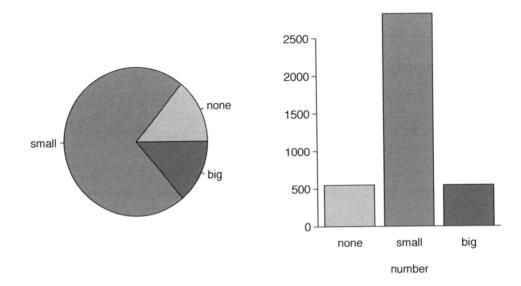

Figure 1.42: A pie chart and bar plot of `number` for the `email` data set.

1.7.5 Comparing numerical data across groups

Some of the more interesting investigations can be considered by examining numerical data across groups. The methods required here aren't really new. All that is required is to make a numerical plot for each group. Here two convenient methods are introduced: side-by-side box plots and hollow histograms.

We will take a look again at the `county` data set and compare the median household income for counties that gained population from 2000 to 2010 versus counties that had no gain. While we might like to make a causal connection here, remember that these are observational data and so such an interpretation would be unjustified.

There were 2,041 counties where the population increased from 2000 to 2010, and there were 1,099 counties with no gain (all but one were a loss). A random sample of 100 counties from the first group and 50 from the second group are shown in Table 1.43 to give a better sense of some of the raw data.

The **side-by-side box plot** is a traditional tool for comparing across groups. An example is shown in the left panel of Figure 1.44, where there are two box plots, one for each group, placed into one plotting window and drawn on the same scale.

Another useful plotting method uses **hollow histograms** to compare numerical data across groups. These are just the outlines of histograms of each group put on the same plot, as shown in the right panel of Figure 1.44.

population gain						no gain		
41.2	33.1	30.4	37.3	79.1	34.5	40.3	33.5	34.8
22.9	39.9	31.4	45.1	50.6	59.4	29.5	31.8	41.3
47.9	36.4	42.2	43.2	31.8	36.9	28	39.1	42.8
50.1	27.3	37.5	53.5	26.1	57.2	38.1	39.5	22.3
57.4	42.6	40.6	48.8	28.1	29.4	43.3	37.5	47.1
43.8	26	33.8	35.7	38.5	42.3	43.7	36.7	36
41.3	40.5	68.3	31	46.7	30.5	35.8	38.7	39.8
68.3	48.3	38.7	62	37.6	32.2	46	42.3	48.2
42.6	53.6	50.7	35.1	30.6	56.8	38.6	31.9	31.1
66.4	41.4	34.3	38.9	37.3	41.7	37.6	29.3	30.1
51.9	83.3	46.3	48.4	40.8	42.6	57.5	32.6	31.1
44.5	34	48.7	45.2	34.7	32.2	46.2	26.5	40.1
39.4	38.6	40	57.3	45.2	33.1	38.4	46.7	25.9
43.8	71.7	45.1	32.2	63.3	54.7	36.4	41.5	45.7
71.3	36.3	36.4	41	37	66.7	39.7	37	37.7
50.2	45.8	45.7	60.2	53.1		21.4	29.3	50.1
35.8	40.4	51.5	66.4	36.1		43.6	39.8	

Table 1.43: In this table, median household income (in \$1000s) from a random sample of 100 counties that gained population over 2000-2010 are shown on the left. Median incomes from a random sample of 50 counties that had no population gain are shown on the right.

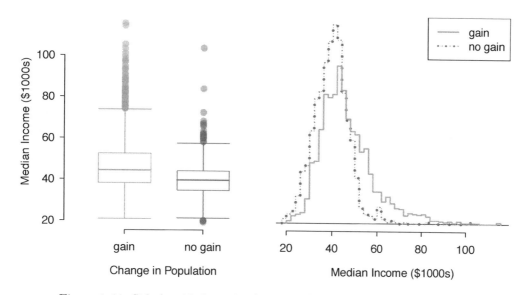

Figure 1.44: Side-by-side box plot (left panel) and hollow histograms (right panel) for med_income, where the counties are split by whether there was a population gain or loss from 2000 to 2010. The income data were collected between 2006 and 2010.

⊙ **Guided Practice 1.44** Use the plots in Figure 1.44 to compare the incomes for counties across the two groups. What do you notice about the approximate center of each group? What do you notice about the variability between groups? Is the shape relatively consistent between groups? How many *prominent* modes are there for each group?[44]

⊙ **Guided Practice 1.45** What components of each plot in Figure 1.44 do you find most useful?[45]

1.8 Case study: gender discrimination 📷📹 (special topic)

● **Example 1.46** Suppose your professor splits the students in class into two groups: students on the left and students on the right. If $\hat{p}_L$ and $\hat{p}_R$ represent the proportion of students who own an Apple product on the left and right, respectively, would you be surprised if $\hat{p}_L$ did not exactly equal $\hat{p}_R$?

While the proportions would probably be close to each other, it would be unusual for them to be exactly the same. We would probably observe a small difference due to chance.

⊙ **Guided Practice 1.47** If we don't think the side of the room a person sits on in class is related to whether the person owns an Apple product, what assumption are we making about the relationship between these two variables?[46]

1.8.1 Variability within data

We consider a study investigating gender discrimination in the 1970s, which is set in the context of personnel decisions within a bank.[47] The research question we hope to answer is, "Are females unfairly discriminated against in promotion decisions made by male managers?"

The participants in this study are 48 male bank supervisors attending a management institute at the University of North Carolina in 1972. They were asked to assume the role of the personnel director of a bank and were given a personnel file to judge whether the person should be promoted to a branch manager position. The files given to the participants were identical, except that half of them indicated the candidate was male and the other half indicated the candidate was female. These files were randomly assigned to the subjects.

---

[44]Answers may vary a little. The counties with population gains tend to have higher income (median of about $45,000) versus counties without a gain (median of about $40,000). The variability is also slightly larger for the population gain group. This is evident in the IQR, which is about 50% bigger in the *gain* group. Both distributions show slight to moderate right skew and are unimodal. There is a secondary small bump at about $60,000 for the *no gain* group, visible in the hollow histogram plot, that seems out of place. (Looking into the data set, we would find that 8 of these 15 counties are in Alaska and Texas.) The box plots indicate there are many observations far above the median in each group, though we should anticipate that many observations will fall beyond the whiskers when using such a large data set.

[45]Answers will vary. The side-by-side box plots are especially useful for comparing centers and spreads, while the hollow histograms are more useful for seeing distribution shape, skew, and groups of anomalies.

[46]We would be assuming that these two variables are independent.

[47]Rosen B and Jerdee T. 1974. Influence of sex role stereotypes on personnel decisions. Journal of Applied Psychology 59(1):9-14.

⊙ **Guided Practice 1.48**   Is this an observational study or an experiment? What implications does the study type have on what can be inferred from the results?[48]

For each supervisor we record the gender associated with the assigned file and the promotion decision. Using the results of the study summarized in Table 1.45, we would like to evaluate if females are unfairly discriminated against in promotion decisions. In this study, a smaller proportion of females are promoted than males (0.583 versus 0.875), but it is unclear whether the difference provides *convincing evidence* that females are unfairly discriminated against.

		decision		
		promoted	not promoted	Total
gender	male	21	3	24
	female	14	10	24
	Total	35	13	48

Table 1.45: Summary results for the gender discrimination study.

● **Example 1.49**   Statisticians are sometimes called upon to evaluate the strength of evidence. When looking at the rates of promotion for males and females in this study, what comes to mind as we try to determine whether the data show convincing evidence of a real difference?

The observed promotion rates (58.3% for females versus 87.5% for males) suggest there might be discrimination against women in promotion decisions. However, we cannot be sure if the observed difference represents discrimination or is just from random chance. Generally there is a little bit of fluctuation in sample data, and we wouldn't expect the sample proportions to be *exactly* equal, even if the truth was that the promotion decisions were independent of gender.

Example 1.49 is a reminder that the observed outcomes in the sample may not perfectly reflect the true relationships between variables in the underlying population. Table 1.45 shows there were 7 fewer promotions in the female group than in the male group, a difference in promotion rates of 29.2% $\left(\frac{21}{24} - \frac{14}{24} = 0.292\right)$. This difference is large, but the sample size for the study is small, making it unclear if this observed difference represents discrimination or whether it is simply due to chance. We label these two competing claims, H_0 and H_A:

H_0: **Independence model.** The variables gender and decision are independent. They have no relationship, and the observed difference between the proportion of males and females who were promoted, 29.2%, was due to chance.

H_A: **Alternative model.** The variables gender and decision are *not* independent. The difference in promotion rates of 29.2% was not due to chance, and equally qualified females are less likely to be promoted than males.

What would it mean if the independence model, which says the variables gender and decision are unrelated, is true? It would mean each banker was going to decide whether to promote the candidate without regard to the gender indicated on the file. That is,

---

[48]The study is an experiment, as subjects were randomly assigned a male file or a female file. Since this is an experiment, the results can be used to evaluate a causal relationship between gender of a candidate and the promotion decision.

the difference in the promotion percentages was due to the way the files were randomly divided to the bankers, and the randomization just happened to give rise to a relatively large difference of 29.2%.

Consider the alternative model: bankers were influenced by which gender was listed on the personnel file. If this was true, and especially if this influence was substantial, we would expect to see some difference in the promotion rates of male and female candidates. If this gender bias was against females, we would expect a smaller fraction of promotion decisions for female personnel files relative to the male files.

We choose between these two competing claims by assessing if the data conflict so much with H_0 that the independence model cannot be deemed reasonable. If this is the case, and the data support H_A, then we will reject the notion of independence and conclude there was discrimination.

1.8.2   Simulating the study

Table 1.45 shows that 35 bank supervisors recommended promotion and 13 did not. Now, suppose the bankers' decisions were independent of gender. Then, if we conducted the experiment again with a different random arrangement of files, differences in promotion rates would be based only on random fluctuation. We can actually perform this **randomization**, which simulates what would have happened if the bankers' decisions had been independent of gender but we had distributed the files differently.

In this **simulation**, we thoroughly shuffle 48 personnel files, 24 labeled `male_sim` and 24 labeled `female_sim`, and deal these files into two stacks. We will deal 35 files into the first stack, which will represent the 35 supervisors who recommended promotion. The second stack will have 13 files, and it will represent the 13 supervisors who recommended against promotion. Then, as we did with the original data, we tabulate the results and determine the fraction of `male_sim` and `female_sim` who were promoted. The randomization of files in this simulation is independent of the promotion decisions, which means any difference in the two fractions is entirely due to chance. Table 1.46 show the results of such a simulation.

| | | decision | | |
		promoted	not promoted	Total
	`male_sim`	18	6	24
`gender_sim`	`female_sim`	17	7	24
	Total	35	13	48

Table 1.46: Simulation results, where any difference in promotion rates between `male_sim` and `female_sim` is purely due to chance.

⊙ **Guided Practice 1.50**   What is the difference in promotion rates between the two simulated groups in Table 1.46? How does this compare to the observed 29.2% in the actual groups?[49]

---

[49]$18/24 - 17/24 = 0.042$ or about 4.2% in favor of the men. This difference due to chance is much smaller than the difference observed in the actual groups.

1.8.3 Checking for independence

We computed one possible difference under the independence model in Guided Practice 1.50, which represents one difference due to chance. While in this first simulation, we physically dealt out files, it is more efficient to perform this simulation using a computer. Repeating the simulation on a computer, we get another difference due to chance: -0.042. And another: 0.208. And so on until we repeat the simulation enough times that we have a good idea of what represents the *distribution of differences from chance alone.* Figure 1.47 shows a plot of the differences found from 100 simulations, where each dot represents a simulated difference between the proportions of male and female files that were recommended for promotion.

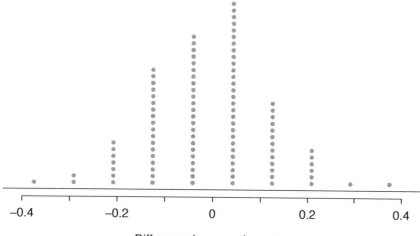

Figure 1.47: A stacked dot plot of differences from 100 simulations produced under the independence model, H_0, where gender_sim and decision are independent. Two of the 100 simulations had a difference of at least 29.2%, the difference observed in the study.

Note that the distribution of these simulated differences is centered around 0. We simulated these differences assuming that the independence model was true, and under this condition, we expect the difference to be zero with some random fluctuation. We would generally be surprised to see a difference of *exactly* 0: sometimes, just by chance, the difference is higher than 0, and other times it is lower than zero.

● **Example 1.51** How often would you observe a difference of at least 29.2% (0.292) according to Figure 1.47? Often, sometimes, rarely, or never?

It appears that a difference of at least 29.2% due to chance alone would only happen about 2% of the time according to Figure 1.47. Such a low probability indicates a rare event.

The difference of 29.2% being a rare event suggests two possible interpretations of the results of the study:

H_0 **Independence model.** Gender has no effect on promotion decision, and we observed a difference that would only happen rarely.

H_A **Alternative model.** Gender has an effect on promotion decision, and what we observed was actually due to equally qualified women being discriminated against in promotion decisions, which explains the large difference of 29.2%.

Based on the simulations, we have two options. (1) We conclude that the study results do not provide strong evidence against the independence model. That is, we do not have sufficiently strong evidence to conclude there was gender discrimination. (2) We conclude the evidence is sufficiently strong to reject H_0 and assert that there was gender discrimination. When we conduct formal studies, usually we reject the notion that we just happened to observe a rare event.[50] So in this case, we reject the independence model in favor of the alternative. That is, we are concluding the data provide strong evidence of gender discrimination against women by the supervisors.

One field of statistics, statistical inference, is built on evaluating whether such differences are due to chance. In statistical inference, statisticians evaluate which model is most reasonable given the data. Errors do occur, just like rare events, and we might choose the wrong model. While we do not always choose correctly, statistical inference gives us tools to control and evaluate how often these errors occur. In Chapter 4, we give a formal introduction to the problem of model selection. We spend the next two chapters building a foundation of probability and theory necessary to make that discussion rigorous.

---

[50]This reasoning does not generally extend to anecdotal observations. Each of us observes incredibly rare events every day, events we could not possibly hope to predict. However, in the non-rigorous setting of anecdotal evidence, almost anything may appear to be a rare event, so the idea of looking for rare events in day-to-day activities is treacherous. For example, we might look at the lottery: there was only a 1 in 176 million chance that the Mega Millions numbers for the largest jackpot in history (March 30, 2012) would be (2, 4, 23, 38, 46) with a Mega ball of (23), but nonetheless those numbers came up! However, no matter what numbers had turned up, they would have had the same incredibly rare odds. That is, *any set of numbers we could have observed would ultimately be incredibly rare.* This type of situation is typical of our daily lives: each possible event in itself seems incredibly rare, but if we consider every alternative, those outcomes are also incredibly rare. We should be cautious not to misinterpret such anecdotal evidence.

1.9 Exercises

1.9.1 Case study: using stents to prevent strokes

1.1 Migraine and acupuncture. A migraine is a particularly painful type of headache, which patients sometimes wish to treat with acupuncture. To determine whether acupuncture relieves migraine pain, researchers conducted a randomized controlled study where 89 females diagnosed with migraine headaches were randomly assigned to one of two groups: treatment or control. 43 patients in the treatment group received acupuncture that is specifically designed to treat migraines. 46 patients in the control group received placebo acupuncture (needle insertion at non-acupoint locations). 24 hours after patients received acupuncture, they were asked if they were pain free. Results are summarized in the contingency table below.[51]

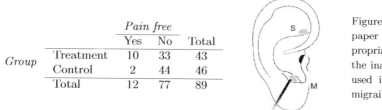

		Pain free		
		Yes	No	Total
Group	Treatment	10	33	43
	Control	2	44	46
	Total	12	77	89

Figure from the original paper displaying the appropriate area (M) versus the inappropriate area (S) used in the treatment of migraine attacks.

(a) What percent of patients in the treatment group were pain free 24 hours after receiving acupuncture? What percent in the control group?

(b) At first glance, does acupuncture appear to be an effective treatment for migraines? Explain your reasoning.

(c) Do the data provide convincing evidence that there is a real pain reduction for those patients in the treatment group? Or do you think that the observed difference might just be due to chance?

1.2 Sinusitis and antibiotics. Researchers studying the effect of antibiotic treatment for acute sinusitis compared to symptomatic treatments randomly assigned 166 adults diagnosed with acute sinusitis to one of two groups: treatment or control. Study participants received either a 10-day course of amoxicillin (an antibiotic) or a placebo similar in appearance and taste. The placebo consisted of symptomatic treatments such as acetaminophen, nasal decongestants, etc. At the end of the 10-day period patients were asked if they experienced significant improvement in symptoms. The distribution of responses is summarized below.[52]

		Self-reported significant improvement in symptoms		
		Yes	No	Total
Group	Treatment	66	19	85
	Control	65	16	81
	Total	131	35	166

(a) What percent of patients in the treatment group experienced a significant improvement in symptoms? What percent in the control group?

(b) Based on your findings in part (a), which treatment appears to be more effective for sinusitis?

(c) Do the data provide convincing evidence that there is a difference in the improvement rates of sinusitis symptoms? Or do you think that the observed difference might just be due to chance?

---

[51]G. Allais et al. "Ear acupuncture in the treatment of migraine attacks: a randomized trial on the efficacy of appropriate versus inappropriate acupoints". In: *Neurological Sci.* 32.1 (2011), pp. 173–175.

[52]J.M. Garbutt et al. "Amoxicillin for Acute Rhinosinusitis: A Randomized Controlled Trial". In: *JAMA: The Journal of the American Medical Association* 307.7 (2012), pp. 685–692.

1.9.2   Data basics

1.3   Air pollution and birth outcomes, study components. Researchers collected data to examine the relationship between air pollutants and preterm births in Southern California. During the study air pollution levels were measured by air quality monitoring stations. Specifically, levels of carbon monoxide were recorded in parts per million, nitrogen dioxide and ozone in parts per hundred million, and coarse particulate matter (PM_{10}) in $\mu g/m^3$. Length of gestation data were collected on 143,196 births between the years 1989 and 1993, and air pollution exposure during gestation was calculated for each birth. The analysis suggested that increased ambient PM_{10} and, to a lesser degree, CO concentrations may be associated with the occurrence of preterm births.[53]. Identify

(a) the cases,

(b) the variables and their types, and

(c) the main research question

in this study.

1.4   Buteyko method, study components. The Buteyko method is a shallow breathing technique developed by Konstantin Buteyko, a Russian doctor, in 1952. Anecdotal evidence suggests that the Buteyko method can reduce asthma symptoms and improve quality of life. In a scientific study to determine the effectiveness of this method, researchers recruited 600 asthma patients aged 18-69 who relied on medication for asthma treatment. These patients were split into two research groups: one practiced the Buteyko method and the other did not. Patients were scored on quality of life, activity, asthma symptoms, and medication reduction on a scale from 0 to 10. On average, the participants in the Buteyko group experienced a significant reduction in asthma symptoms and an improvement in quality of life.[54]. Identify

(a) the cases,

(b) the variables and their types, and

(c) the main research question

in this study.

1.5   Cheaters, study components. Researchers studying the relationship between honesty, age and self-control conducted an experiment on 160 children between the ages of 5 and 15. Participants reported their age, sex, and whether they were an only child or not. The researchers asked each child to toss a fair coin in private and to record the outcome (white or black) on a paper sheet, and said they would only reward children who report white. Half the students were explicitly told not to cheat and the others were not given any explicit instructions. In the no instruction group probability of cheating was found to be uniform across groups based on child's characteristics. In the group that was explicitly told to not cheat, girls were less likely to cheat, and while rate of cheating didn't vary by age for boys, it decreased with age for girls.[55] Identify

(a) the cases,

(b) the variables and their types, and

(c) the main research question

in this study.

---

[53]B. Ritz et al. "Effect of air pollution on preterm birth among children born in Southern California between 1989 and 1993". In: *Epidemiology* 11.5 (2000), pp. 502–511.

[54]J. McGowan. "Health Education: Does the Buteyko Institute Method make a difference?" In: *Thorax* 58 (2003).

[55]Alessandro Bucciol and Marco Piovesan. "Luck or cheating? A field experiment on honesty with children". In: *Journal of Economic Psychology* 32.1 (2011), pp. 73–78.

1.6 Stealers, study components. In a study of the relationship between socio-economic class and unethical behavior, 129 University of California undergraduates at Berkeley were asked to identify themselves as having low or high social-class by comparing themselves to others with the most (least) money, most (least) education, and most (least) respected jobs. They were also presented with a jar of individually wrapped candies and informed that the candies were for children in a nearby laboratory, but that they could take some if they wanted. After completing some unrelated tasks, participants reported the number of candies they had taken. It was found that those who were identified as upper-class took more candy than others.[56] Identify

(a) the cases,

(b) the variables and their types, and

(c) the main research question

in this study.

1.7 Fisher's irises. Sir Ronald Aylmer Fisher was an English statistician, evolutionary biologist, and geneticist who worked on a data set that contained sepal length and width, and petal length and width from three species of iris flowers (*setosa*, *versicolor* and *virginica*). There were 50 flowers from each species in the data set.[57]

(a) How many cases were included in the data?

(b) How many numerical variables are included in the data? Indicate what they are, and if they are continuous or discrete.

(c) How many categorical variables are included in the data, and what are they? List the corresponding levels (categories).

Photo by Ryan Claussen (http://flic.kr/p/6QTcuX) CC BY-SA 2.0 license

1.8 Smoking habits of UK residents. A survey was conducted to study the smoking habits of UK residents. Below is a data matrix displaying a portion of the data collected in this survey. Note that "£" stands for British Pounds Sterling, "cig" stands for cigarettes, and "N/A" refers to a missing component of the data.[58]

	sex	age	marital	grossIncome	smoke	amtWeekends	amtWeekdays
1	Female	42	Single	Under £2,600	Yes	12 cig/day	12 cig/day
2	Male	44	Single	£10,400 to £15,600	No	N/A	N/A
3	Male	53	Married	Above £36,400	Yes	6 cig/day	6 cig/day
⋮	⋮	⋮	⋮	⋮	⋮	⋮	⋮
1691	Male	40	Single	£2,600 to £5,200	Yes	8 cig/day	8 cig/day

(a) What does each row of the data matrix represent?

(b) How many participants were included in the survey?

(c) Indicate whether each variable in the study is numerical or categorical. If numerical, identify as continuous or discrete. If categorical, indicate if the variable is ordinal.

---

[56] P.K. Piff et al. "Higher social class predicts increased unethical behavior". In: *Proceedings of the National Academy of Sciences* (2012).

[57] R.A Fisher. "The Use of Multiple Measurements in Taxonomic Problems". In: *Annals of Eugenics* 7 (1936), pp. 179–188.

[58] National STEM Centre, Large Datasets from stats4schools.

1.9.3   Overview of data collection principles

1.9   Air pollution and birth outcomes, scope of inference. Exercise 1.3 introduces a study where researchers collected data to examine the relationship between air pollutants and preterm births in Southern California. During the study air pollution levels were measured by air quality monitoring stations. Length of gestation data were collected on 143,196 births between the years 1989 and 1993, and air pollution exposure during gestation was calculated for each birth.

(a) Identify the population of interest and the sample in this study.

(b) Comment on whether or not the results of the study can be generalized to the population, and if the findings of the study can be used to establish causal relationships.

1.10   Cheaters, scope of inference. Exercise 1.5 introduces a study where researchers studying the relationship between honesty, age, and self-control conducted an experiment on 160 children between the ages of 5 and 15. The researchers asked each child to toss a fair coin in private and to record the outcome (white or black) on a paper sheet, and said they would only reward children who report white. Half the students were explicitly told not to cheat and the others were not given any explicit instructions. Differences were observed in the cheating rates in the instruction and no instruction groups, as well as some differences across children's characteristics within each group.

(a) Identify the population of interest and the sample in this study.

(b) Comment on whether or not the results of the study can be generalized to the population, and if the findings of the study can be used to establish causal relationships.

1.11   Buteyko method, scope of inference. Exercise 1.4 introduces a study on using the Buteyko shallow breathing technique to reduce asthma symptoms and improve quality of life. As part of this study 600 asthma patients aged 18-69 who relied on medication for asthma treatment were recruited and randomly assigned to two groups: one practiced the Buteyko method and the other did not. Those in the Buteyko group experienced, on average, a significant reduction in asthma symptoms and an improvement in quality of life.

(a) Identify the population of interest and the sample in this study.

(b) Comment on whether or not the results of the study can be generalized to the population, and if the findings of the study can be used to establish causal relationships.

1.12   Stealers, scope of inference. Exercise 1.6 introduces a study on the relationship between socio-economic class and unethical behavior. As part of this study 129 University of California Berkeley undergraduates were asked to identify themselves as having low or high social-class by comparing themselves to others with the most (least) money, most (least) education, and most (least) respected jobs. They were also presented with a jar of individually wrapped candies and informed that the candies were for children in a nearby laboratory, but that they could take some if they wanted. After completing some unrelated tasks, participants reported the number of candies they had taken. It was found that those who were identified as upper-class took more candy than others.

(a) Identify the population of interest and the sample in this study.

(b) Comment on whether or not the results of the study can be generalized to the population, and if the findings of the study can be used to establish causal relationships.

1.13 Relaxing after work. The 2010 General Social Survey asked the question, "After an average work day, about how many hours do you have to relax or pursue activities that you enjoy?" to a random sample of 1,155 Americans. The average relaxing time was found to be 1.65 hours. Determine which of the following is an observation, a variable, a sample statistic, or a population parameter.

(a) An American in the sample.

(b) Number of hours spent relaxing after an average work day.

(c) 1.65.

(d) Average number of hours all Americans spend relaxing after an average work day.

1.14 Cats on YouTube. Suppose you want to estimate the percentage of videos on YouTube that are cat videos. It is impossible for you to watch all videos on YouTube so you use a random video picker to select 1000 videos for you. You find that 2% of these videos are cat videos. Determine which of the following is an observation, a variable, a sample statistic, or a population parameter.

(a) Percentage of all videos on YouTube that are cat videos.

(b) 2%.

(c) A video in your sample.

(d) Whether or not a video is a cat video.

1.15 GPA and study hours. A survey was conducted on 193 Duke University undergraduates who took an introductory statistics course in 2012. Among many other questions, this survey asked them about their GPA, which can range between 0 and 4 points, and the number of hours they spent studying per week. The scatterplot below displays the relationship between these two variables.

(a) What is the explanatory variable and what is the response variable?

(b) Describe the relationship between the two variables. Make sure to discuss unusual observations, if any.

(c) Is this an experiment or an observational study?

(d) Can we conclude that studying longer hours leads to higher GPAs?

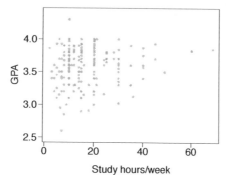

1.16 Income and education in US counties. The scatterplot below shows the relationship between per capita income (in thousands of dollars) and percent of population with a bachelor's degree in 3,143 counties in the US in 2010.

(a) What are the explanatory and response variables?

(b) Describe the relationship between the two variables. Make sure to discuss unusual observations, if any.

(c) Can we conclude that having a bachelor's degree increases one's income?

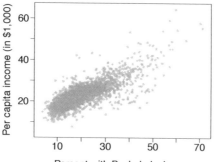

1.9.4   Observational studies and sampling strategies

1.17   Course satisfaction across sections. A large college class has 160 students. All 160 students attend the lectures together, but the students are divided into 4 groups, each of 40 students, for lab sections administered by different teaching assistants. The professor wants to conduct a survey about how satisfied the students are with the course, and he believes that the lab section a student is in might affect the student's overall satisfaction with the course.

(a) What type of study is this?

(b) Suggest a sampling strategy for carrying out this study.

1.18   Housing proposal across dorms. On a large college campus first-year students and sophomores live in dorms located on the eastern part of the campus and juniors and seniors live in dorms located on the western part of the campus. Suppose you want to collect student opinions on a new housing structure the college administration is proposing and you want to make sure your survey equally represents opinions from students from all years.

(a) What type of study is this?

(b) Suggest a sampling strategy for carrying out this study.

1.19   Internet use and life expectancy. The following scatterplot was created as part of a study evaluating the relationship between estimated life expectancy at birth (as of 2014) and percentage of internet users (as of 2009) in 208 countries for which such data were available.[59]

(a) Describe the relationship between life expectancy and percentage of internet users.

(b) What type of study is this?

(c) State a possible confounding variable that might explain this relationship and describe its potential effect.

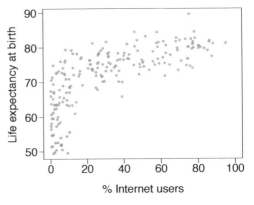

1.20   Stressed out, Part I. A study that surveyed a random sample of otherwise healthy high school students found that they are more likely to get muscle cramps when they are stressed. The study also noted that students drink more coffee and sleep less when they are stressed.

(a) What type of study is this?

(b) Can this study be used to conclude a causal relationship between increased stress and muscle cramps?

(c) State possible confounding variables that might explain the observed relationship between increased stress and muscle cramps.

1.21   Evaluate sampling methods. A university wants to determine what fraction of its undergraduate student body support a new $25 annual fee to improve the student union. For each proposed method below, indicate whether the method is reasonable or not.

(a) Survey a simple random sample of 500 students.

(b) Stratify students by their field of study, then sample 10% of students from each stratum.

(c) Cluster students by their ages (e.g. 18 years old in one cluster, 19 years old in one cluster, etc.), then randomly sample three clusters and survey all students in those clusters.

---

[59]CIA Factbook, Country Comparisons, 2014.

1.22 Random digit dialing. The Gallup Poll uses a procedure called random digit dialing, which creates phone numbers based on a list of all area codes in America in conjunction with the associated number of residential households in each area code. Give a possible reason the Gallup Poll chooses to use random digit dialing instead of picking phone numbers from the phone book.

1.23 Haters are gonna hate, study confirms. A study published in the *Journal of Personality and Social Psychology* asked a group of 200 randomly sampled men and women to evaluate how they felt about various subjects, such as camping, health care, architecture, taxidermy, crossword puzzles, and Japan in order to measure their dispositional attitude towards mostly independent stimuli. Then, they presented the participants with information about a new product: a microwave oven. This microwave oven does not exist, but the participants didn't know this, and were given three positive and three negative fake reviews. People who reacted positively to the subjects on the dispositional attitude measurement also tended to react positively to the microwave oven, and those who reacted negatively also tended to react negatively to it. Researchers concluded that "some people tend to like things, whereas others tend to dislike things, and a more thorough understanding of this tendency will lead to a more thorough understanding of the psychology of attitudes."[60]

(a) What are the cases?

(b) What is (are) the response variable(s) in this study?

(c) What is (are) the explanatory variable(s) in this study?

(d) Does the study employ random sampling?

(e) Is this an observational study or an experiment? Explain your reasoning.

(f) Can we establish a causal link between the explanatory and response variables?

(g) Can the results of the study be generalized to the population at large?

1.24 Family size. Suppose we want to estimate household size, where a "household" is defined as people living together in the same dwelling, and sharing living accommodations. If we select students at random at an elementary school and ask them what their family size is, will this be a good measure of household size? Or will our average be biased? If so, will it overestimate or underestimate the true value?

1.25 Flawed reasoning. Identify the flaw(s) in reasoning in the following scenarios. Explain what the individuals in the study should have done differently if they wanted to make such strong conclusions.

(a) Students at an elementary school are given a questionnaire that they are asked to return after their parents have completed it. One of the questions asked is, "Do you find that your work schedule makes it difficult for you to spend time with your kids after school?" Of the parents who replied, 85% said "no". Based on these results, the school officials conclude that a great majority of the parents have no difficulty spending time with their kids after school.

(b) A survey is conducted on a simple random sample of 1,000 women who recently gave birth, asking them about whether or not they smoked during pregnancy. A follow-up survey asking if the children have respiratory problems is conducted 3 years later, however, only 567 of these women are reached at the same address. The researcher reports that these 567 women are representative of all mothers.

(c) An orthopedist administers a questionnaire to 30 of his patients who do not have any joint problems and finds that 20 of them regularly go running. He concludes that running decreases the risk of joint problems.

---

[60]Justin Hepler and Dolores Albarracín. "Attitudes without objects - Evidence for a dispositional attitude, its measurement, and its consequences". In: *Journal of personality and social psychology* 104.6 (2013), p. 1060.

1.26 City council survey. A city council has requested a household survey be conducted in a suburban area of their city. The area is broken into many distinct and unique neighborhoods, some including large homes, some with only apartments, and others a diverse mixture of housing structures. Identify the sampling methods described below, and comment on whether or not you think they would be effective in this setting.

(a) Randomly sample 50 households from the city.

(b) Divide the city into neighborhoods, and sample 20 households from each neighborhood.

(c) Divide the city into neighborhoods, randomly sample 10 neighborhoods, and sample all households from those neighborhoods.

(d) Divide the city into neighborhoods, randomly sample 10 neighborhoods, and then randomly sample 20 households from those neighborhoods.

(e) Sample the 200 households closest to the city council offices.

1.27 Sampling strategies. A statistics student who is curious about the relationship between the amount of time students spend on social networking sites and their performance at school decides to conduct a survey. Various research strategies for collecting data are described below. In each, name the sampling method proposed and any bias you might expect.

(a) He randomly samples 40 students from the study's population, gives them the survey, asks them to fill it out and bring it back the next day.

(b) He gives out the survey only to his friends, making sure each one of them fills out the survey.

(c) He posts a link to an online survey on Facebook and asks his friends to fill out the survey.

(d) He randomly samples 5 classes and asks a random sample of students from those classes to fill out the survey.

1.28 Reading the paper. Below are excerpts from two articles published in the *NY Times*:

(a) An article titled *Risks: Smokers Found More Prone to Dementia* states the following:[61]

> "Researchers analyzed data from 23,123 health plan members who participated in a voluntary exam and health behavior survey from 1978 to 1985, when they were 50-60 years old. 23 years later, about 25% of the group had dementia, including 1,136 with Alzheimer's disease and 416 with vascular dementia. After adjusting for other factors, the researchers concluded that pack-a-day smokers were 37% more likely than nonsmokers to develop dementia, and the risks went up with increased smoking; 44% for one to two packs a day; and twice the risk for more than two packs."

Based on this study, can we conclude that smoking causes dementia later in life? Explain your reasoning.

(b) Another article titled *The School Bully Is Sleepy* states the following:[62]

> "The University of Michigan study, collected survey data from parents on each child's sleep habits and asked both parents and teachers to assess behavioral concerns. About a third of the students studied were identified by parents or teachers as having problems with disruptive behavior or bullying. The researchers found that children who had behavioral issues and those who were identified as bullies were twice as likely to have shown symptoms of sleep disorders."

A friend of yours who read the article says, "The study shows that sleep disorders lead to bullying in school children." Is this statement justified? If not, how best can you describe the conclusion that can be drawn from this study?

---

[61] R.C. Rabin. "Risks: Smokers Found More Prone to Dementia". In: *New York Times* (2010).
[62] T. Parker-Pope. "The School Bully Is Sleepy". In: *New York Times* (2011).

1.29 Shyness on Facebook. Given the anonymity afforded to individuals in online interactions, researchers hypothesized that shy individuals might have more favorable attitudes toward Facebook, and that shyness might be positively correlated with time spent on Facebook. They also hypothesized that shy individuals might have fewer Facebook "friends" as they tend to have fewer friends than non-shy individuals have in the offline world. 103 undergraduate students at an Ontario university were surveyed via online questionnaires. The study states "Participants were recruited through the university's psychology participation pool. After indicating an interest in the study, participants were sent an e-mail containing the study's URL." Are the results of this study generalizable to the population of all Facebook users?[63]

1.9.5 Experiments

1.30 Stressed out, Part II. In a study evaluating the relationship between stress and muscle cramps, half the subjects are randomly assigned to be exposed to increased stress by being placed into an elevator that falls rapidly and stops abruptly and the other half are left at no or baseline stress.

(a) What type of study is this?

(b) Can this study be used to conclude a causal relationship between increased stress and muscle cramps?

1.31 Light and exam performance. A study is designed to test the effect of light level on exam performance of students. The researcher believes that light levels might have different effects on males and females, so wants to make sure both are equally represented in each treatment. The treatments are fluorescent overhead lighting, yellow overhead lighting, no overhead lighting (only desk lamps).

(a) What is the response variable?

(b) What is the explanatory variable? What are its levels?

(c) What is the blocking variable? What are its levels?

1.32 Vitamin supplements. In order to assess the effectiveness of taking large doses of vitamin C in reducing the duration of the common cold, researchers recruited 400 healthy volunteers from staff and students at a university. A quarter of the patients were assigned a placebo, and the rest were evenly divided between 1g Vitamin C, 3g Vitamin C, or 3g Vitamin C plus additives to be taken at onset of a cold for the following two days. All tablets had identical appearance and packaging. The nurses who handed the prescribed pills to the patients knew which patient received which treatment, but the researchers assessing the patients when they were sick did not. No significant differences were observed in any measure of cold duration or severity between the four medication groups, and the placebo group had the shortest duration of symptoms.[64]

(a) Was this an experiment or an observational study? Why?

(b) What are the explanatory and response variables in this study?

(c) Were the patients blinded to their treatment?

(d) Was this study double-blind?

(e) Participants are ultimately able to choose whether or not to use the pills prescribed to them. We might expect that not all of them will adhere and take their pills. Does this introduce a confounding variable to the study? Explain your reasoning.

---

[63]E.S. Orr et al. "The influence of shyness on the use of Facebook in an undergraduate sample". In: *CyberPsychology & Behavior* 12.3 (2009), pp. 337–340.

[64]C. Audera et al. "Mega-dose vitamin C in treatment of the common cold: a randomised controlled trial". In: *Medical Journal of Australia* 175.7 (2001), pp. 359–362.

1.33 Light, noise, and exam performance. A study is designed to test the effect of light level and noise level on exam performance of students. The researcher believes that light and noise levels might have different effects on males and females, so wants to make sure both are equally represented in each treatment. The light treatments considered are fluorescent overhead lighting, yellow overhead lighting, no overhead lighting (only desk lamps). The noise treatments considered are no noise, construction noise, and human chatter noise.

(a) What is the response variable?

(b) How many factors are considered in this study? Identify them, and describe their levels.

(c) What is the role of the sex variable in this study?

1.34 Music and learning. You would like to conduct an experiment in class to see if students learn better if they study without any music, with music that has no lyrics (instrumental), or with music that has lyrics. Briefly outline a design for this study.

1.35 Soda preference. You would like to conduct an experiment in class to see if your class-mates prefer the taste of regular Coke or Diet Coke. Briefly outline a design for this study.

1.36 Exercise and mental health. A researcher is interested in the effects of exercise on mental health and he proposes the following study: Use stratified random sampling to ensure representative proportions of 18-30, 31-40 and 41- 55 year olds from the population. Next, randomly assign half the subjects from each age group to exercise twice a week, and instruct the rest not to exercise. Conduct a mental health exam at the beginning and at the end of the study, and compare the results.

(a) What type of study is this?

(b) What are the treatment and control groups in this study?

(c) Does this study make use of blocking? If so, what is the blocking variable?

(d) Does this study make use of blinding?

(e) Comment on whether or not the results of the study can be used to establish a causal relationship between exercise and mental health, and indicate whether or not the conclusions can be generalized to the population at large.

(f) Suppose you are given the task of determining if this proposed study should get funding. Would you have any reservations about the study proposal?

1.37 Chia seeds and weight loss. Chia Pets – those terra-cotta figurines that sprout fuzzy green hair – made the chia plant a household name. But chia has gained an entirely new reputation as a diet supplement. In one 2009 study, a team of researchers recruited 38 men and divided them randomly into two groups: treatment or control. They also recruited 38 women, and they randomly placed half of these participants into the treatment group and the other half into the control group. One group was given 25 grams of chia seeds twice a day, and the other was given a placebo. The subjects volunteered to be a part of the study. After 12 weeks, the scientists found no significant difference between the groups in appetite or weight loss.[65]

(a) What type of study is this?

(b) What are the experimental and control treatments in this study?

(c) Has blocking been used in this study? If so, what is the blocking variable?

(d) Has blinding been used in this study?

(e) Comment on whether or not we can make a causal statement, and indicate whether or not we can generalize the conclusion to the population at large.

---

[65]D.C. Nieman et al. "Chia seed does not promote weight loss or alter disease risk factors in overweight adults". In: *Nutrition Research* 29.6 (2009), pp. 414–418.

1.9.6 Examining numerical data

1.38 Mammal life spans. Data were collected on life spans (in years) and gestation lengths (in days) for 62 mammals. A scatterplot of life span versus length of gestation is shown below.[66]

(a) What type of an association is apparent between life span and length of gestation?

(b) What type of an association would you expect to see if the axes of the plot were reversed, i.e. if we plotted length of gestation versus life span?

(c) Are life span and length of gestation independent? Explain your reasoning.

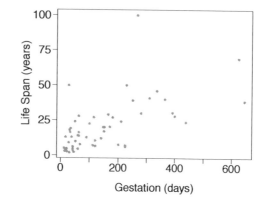

1.39 Associations. Indicate which of the plots show a

(a) positive association

(b) negative association

(c) no association

Also determine if the positive and negative associations are linear or nonlinear. Each part may refer to more than one plot.

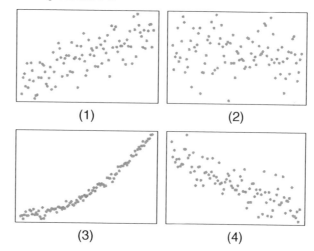

1.40 Office productivity. Office productivity is relatively low when the employees feel no stress about their work or job security. However, high levels of stress can also lead to reduced employee productivity. Sketch a plot to represent the relationship between stress and productivity.

1.41 Reproducing bacteria. Suppose that there is only sufficient space and nutrients to support one million bacterial cells in a petri dish. You place a few bacterial cells in this petri dish, allow them to reproduce freely, and record the number of bacterial cells in the dish over time. Sketch a plot representing the relationship between number of bacterial cells and time.

1.42 Sleeping in college. A recent article in a college newspaper stated that college students get an average of 5.5 hrs of sleep each night. A student who was skeptical about this value decided to conduct a survey by randomly sampling 25 students. On average, the sampled students slept 6.25 hours per night. Identify which value represents the sample mean and which value represents the claimed population mean.

---

[66]T. Allison and D.V. Cicchetti. "Sleep in mammals: ecological and constitutional correlates". In: *Arch. Hydrobiol* 75 (1975), p. 442.

1.43 Parameters and statistics. Identify which value represents the sample mean and which value represents the claimed population mean.

(a) American households spent an average of about $52 in 2007 on Halloween merchandise such as costumes, decorations and candy. To see if this number had changed, researchers conducted a new survey in 2008 before industry numbers were reported. The survey included 1,500 households and found that average Halloween spending was $58 per household.

(b) The average GPA of students in 2001 at a private university was 3.37. A survey on a sample of 203 students from this university yielded an average GPA of 3.59 in Spring semester of 2012.

1.44 Make-up exam. In a class of 25 students, 24 of them took an exam in class and 1 student took a make-up exam the following day. The professor graded the first batch of 24 exams and found an average score of 74 points with a standard deviation of 8.9 points. The student who took the make-up the following day scored 64 points on the exam.

(a) Does the new student's score increase or decrease the average score?

(b) What is the new average?

(c) Does the new student's score increase or decrease the standard deviation of the scores?

1.45 Days off at a mining plant. Workers at a particular mining site receive an average of 35 days paid vacation, which is lower than the national average. The manager of this plant is under pressure from a local union to increase the amount of paid time off. However, he does not want to give more days off to the workers because that would be costly. Instead he decides he should fire 10 employees in such a way as to raise the average number of days off that are reported by his employees. In order to achieve this goal, should he fire employees who have the most number of days off, least number of days off, or those who have about the average number of days off?

1.46 Medians and IQRs. For each part, compare distributions (1) and (2) based on their medians and IQRs. You do not need to calculate these statistics; simply state how the medians and IQRs compare. Make sure to explain your reasoning.

(a) (1) 3, 5, 6, 7, 9
     (2) 3, 5, 6, 7, 20

(b) (1) 3, 5, 6, 7, 9
     (2) 3, 5, 7, 8, 9

(c) (1) 1, 2, 3, 4, 5
     (2) 6, 7, 8, 9, 10

(d) (1) 0, 10, 50, 60, 100
     (2) 0, 100, 500, 600, 1000

1.47 Means and SDs. For each part, compare distributions (1) and (2) based on their means and standard deviations. You do not need to calculate these statistics; simply state how the means and the standard deviations compare. Make sure to explain your reasoning. *Hint:* It may be useful to sketch dot plots of the distributions.

(a) (1) 3, 5, 5, 5, 8, 11, 11, 11, 13
     (2) 3, 5, 5, 5, 8, 11, 11, 11, 20

(b) (1) -20, 0, 0, 0, 15, 25, 30, 30
     (2) -40, 0, 0, 0, 15, 25, 30, 30

(c) (1) 0, 2, 4, 6, 8, 10
     (2) 20, 22, 24, 26, 28, 30

(d) (1) 100, 200, 300, 400, 500
     (2) 0, 50, 300, 550, 600

1.48 Stats scores. Below are the final exam scores of twenty introductory statistics students.

57, 66, 69, 71, 72, 73, 74, 77, 78, 78, 79, 79, 81, 81, 82, 83, 83, 88, 89, 94

Create a box plot of the distribution of these scores. The five number summary provided below may be useful.

Min	Q1	Q2 (Median)	Q3	Max
57	72.5	78.5	82.5	94

1.49 Infant mortality. The infant mortality rate is defined as the number of infant deaths per 1,000 live births. This rate is often used as an indicator of the level of health in a country. The relative frequency histogram below shows the distribution of estimated infant death rates for 224 countries for which such data were available in 2014.[67]

(a) Estimate Q1, the median, and Q3 from the histogram.

(b) Would you expect the mean of this data set to be smaller or larger than the median? Explain your reasoning.

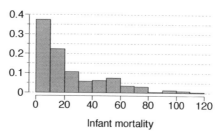

1.50 Mix-and-match. Describe the distribution in the histograms below and match them to the box plots.

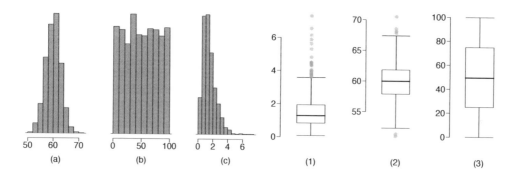

1.51 Air quality. Daily air quality is measured by the air quality index (AQI) reported by the Environmental Protection Agency. This index reports the pollution level and what associated health effects might be a concern. The index is calculated for five major air pollutants regulated by the Clean Air Act and takes values from 0 to 300, where a higher value indicates lower air quality. AQI was reported for a sample of 91 days in 2011 in Durham, NC. The relative frequency histogram below shows the distribution of the AQI values on these days.[68]

(a) Estimate the median AQI value of this sample.

(b) Would you expect the mean AQI value of this sample to be higher or lower than the median? Explain your reasoning.

(c) Estimate Q1, Q3, and IQR for the distribution.

(d) Would any of the days in this sample be considered to have an unusually low or high AQI? Explain your reasoning.

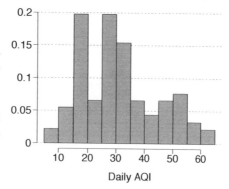

---

[67]CIA Factbook, Country Comparisons, 2014.
[68]US Environmental Protection Agency, AirData, 2011.

1.52 Median vs. mean. Estimate the median for the 400 observations shown in the histogram, and note whether you expect the mean to be higher or lower than the median.

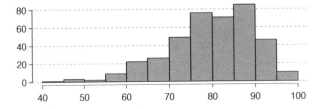

1.53 Histograms vs. box plots. Compare the two plots below. What characteristics of the distribution are apparent in the histogram and not in the box plot? What characteristics are apparent in the box plot but not in the histogram?

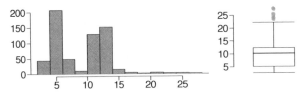

1.54 Marathon winners. The histogram and box plots below show the distribution of finishing times for male and female winners of the New York Marathon between 1970 and 1999.

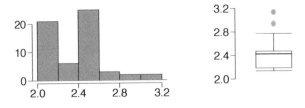

(a) What features of the distribution are apparent in the histogram and not the box plot? What features are apparent in the box plot but not in the histogram?

(b) What may be the reason for the bimodal distribution? Explain.

(c) Compare the distribution of marathon times for men and women based on the box plot shown below.

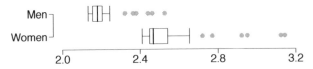

(d) The time series plot shown below is another way to look at these data. Describe what is visible in this plot but not in the others.

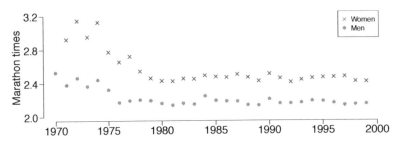

1.55 Distributions and appropriate statistics, Part I. For each of the following, state whether you expect the distribution to be symmetric, right skewed, or left skewed. Also specify whether the mean or median would best represent a typical observation in the data, and whether the variability of observations would be best represented using the standard deviation or IQR. Explain your reasoning.

(a) Number of pets per household.

(b) Distance to work, i.e. number of miles between work and home.

(c) Heights of adult males.

1.56 Distributions and appropriate statistics, Part II . For each of the following, state whether you expect the distribution to be symmetric, right skewed, or left skewed. Also specify whether the mean or median would best represent a typical observation in the data, and whether the variability of observations would be best represented using the standard deviation or IQR. Explain your reasoning.

(a) Housing prices in a country where 25% of the houses cost below $350,000, 50% of the houses cost below $450,000, 75% of the houses cost below $1,000,000 and there are a meaningful number of houses that cost more than $6,000,000.

(b) Housing prices in a country where 25% of the houses cost below $300,000, 50% of the houses cost below $600,000, 75% of the houses cost below $900,000 and very few houses that cost more than $1,200,000.

(c) Number of alcoholic drinks consumed by college students in a given week. Assume that most of these students don't drink since they are under 21 years old, and only a few drink excessively.

(d) Annual salaries of the employees at a Fortune 500 company where only a few high level executives earn much higher salaries than all the other employees.

1.57 TV watchers. Students in an AP Statistics class were asked how many hours of television they watch per week (including online streaming). This sample yielded an average of 4.71 hours, with a standard deviation of 4.18 hours. Is the distribution of number of hours students watch television weekly symmetric? If not, what shape would you expect this distribution to have? Explain your reasoning.

1.58 Exam scores. The average on a history exam (scored out of 100 points) was 85, with a standard deviation of 15. Is the distribution of the scores on this exam symmetric? If not, what shape would you expect this distribution to have? Explain your reasoning.

1.59 Facebook friends. Facebook data indicate that 50% of Facebook users have 100 or more friends, and that the average friend count of users is 190. What do these findings suggest about the shape of the distribution of number of friends of Facebook users?[69]

1.60 A new statistic. The statistic $\frac{\bar{x}}{median}$ can be used as a measure of skewness. Suppose we have a distribution where all observations are greater than 0, $x_i > 0$. What is the expected shape of the distribution under the following conditions? Explain your reasoning.

(a) $\frac{\bar{x}}{median} = 1$

(b) $\frac{\bar{x}}{median} < 1$

(c) $\frac{\bar{x}}{median} > 1$

---

[69]Lars Backstrom. "Anatomy of Facebook". In: *Facebook Data Teams Notes* (2011).

1.61 Income at the coffee shop. The first histogram below shows the distribution of the yearly incomes of 40 patrons at a college coffee shop. Suppose two new people walk into the coffee shop: one making $225,000 and the other $250,000. The second histogram shows the new income distribution. Summary statistics are also provided.

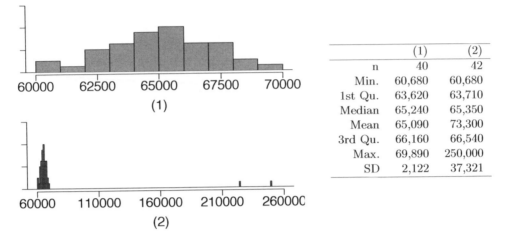

	(1)	(2)
n	40	42
Min.	60,680	60,680
1st Qu.	63,620	63,710
Median	65,240	65,350
Mean	65,090	73,300
3rd Qu.	66,160	66,540
Max.	69,890	250,000
SD	2,122	37,321

(a) Would the mean or the median best represent what we might think of as a typical income for the 42 patrons at this coffee shop? What does this say about the robustness of the two measures?

(b) Would the standard deviation or the IQR best represent the amount of variability in the incomes of the 42 patrons at this coffee shop? What does this say about the robustness of the two measures?

1.62 Midrange. The *midrange* of a distribution is defined as the average of the maximum and the minimum of that distribution. Is this statistic robust to outliers and extreme skew? Explain your reasoning

1.63 Commute times. The US census collects data on time it takes Americans to commute to work, among many other variables. The histogram below shows the distribution of average commute times in 3,143 US counties in 2010. Also shown below is a spatial intensity map of the same data.

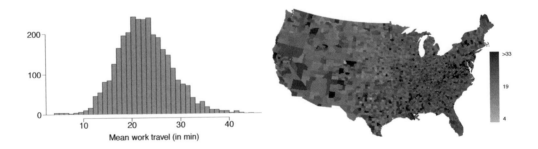

(a) Describe the numerical distribution and comment on whether or not a log transformation may be advisable for these data.

(b) Describe the spatial distribution of commuting times using the map below.

1.64 Hispanic population. The US census collects data on race and ethnicity of Americans, among many other variables. The histogram below shows the distribution of the percentage of the population that is Hispanic in 3,143 counties in the US in 2010. Also shown is a histogram of logs of these values.

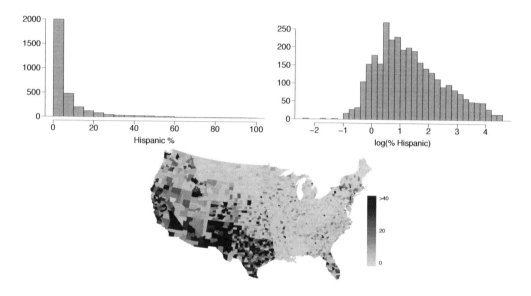

(a) Describe the numerical distribution and comment on why we might want to use log-transformed values in analyzing or modeling these data.

(b) What features of the distribution of the Hispanic population in US counties are apparent in the map but not in the histogram? What features are apparent in the histogram but not the map?

(c) Is one visualization more appropriate or helpful than the other? Explain your reasoning.

1.9.7 Considering categorical data

1.65 Antibiotic use in children. The bar plot and the pie chart below show the distribution of pre-existing medical conditions of children involved in a study on the optimal duration of antibiotic use in treatment of tracheitis, which is an upper respiratory infection.

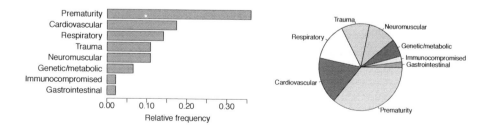

(a) What features are apparent in the bar plot but not in the pie chart?

(b) What features are apparent in the pie chart but not in the bar plot?

(c) Which graph would you prefer to use for displaying these categorical data?

1.66 Views on immigration. 910 randomly sampled registered voters from Tampa, FL were asked if they thought workers who have illegally entered the US should be (i) allowed to keep their jobs and apply for US citizenship, (ii) allowed to keep their jobs as temporary guest workers but not allowed to apply for US citizenship, or (iii) lose their jobs and have to leave the country. The results of the survey by political ideology are shown below.[70]

		Political ideology			
		Conservative	Moderate	Liberal	Total
	(i) Apply for citizenship	57	120	101	278
	(ii) Guest worker	121	113	28	262
Response	(iii) Leave the country	179	126	45	350
	(iv) Not sure	15	4	1	20
	Total	372	363	175	910

(a) What percent of these Tampa, FL voters identify themselves as conservatives?

(b) What percent of these Tampa, FL voters are in favor of the citizenship option?

(c) What percent of these Tampa, FL voters identify themselves as conservatives and are in favor of the citizenship option?

(d) What percent of these Tampa, FL voters who identify themselves as conservatives are also in favor of the citizenship option? What percent of moderates share this view? What percent of liberals share this view?

(e) Do political ideology and views on immigration appear to be independent? Explain your reasoning.

1.67 Views on the DREAM Act. A random sample of registered voters from Tampa, FL were asked if they support the DREAM Act, a proposed law which would provide a path to citizenship for people brought illegally to the US as children. The survey also collected information on the political ideology of the respondents. Based on the mosaic plot shown below, do views on the DREAM Act and political ideology appear to be independent? Explain your reasoning.[71]

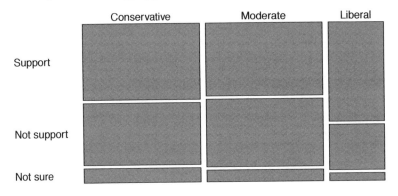

---

[70]SurveyUSA, News Poll #18927, data collected Jan 27-29, 2012.

[71]SurveyUSA, News Poll #18927, data collected Jan 27-29, 2012.

1.68 Raise taxes. A random sample of registered voters nationally were asked whether they think it's better to raise taxes on the rich or raise taxes on the poor. The survey also collected information on the political party affiliation of the respondents. Based on the mosaic plot shown below, do views on raising taxes and political affiliation appear to be independent? Explain your reasoning.[72]

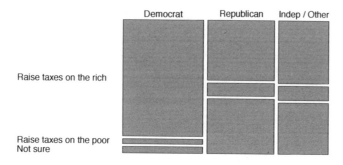

1.9.8 Case study: gender discrimination

1.69 Side effects of Avandia. Rosiglitazone is the active ingredient in the controversial type 2 diabetes medicine Avandia and has been linked to an increased risk of serious cardiovascular problems such as stroke, heart failure, and death. A common alternative treatment is pioglitazone, the active ingredient in a diabetes medicine called Actos. In a nationwide retrospective observational study of 227,571 Medicare beneficiaries aged 65 years or older, it was found that 2,593 of the 67,593 patients using rosiglitazone and 5,386 of the 159,978 using pioglitazone had serious cardiovascular problems. These data are summarized in the contingency table below.[73]

| | | *Cardiovascular problems* | | |
		Yes	No	Total
Treatment	Rosiglitazone	2,593	65,000	67,593
	Pioglitazone	5,386	154,592	159,978
	Total	7,979	219,592	227,571

(a) Determine if each of the following statements is true or false. If false, explain why. *Be careful:* The reasoning may be wrong even if the statement's conclusion is correct. In such cases, the statement should be considered false.

    i. Since more patients on pioglitazone had cardiovascular problems (5,386 vs. 2,593), we can conclude that the rate of cardiovascular problems for those on a pioglitazone treatment is higher.

    ii. The data suggest that diabetic patients who are taking rosiglitazone are more likely to have cardiovascular problems since the rate of incidence was (2,593 / 67,593 = 0.038) 3.8% for patients on this treatment, while it was only (5,386 / 159,978 = 0.034) 3.4% for patients on pioglitazone.

    iii. The fact that the rate of incidence is higher for the rosiglitazone group proves that rosiglitazone causes serious cardiovascular problems.

    iv. Based on the information provided so far, we cannot tell if the difference between the rates of incidences is due to a relationship between the two variables or due to chance.

(See the next page for additional parts to this question.)

---

[72]Public Policy Polling, Americans on College Degrees, Classic Literature, the Seasons, and More, data collected Feb 20-22, 2015.

[73]D.J. Graham et al. "Risk of acute myocardial infarction, stroke, heart failure, and death in elderly Medicare patients treated with rosiglitazone or pioglitazone". In: *JAMA* 304.4 (2010), p. 411. ISSN: 0098-7484.

(b) What proportion of all patients had cardiovascular problems?

(c) If the type of treatment and having cardiovascular problems were independent, about how many patients in the rosiglitazone group would we expect to have had cardiovascular problems?

(d) We can investigate the relationship between outcome and treatment in this study using a randomization technique. While in reality we would carry out the simulations required for randomization using statistical software, suppose we actually simulate using index cards. In order to simulate from the independence model, which states that the outcomes were independent of the treatment, we write whether or not each patient had a cardiovascular problem on cards, shuffled all the cards together, then deal them into two groups of size 67,593 and 159,978. We repeat this simulation 1,000 times and each time record the number of people in the rosiglitazone group who had cardiovascular problems. Use the relative frequency histogram of these counts to answer (i)-(iii).

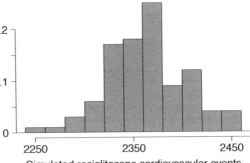

i. What are the claims being tested?

ii. Compared to the number calculated in part (b), which would provide more support for the alternative hypothesis, *more* or *fewer* patients with cardiovascular problems in the rosiglitazone group?

iii. What do the simulation results suggest about the relationship between taking rosiglitazone and having cardiovascular problems in diabetic patients?

1.70 Heart transplants. The Stanford University Heart Transplant Study was conducted to determine whether an experimental heart transplant program increased lifespan. Each patient entering the program was designated an official heart transplant candidate, meaning that he was gravely ill and would most likely benefit from a new heart. Some patients got a transplant and some did not. The variable `transplant` indicates which group the patients were in; patients in the treatment group got a transplant and those in the control group did not. Another variable called `survived` was used to indicate whether or not the patient was alive at the end of the study. Of the 34 patients in the control group, 30 died. Of the 69 people in the treatment group, 45 died.[74]

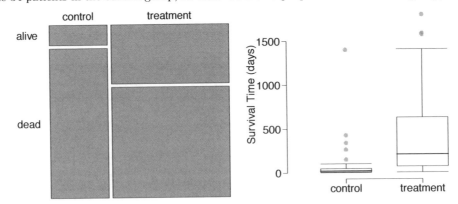

(a) Based on the mosaic plot, is survival independent of whether or not the patient got a transplant? Explain your reasoning.

(See the next page for additional parts to this question.)

---

[74]B. Turnbull et al. "Survivorship of Heart Transplant Data". In: *Journal of the American Statistical Association* 69 (1974), pp. 74–80.

(b) What do the box plots below suggest about the efficacy (effectiveness) of the heart transplant treatment.

(c) What proportion of patients in the treatment group and what proportion of patients in the control group died?

(d) One approach for investigating whether or not the treatment is effective is to use a randomization technique.

 i. What are the claims being tested?

 ii. The paragraph below describes the set up for such approach, if we were to do it without using statistical software. Fill in the blanks with a number or phrase, whichever is appropriate.

 > We write *alive* on _____ cards representing patients who were alive at the end of the study, and *dead* on _____ cards representing patients who were not. Then, we shuffle these cards and split them into two groups: one group of size _____ representing treatment, and another group of size _____ representing control. We calculate the difference between the proportion of *dead* cards in the treatment and control groups (treatment - control) and record this value. We repeat this 100 times to build a distribution centered at _____. Lastly, we calculate the fraction of simulations where the simulated differences in proportions are _____. If this fraction is low, we conclude that it is unlikely to have observed such an outcome by chance and that the null hypothesis should be rejected in favor of the alternative.

 iii. What do the simulation results shown below suggest about the effectiveness of the transplant program?

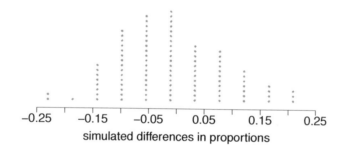

simulated differences in proportions

Chapter 2

Probability (special topic)

Probability forms a foundation for statistics. You might already be familiar with many aspects of probability, however, formalization of the concepts is new for most. This chapter aims to introduce probability on familiar terms using processes most people have seen before.

2.1 Defining probability (special topic)

● **Example 2.1** A "die", the singular of dice, is a cube with six faces numbered 1, 2, 3, 4, 5, and 6. What is the chance of getting 1 when rolling a die?

If the die is fair, then the chance of a 1 is as good as the chance of any other number. Since there are six outcomes, the chance must be 1-in-6 or, equivalently, 1/6.

● **Example 2.2** What is the chance of getting a 1 or 2 in the next roll?

1 and 2 constitute two of the six equally likely possible outcomes, so the chance of getting one of these two outcomes must be $2/6 = 1/3$.

● **Example 2.3** What is the chance of getting either 1, 2, 3, 4, 5, or 6 on the next roll?

100%. The outcome must be one of these numbers.

● **Example 2.4** What is the chance of not rolling a 2?

Since the chance of rolling a 2 is 1/6 or $16.\bar{6}\%$, the chance of not rolling a 2 must be $100\% - 16.\bar{6}\% = 83.\bar{3}\%$ or 5/6.

Alternatively, we could have noticed that not rolling a 2 is the same as getting a 1, 3, 4, 5, or 6, which makes up five of the six equally likely outcomes and has probability 5/6.

● **Example 2.5** Consider rolling two dice. If $1/6^{th}$ of the time the first die is a 1 and $1/6^{th}$ of those times the second die is a 1, what is the chance of getting two 1s?

If $16.\bar{6}\%$ of the time the first die is a 1 and $1/6^{th}$ of *those* times the second die is also a 1, then the chance that both dice are 1 is $(1/6) \times (1/6)$ or 1/36.

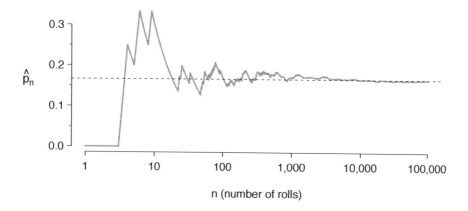

Figure 2.1: The fraction of die rolls that are 1 at each stage in a simulation. The proportion tends to get closer to the probability $1/6 \approx 0.167$ as the number of rolls increases.

2.1.1 Probability

We use probability to build tools to describe and understand apparent randomness. We often frame probability in terms of a **random process** giving rise to an **outcome**.

Roll a die $\rightarrow$ 1, 2, 3, 4, 5, or 6
Flip a coin $\rightarrow$ H or T

Rolling a die or flipping a coin is a seemingly random process and each gives rise to an outcome.

Probability

The **probability** of an outcome is the proportion of times the outcome would occur if we observed the random process an infinite number of times.

Probability is defined as a proportion, and it always takes values between 0 and 1 (inclusively). It may also be displayed as a percentage between 0% and 100%.

Probability can be illustrated by rolling a die many times. Let $\hat{p}_n$ be the proportion of outcomes that are 1 after the first n rolls. As the number of rolls increases, $\hat{p}_n$ will converge to the probability of rolling a 1, $p = 1/6$. Figure 2.1 shows this convergence for 100,000 die rolls. The tendency of $\hat{p}_n$ to stabilize around p is described by the **Law of Large Numbers**.

Law of Large Numbers

As more observations are collected, the proportion $\hat{p}_n$ of occurrences with a particular outcome converges to the probability p of that outcome.

Occasionally the proportion will veer off from the probability and appear to defy the Law of Large Numbers, as $\hat{p}_n$ does many times in Figure 2.1. However, these deviations become smaller as the number of rolls increases.

Above we write p as the probability of rolling a 1. We can also write this probability as

$$P(\text{rolling a 1})$$

As we become more comfortable with this notation, we will abbreviate it further. For instance, if it is clear that the process is "rolling a die", we could abbreviate $P(\text{rolling a 1})$ as $P(1)$.

⊙ **Guided Practice 2.6**   Random processes include rolling a die and flipping a coin. (a) Think of another random process. (b) Describe all the possible outcomes of that process. For instance, rolling a die is a random process with possible outcomes 1, 2, ..., 6.[1]

What we think of as random processes are not necessarily random, but they may just be too difficult to understand exactly. The fourth example in the footnote solution to Guided Practice 2.6 suggests a roommate's behavior is a random process. However, even if a roommate's behavior is not truly random, modeling her behavior as a random process can still be useful.

TIP: Modeling a process as random
It can be helpful to model a process as random even if it is not truly random.

2.1.2   Disjoint or mutually exclusive outcomes

Two outcomes are called **disjoint** or **mutually exclusive** if they cannot both happen. For instance, if we roll a die, the outcomes 1 and 2 are disjoint since they cannot both occur. On the other hand, the outcomes 1 and "rolling an odd number" are not disjoint since both occur if the outcome of the roll is a 1. The terms *disjoint* and *mutually exclusive* are equivalent and interchangeable.

Calculating the probability of disjoint outcomes is easy. When rolling a die, the outcomes 1 and 2 are disjoint, and we compute the probability that one of these outcomes will occur by adding their separate probabilities:

$$P(1 \text{ or } 2) = P(1) + P(2) = 1/6 + 1/6 = 1/3$$

What about the probability of rolling a 1, 2, 3, 4, 5, or 6? Here again, all of the outcomes are disjoint so we add the probabilities:

$$P(1 \text{ or } 2 \text{ or } 3 \text{ or } 4 \text{ or } 5 \text{ or } 6)$$
$$= P(1) + P(2) + P(3) + P(4) + P(5) + P(6)$$
$$= 1/6 + 1/6 + 1/6 + 1/6 + 1/6 + 1/6 = 1.$$

---

[1]Here are four examples. (i) Whether someone gets sick in the next month or not is an apparently random process with outcomes sick and not. (ii) We can *generate* a random process by randomly picking a person and measuring that person's height. The outcome of this process will be a positive number. (iii) Whether the stock market goes up or down next week is a seemingly random process with possible outcomes up, down, and no_change. Alternatively, we could have used the percent change in the stock market as a numerical outcome. (iv) Whether your roommate cleans her dishes tonight probably seems like a random process with possible outcomes cleans_dishes and leaves_dishes.

The **Addition Rule** guarantees the accuracy of this approach when the outcomes are disjoint.

Addition Rule of disjoint outcomes

If A_1 and A_2 represent two disjoint outcomes, then the probability that one of them occurs is given by

$$P(A_1 \text{ or } A_2) = P(A_1) + P(A_2)$$

If there are many disjoint outcomes A_1, ..., A_k, then the probability that one of these outcomes will occur is

$$P(A_1) + P(A_2) + \cdots + P(A_k) \tag{2.7}$$

⊙ **Guided Practice 2.8** We are interested in the probability of rolling a 1, 4, or 5. (a) Explain why the outcomes 1, 4, and 5 are disjoint. (b) Apply the Addition Rule for disjoint outcomes to determine $P(1 \text{ or } 4 \text{ or } 5)$.[2]

⊙ **Guided Practice 2.9** In the `email` data set in Chapter 1, the `number` variable described whether no number (labeled `none`), only one or more small numbers (`small`), or whether at least one big number appeared in an email (`big`). Of the 3,921 emails, 549 had no numbers, 2,827 had only one or more small numbers, and 545 had at least one big number. (a) Are the outcomes `none`, `small`, and `big` disjoint? (b) Determine the proportion of emails with value `small` and `big` separately. (c) Use the Addition Rule for disjoint outcomes to compute the probability a randomly selected email from the data set has a number in it, small or big.[3]

Statisticians rarely work with individual outcomes and instead consider *sets* or *collections* of outcomes. Let A represent the event where a die roll results in 1 or 2 and B represent the event that the die roll is a 4 or a 6. We write A as the set of outcomes $\{1, 2\}$ and $B = \{4, 6\}$. These sets are commonly called **events**. Because A and B have no elements in common, they are disjoint events. A and B are represented in Figure 2.2.

Figure 2.2: Three events, A, B, and D, consist of outcomes from rolling a die. A and B are disjoint since they do not have any outcomes in common.

---

[2](a) The random process is a die roll, and at most one of these outcomes can come up. This means they are disjoint outcomes. (b) $P(1 \text{ or } 4 \text{ or } 5) = P(1) + P(4) + P(5) = \frac{1}{6} + \frac{1}{6} + \frac{1}{6} = \frac{3}{6} = \frac{1}{2}$

[3](a) Yes. Each email is categorized in only one level of `number`. (b) Small: $\frac{2827}{3921} = 0.721$. Big: $\frac{545}{3921} = 0.139$. (c) $P(\text{small or big}) = P(\text{small}) + P(\text{big}) = 0.721 + 0.139 = 0.860$.

The Addition Rule applies to both disjoint outcomes and disjoint events. The probability that one of the disjoint events A or B occurs is the sum of the separate probabilities:

$$P(A \text{ or } B) = P(A) + P(B) = 1/3 + 1/3 = 2/3$$

⊙ **Guided Practice 2.10** (a) Verify the probability of event A, $P(A)$, is 1/3 using the Addition Rule. (b) Do the same for event B.[4]

⊙ **Guided Practice 2.11** (a) Using Figure 2.2 as a reference, what outcomes are represented by event D? (b) Are events B and D disjoint? (c) Are events A and D disjoint?[5]

⊙ **Guided Practice 2.12** In Guided Practice 2.11, you confirmed B and D from Figure 2.2 are disjoint. Compute the probability that event B or event D occurs.[6]

2.1.3 Probabilities when events are not disjoint

Let's consider calculations for two events that are not disjoint in the context of a regular deck of 52 cards, represented in Table 2.3. If you are unfamiliar with the cards in a regular deck, please see the footnote.[7]

2♣	3♣	4♣	5♣	6♣	7♣	8♣	9♣	10♣	J♣	Q♣	K♣	A♣
2◇	3◇	4◇	5◇	6◇	7◇	8◇	9◇	10◇	J◇	Q◇	K◇	A◇
2♡	3♡	4♡	5♡	6♡	7♡	8♡	9♡	10♡	J♡	Q♡	K♡	A♡
2♠	3♠	4♠	5♠	6♠	7♠	8♠	9♠	10♠	J♠	Q♠	K♠	A♠

Table 2.3: Representations of the 52 unique cards in a deck.

⊙ **Guided Practice 2.13** (a) What is the probability that a randomly selected card is a diamond? (b) What is the probability that a randomly selected card is a face card?[8]

Venn diagrams are useful when outcomes can be categorized as "in" or "out" for two or three variables, attributes, or random processes. The Venn diagram in Figure 2.4 uses a circle to represent diamonds and another to represent face cards. If a card is both a diamond and a face card, it falls into the intersection of the circles. If it is a diamond but not a face card, it will be in part of the left circle that is not in the right circle (and so on). The total number of cards that are diamonds is given by the total number of cards in the diamonds circle: $10 + 3 = 13$. The probabilities are also shown (e.g. $10/52 = 0.1923$).

---

[4](a) $P(A) = P(1 \text{ or } 2) = P(1) + P(2) = \frac{1}{6} + \frac{1}{6} = \frac{2}{6} = \frac{1}{3}$. (b) Similarly, $P(B) = 1/3$.

[5](a) Outcomes 2 and 3. (b) Yes, events B and D are disjoint because they share no outcomes. (c) The events A and D share an outcome in common, 2, and so are not disjoint.

[6]Since B and D are disjoint events, use the Addition Rule: $P(B \text{ or } D) = P(B) + P(D) = \frac{1}{3} + \frac{1}{3} = \frac{2}{3}$.

[7]The 52 cards are split into four **suits**: ♣ (club), ◇ (diamond), ♡ (heart), ♠ (spade). Each suit has its 13 cards labeled: 2, 3, ..., 10, J (jack), Q (queen), K (king), and A (ace). Thus, each card is a unique combination of a suit and a label, e.g. 4♡ and J♣. The 12 cards represented by the jacks, queens, and kings are called **face cards**. The cards that are ◇ or ♡ are typically colored red while the other two suits are typically colored black.

[8](a) There are 52 cards and 13 diamonds. If the cards are thoroughly shuffled, each card has an equal chance of being drawn, so the probability that a randomly selected card is a diamond is $P(◇) = \frac{13}{52} = 0.250$. (b) Likewise, there are 12 face cards, so $P(\text{face card}) = \frac{12}{52} = \frac{3}{13} = 0.231$.

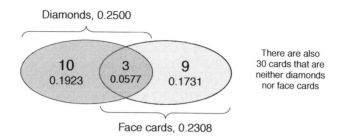

Figure 2.4: A Venn diagram for diamonds and face cards.

Let A represent the event that a randomly selected card is a diamond and B represent the event that it is a face card. How do we compute $P(A \text{ or } B)$? Events A and B are not disjoint – the cards $J\diamond$, $Q\diamond$, and $K\diamond$ fall into both categories – so we cannot use the Addition Rule for disjoint events. Instead we use the Venn diagram. We start by adding the probabilities of the two events:

$$P(A) + P(B) = P(\diamond) + P(\text{face card}) = 13/52 + 12/52$$

However, the three cards that are in both events were counted twice, once in each probability. We must correct this double counting:

$$
\begin{aligned}
P(A \text{ or } B) &= P(\diamond \text{ or face card}) \\
&= P(\diamond) + P(\text{face card}) - P(\diamond \text{ and face card}) \qquad (2.14)\\
&= 13/52 + 12/52 - 3/52 \\
&= 22/52 = 11/26
\end{aligned}
$$

Equation (2.14) is an example of the **General Addition Rule**.

General Addition Rule

If A and B are any two events, disjoint or not, then the probability that at least one of them will occur is

$$P(A \text{ or } B) = P(A) + P(B) - P(A \text{ and } B) \qquad (2.15)$$

where $P(A \text{ and } B)$ is the probability that both events occur.

TIP: "or" is inclusive
When we write "or" in statistics, we mean "and/or" unless we explicitly state otherwise. Thus, A or B occurs means A, B, or both A and B occur.

⊙ **Guided Practice 2.16** (a) If A and B are disjoint, describe why this implies $P(A$ and $B) = 0$. (b) Using part (a), verify that the General Addition Rule simplifies to the simpler Addition Rule for disjoint events if A and B are disjoint.[9]

---

[9](a) If A and B are disjoint, A and B can never occur simultaneously. (b) If A and B are disjoint, then the last term of Equation (2.15) is 0 (see part (a)) and we are left with the Addition Rule for disjoint events.

⊙ **Guided Practice 2.17** In the email data set with 3,921 emails, 367 were spam, 2,827 contained some small numbers but no big numbers, and 168 had both characteristics. Create a Venn diagram for this setup.[10]

⊙ **Guided Practice 2.18** (a) Use your Venn diagram from Guided Practice 2.17 to determine the probability a randomly drawn email from the email data set is spam and had small numbers (but not big numbers). (b) What is the probability that the email had either of these attributes?[11]

2.1.4 Probability distributions

A **probability distribution** is a table of all disjoint outcomes and their associated probabilities. Table 2.5 shows the probability distribution for the sum of two dice.

Dice sum	2	3	4	5	6	7	8	9	10	11	12
Probability	$\frac{1}{36}$	$\frac{2}{36}$	$\frac{3}{36}$	$\frac{4}{36}$	$\frac{5}{36}$	$\frac{6}{36}$	$\frac{5}{36}$	$\frac{4}{36}$	$\frac{3}{36}$	$\frac{2}{36}$	$\frac{1}{36}$

Table 2.5: Probability distribution for the sum of two dice.

Rules for probability distributions

A probability distribution is a list of the possible outcomes with corresponding probabilities that satisfies three rules:

1. The outcomes listed must be disjoint.

2. Each probability must be between 0 and 1.

3. The probabilities must total 1.

⊙ **Guided Practice 2.19** Table 2.6 suggests three distributions for household income in the United States. Only one is correct. Which one must it be? What is wrong with the other two?[12]

Chapter 1 emphasized the importance of plotting data to provide quick summaries. Probability distributions can also be summarized in a bar plot. For instance, the distribution of US household incomes is shown in Figure 2.7 as a bar plot. The probability distribution for the sum of two dice is shown in Table 2.5 and plotted in Figure 2.8.

In these bar plots, the bar heights represent the probabilities of outcomes. If the outcomes are numerical and discrete, it is usually (visually) convenient to make a bar plot that resembles a histogram, as in the case of the sum of two dice. Another example of plotting the bars at their respective locations is shown in Figure 2.20 on page 104.

---

[10]Both the counts and corresponding probabilities (e.g. 2659/3921 = 0.678) are shown. Notice that the number of emails represented in the left circle corresponds to 2659 + 168 = 2827, and the number represented in the right circle is 168 + 199 = 367.

[11](a) The solution is represented by the intersection of the two circles: 0.043. (b) This is the sum of the three disjoint probabilities shown in the circles: 0.678 + 0.043 + 0.051 = 0.772.

[12]The probabilities of (a) do not sum to 1. The second probability in (b) is negative. This leaves (c), which sure enough satisfies the requirements of a distribution. One of the three was said to be the actual distribution of US household incomes, so it must be (c).

Income range ($1000s)	0-25	25-50	50-100	100+
(a)	0.18	0.39	0.33	0.16
(b)	0.38	-0.27	0.52	0.37
(c)	0.28	0.27	0.29	0.16

Table 2.6: Proposed distributions of US household incomes (Guided Practice 2.19).

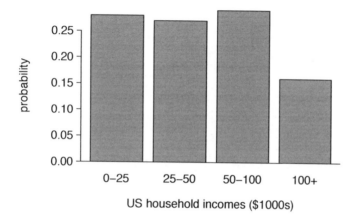

Figure 2.7: The probability distribution of US household income.

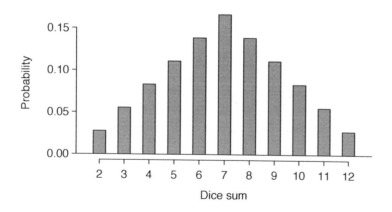

Figure 2.8: The probability distribution of the sum of two dice.

2.1.5   Complement of an event

S
Sample space

Rolling a die produces a value in the set $\{1, 2, 3, 4, 5, 6\}$. This set of all possible outcomes is called the **sample space** (S) for rolling a die. We often use the sample space to examine the scenario where an event does not occur.

A^c
Complement
of outcome A

Let $D = \{2, 3\}$ represent the event that the outcome of a die roll is 2 or 3. Then the **complement** of D represents all outcomes in our sample space that are not in D, which is denoted by $D^c = \{1, 4, 5, 6\}$. That is, D^c is the set of all possible outcomes not already included in D. Figure 2.9 shows the relationship between D, D^c, and the sample space S.

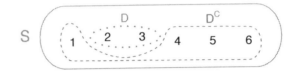

Figure 2.9: Event $D = \{2, 3\}$ and its complement, $D^c = \{1, 4, 5, 6\}$.
S represents the sample space, which is the set of all possible events.

⊙ **Guided Practice 2.20**   (a) Compute $P(D^c) = P$(rolling a 1, 4, 5, or 6). (b) What is $P(D) + P(D^c)$?[13]

⊙ **Guided Practice 2.21**   Events $A = \{1, 2\}$ and $B = \{4, 6\}$ are shown in Figure 2.2 on page 79. (a) Write out what A^c and B^c represent. (b) Compute $P(A^c)$ and $P(B^c)$. (c) Compute $P(A) + P(A^c)$ and $P(B) + P(B^c)$.[14]

A complement of an event A is constructed to have two very important properties: (i) every possible outcome not in A is in A^c, and (ii) A and A^c are disjoint. Property (i) implies

$$P(A \text{ or } A^c) = 1 \qquad (2.22)$$

That is, if the outcome is not in A, it must be represented in A^c. We use the Addition Rule for disjoint events to apply Property (ii):

$$P(A \text{ or } A^c) = P(A) + P(A^c) \qquad (2.23)$$

Combining Equations (2.22) and (2.23) yields a very useful relationship between the probability of an event and its complement.

Complement

The complement of event A is denoted A^c, and A^c represents all outcomes not in A. A and A^c are mathematically related:

$$P(A) + P(A^c) = 1, \quad \text{i.e.} \quad P(A) = 1 - P(A^c) \qquad (2.24)$$

---

[13](a) The outcomes are disjoint and each has probability 1/6, so the total probability is $4/6 = 2/3$. (b) We can also see that $P(D) = \frac{1}{6} + \frac{1}{6} = 1/3$. Since D and D^c are disjoint, $P(D) + P(D^c) = 1$.

[14]Brief solutions: (a) $A^c = \{3, 4, 5, 6\}$ and $B^c = \{1, 2, 3, 5\}$. (b) Noting that each outcome is disjoint, add the individual outcome probabilities to get $P(A^c) = 2/3$ and $P(B^c) = 2/3$. (c) A and A^c are disjoint, and the same is true of B and B^c. Therefore, $P(A) + P(A^c) = 1$ and $P(B) + P(B^c) = 1$.

In simple examples, computing A or A^c is feasible in a few steps. However, using the complement can save a lot of time as problems grow in complexity.

⊙ **Guided Practice 2.25** Let A represent the event where we roll two dice and their total is less than 12. (a) What does the event A^c represent? (b) Determine $P(A^c)$ from Table 2.5 on page 82. (c) Determine $P(A)$.[15]

⊙ **Guided Practice 2.26** Consider again the probabilities from Table 2.5 and rolling two dice. Find the following probabilities: (a) The sum of the dice is *not* 6. (b) The sum is at least 4. That is, determine the probability of the event $B = \{4, 5, ..., 12\}$. (c) The sum is no more than 10. That is, determine the probability of the event $D = \{2, 3, ..., 10\}$.[16]

2.1.6 Independence

Just as variables and observations can be independent, random processes can be independent, too. Two processes are **independent** if knowing the outcome of one provides no useful information about the outcome of the other. For instance, flipping a coin and rolling a die are two independent processes – knowing the coin was heads does not help determine the outcome of a die roll. On the other hand, stock prices usually move up or down together, so they are not independent.

Example 2.5 provides a basic example of two independent processes: rolling two dice. We want to determine the probability that both will be 1. Suppose one of the dice is red and the other white. If the outcome of the red die is a 1, it provides no information about the outcome of the white die. We first encountered this same question in Example 2.5 (page 76), where we calculated the probability using the following reasoning: $1/6^{th}$ of the time the red die is a 1, and $1/6^{th}$ of *those* times the white die will also be 1. This is illustrated in Figure 2.10. Because the rolls are independent, the probabilities of the corresponding outcomes can be multiplied to get the final answer: $(1/6) \times (1/6) = 1/36$. This can be generalized to many independent processes.

● **Example 2.27** What if there was also a blue die independent of the other two? What is the probability of rolling the three dice and getting all 1s?

The same logic applies from Example 2.5. If $1/36^{th}$ of the time the white and red dice are both 1, then $1/6^{th}$ of *those* times the blue die will also be 1, so multiply:

$$P(white = 1 \text{ and } red = 1 \text{ and } blue = 1) = P(white = 1) \times P(red = 1) \times P(blue = 1)$$
$$= (1/6) \times (1/6) \times (1/6) = 1/216$$

Example 2.27 illustrates what is called the Multiplication Rule for independent processes.

---

[15](a) The complement of A: when the total is equal to 12. (b) $P(A^c) = 1/36$. (c) Use the probability of the complement from part (b), $P(A^c) = 1/36$, and Equation (2.24): $P(\text{less than 12}) = 1 - P(12) = 1 - 1/36 = 35/36$.

[16](a) First find $P(6) = 5/36$, then use the complement: $P(\text{not } 6) = 1 - P(6) = 31/36$.

(b) First find the complement, which requires much less effort: $P(2 \text{ or } 3) = 1/36 + 2/36 = 1/12$. Then calculate $P(B) = 1 - P(B^c) = 1 - 1/12 = 11/12$.

(c) As before, finding the complement is the clever way to determine $P(D)$. First find $P(D^c) = P(11 \text{ or } 12) = 2/36 + 1/36 = 1/12$. Then calculate $P(D) = 1 - P(D^c) = 11/12$.

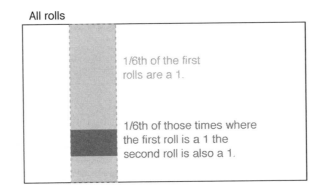

Figure 2.10: $1/6^{th}$ of the time, the first roll is a 1. Then $1/6^{th}$ of *those* times, the second roll will also be a 1.

Multiplication Rule for independent processes

If A and B represent events from two different and independent processes, then the probability that both A and B occur can be calculated as the product of their separate probabilities:

$$P(A \text{ and } B) = P(A) \times P(B) \tag{2.28}$$

Similarly, if there are k events A_1, ..., A_k from k independent processes, then the probability they all occur is

$$P(A_1) \times P(A_2) \times \cdots \times P(A_k)$$

⊙ **Guided Practice 2.29**   About 9% of people are left-handed. Suppose 2 people are selected at random from the U.S. population. Because the sample size of 2 is very small relative to the population, it is reasonable to assume these two people are independent. (a) What is the probability that both are left-handed? (b) What is the probability that both are right-handed?[17]

---

[17](a) The probability the first person is left-handed is 0.09, which is the same for the second person. We apply the Multiplication Rule for independent processes to determine the probability that both will be left-handed: $0.09 \times 0.09 = 0.0081$.

(b) It is reasonable to assume the proportion of people who are ambidextrous (both right and left handed) is nearly 0, which results in $P(\text{right-handed}) = 1 - 0.09 = 0.91$. Using the same reasoning as in part (a), the probability that both will be right-handed is $0.91 \times 0.91 = 0.8281$.

⊙ **Guided Practice 2.30**   Suppose 5 people are selected at random.[18]

(a) What is the probability that all are right-handed?

(b) What is the probability that all are left-handed?

(c) What is the probability that not all of the people are right-handed?

Suppose the variables **handedness** and **gender** are independent, i.e. knowing someone's **gender** provides no useful information about their **handedness** and vice-versa. Then we can compute whether a randomly selected person is right-handed and female[19] using the Multiplication Rule:

$$
\begin{aligned}
P(\text{right-handed and female}) &= P(\text{right-handed}) \times P(\text{female}) \\
&= 0.91 \times 0.50 = 0.455
\end{aligned}
$$

⊙ **Guided Practice 2.31**   Three people are selected at random.[20]

(a) What is the probability that the first person is male and right-handed?

(b) What is the probability that the first two people are male and right-handed?.

(c) What is the probability that the third person is female and left-handed?

(d) What is the probability that the first two people are male and right-handed and the third person is female and left-handed?

Sometimes we wonder if one outcome provides useful information about another outcome. The question we are asking is, are the occurrences of the two events independent? We say that two events A and B are independent if they satisfy Equation (2.28).

● **Example 2.32**   If we shuffle up a deck of cards and draw one, is the event that the card is a heart independent of the event that the card is an ace?

The probability the card is a heart is $1/4$ and the probability that it is an ace is $1/13$. The probability the card is the ace of hearts is $1/52$. We check whether Equation 2.28 is satisfied:

$$
P(\heartsuit) \times P(\text{ace}) = \frac{1}{4} \times \frac{1}{13} = \frac{1}{52} = P(\heartsuit \text{ and ace})
$$

Because the equation holds, the event that the card is a heart and the event that the card is an ace are independent events.

---

[18](a) The abbreviations RH and LH are used for right-handed and left-handed, respectively. Since each are independent, we apply the Multiplication Rule for independent processes:

$$
\begin{aligned}
P(\text{all five are RH}) &= P(\text{first} = \text{RH, second} = \text{RH, ..., fifth} = \text{RH}) \\
&= P(\text{first} = \text{RH}) \times P(\text{second} = \text{RH}) \times \cdots \times P(\text{fifth} = \text{RH}) \\
&= 0.91 \times 0.91 \times 0.91 \times 0.91 \times 0.91 = 0.624
\end{aligned}
$$

(b) Using the same reasoning as in (a), $0.09 \times 0.09 \times 0.09 \times 0.09 \times 0.09 = 0.0000059$

(c) Use the complement, $P(\text{all five are RH})$, to answer this question:

$$
P(\text{not all RH}) = 1 - P(\text{all RH}) = 1 - 0.624 = 0.376
$$

[19]The actual proportion of the U.S. population that is **female** is about 50%, and so we use 0.5 for the probability of sampling a woman. However, this probability does differ in other countries.

[20]Brief answers are provided. (a) This can be written in probability notation as P(a randomly selected person is male and right-handed) = 0.455. (b) 0.207. (c) 0.045. (d) 0.0093.

2.2 Conditional probability (special topic)

The family_college data set contains a sample of 792 cases with two variables, teen and parents, and is summarized in Table 2.11.[21] The teen variable is either college or not, where the college label means the teen went to college immediately after high school. The parents variable takes the value degree if at least one parent of the teenager completed a college degree.

		parents		Total
		degree	not	
teen	college	231	214	445
	not	49	298	347
	Total	280	512	792

Table 2.11: Contingency table summarizing the family_college data set.

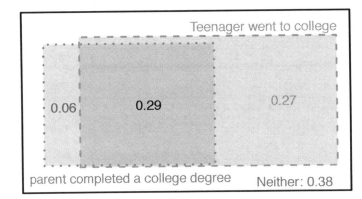

Figure 2.12: A Venn diagram using boxes for the family_college data set.

● **Example 2.33** If at least one parent of a teenager completed a college degree, what is the chance the teenager attended college right after high school?

We can estimate this probability using the data. Of the 280 cases in this data set where parents takes value degree, 231 represent cases where the teen variable takes value college:

$$P(\text{teen college given parents degree}) = \frac{231}{280} = 0.825$$

● **Example 2.34** A teenager is randomly selected from the sample and she did not attend college right after high school. What is the probability that at least one of her parents has a college degree?

If the teenager did not attend, then she is one of the 347 teens in the second row. Of these 347 teens, 49 had at least one parent who got a college degree:

$$P(\text{parents degree given teen not}) = \frac{49}{347} = 0.141$$

[21]A simulated data set based on real population summaries at nces.ed.gov/pubs2001/2001126.pdf.

2.2.1   Marginal and joint probabilities

Table 2.11 includes row and column totals for each variable separately in the `family_college` data set. These totals represent **marginal probabilities** for the sample, which are the probabilities based on a single variable without regard to any other variables. For instance, a probability based solely on the `teen` variable is a marginal probability:

$$P(\texttt{teen college}) = \frac{445}{792} = 0.56$$

A probability of outcomes for two or more variables or processes is called a **joint probability**:

$$P(\texttt{teen college and parents not}) = \frac{214}{792} = 0.27$$

It is common to substitute a comma for "and" in a joint probability, although either is acceptable. That is,

$$P(\texttt{teen college, parents not})$$

means the same thing as

$$P(\texttt{teen college and parents not})$$

Marginal and joint probabilities

If a probability is based on a single variable, it is a *marginal probability*. The probability of outcomes for two or more variables or processes is called a *joint probability*.

We use **table proportions** to summarize joint probabilities for the `family_college` sample. These proportions are computed by dividing each count in Table 2.11 by the table's total, 792, to obtain the proportions in Table 2.13. The joint probability distribution of the `parents` and `teen` variables is shown in Table 2.14.

	parents: degree	parents: not	Total
teen: college	0.29	0.27	0.56
teen: not	0.06	0.38	0.44
Total	0.35	0.65	1.00

Table 2.13: Probability table summarizing whether at least one parent had a college degree and the teenager attended college.

Joint outcome	Probability
parents degree and teen college	0.29
parents degree and teen not	0.06
parents not and teen college	0.27
parents not and teen not	0.38
Total	1.00

Table 2.14: Joint probability distribution for the `family_college` data set.

⊙ **Guided Practice 2.35**   Verify Table 2.14 represents a probability distribution: events are disjoint, all probabilities are non-negative, and the probabilities sum to 1.[22]

We can compute marginal probabilities using joint probabilities in simple cases. For example, the probability a random teenager from the study went to college is found by summing the outcomes where `teen` takes value `college`:

$$P(\underline{\text{teen college}}) = P(\text{parents degree and } \underline{\text{teen college}})$$
$$+ P(\text{parents not and } \underline{\text{teen college}})$$
$$= 0.29 + 0.27$$
$$= 0.56$$

2.2.2   Defining conditional probability

There is some connection between education level of parents and of the teenager: a college degree by a parent is associated with college attendance of the teenager. In this section, we discuss how to use information about associations between two variables to improve probability estimation.

The probability that a random teenager from the study attended college is 0.56. Could we update this probability if we knew that one of the teen's parents has a college degree? Absolutely. To do so, we limit our view to only those 280 cases where a parent has a college degree and look at the fraction where the teenager attended college:

$$P(\text{teen college given parents degree}) = \frac{231}{280} = 0.825$$

We call this a **conditional probability** because we computed the probability under a condition: a parent has a college degree. There are two parts to a conditional probability, the **outcome of interest** and the **condition**. It is useful to think of the condition as information we know to be true, and this information usually can be described as a known outcome or event.

We separate the text inside our probability notation into the outcome of interest and the condition:

$$P(\text{teen college given parents degree})$$

$P(A|B)$

Probability of outcome A given B

$$= P(\text{teen college} \mid \text{parents degree}) = \frac{231}{280} = 0.825 \qquad (2.36)$$

The vertical bar "|" is read as *given*.

In Equation (2.36), we computed the probability a teen attended college based on the condition that at least one parent has a college degree as a fraction:

$$P(\text{teen college} \mid \text{parents degree})$$
$$= \frac{\text{\# cases where } \underline{\text{teen college}} \text{ and } \underline{\text{parents degree}}}{\text{\# cases where } \underline{\text{parents degree}}} \qquad (2.37)$$
$$= \frac{231}{280} = 0.825$$

---

[22]Each of the four outcome combination are disjoint, all probabilities are indeed non-negative, and the sum of the probabilities is $0.29 + 0.06 + 0.27 + 0.38 = 1.00$.

We considered only those cases that met the condition, parents degree, and then we computed the ratio of those cases that satisfied our outcome of interest, the teenager attended college.

Frequently, marginal and joint probabilities are provided instead of count data. For example, disease rates are commonly listed in percentages rather than in a count format. We would like to be able to compute conditional probabilities even when no counts are available, and we use Equation (2.37) as a template to understand this technique.

We considered only those cases that satisfied the condition, parents degree. Of these cases, the conditional probability was the fraction who represented the outcome of interest, teen college. Suppose we were provided only the information in Table 2.13, i.e. only probability data. Then if we took a sample of 1000 people, we would anticipate about 35% or $0.35 \times 1000 = 350$ would meet the information criterion (parents degree). Similarly, we would expect about 29% or $0.29 \times 1000 = 290$ to meet both the information criteria and represent our outcome of interest. Then the conditional probability can be computed as

$$P(\text{teen college} \mid \text{parents degree})$$
$$= \frac{\#\ (\text{teen college and } \underline{\text{parents degree}})}{\#\ (\text{parents degree})}$$
$$= \frac{290}{350} = \frac{0.29}{0.35} = 0.829 \quad \text{(different from 0.825 due to rounding error)} \tag{2.38}$$

In Equation (2.38), we examine exactly the fraction of two probabilities, 0.29 and 0.35, which we can write as

$$P(\text{teen college and parents degree}) \quad \text{and} \quad P(\text{parents degree}).$$

The fraction of these probabilities is an example of the general formula for conditional probability.

Conditional probability

The conditional probability of the outcome of interest A given condition B is computed as the following:

$$P(A|B) = \frac{P(A \text{ and } B)}{P(B)} \tag{2.39}$$

⊙ **Guided Practice 2.40** (a) Write out the following statement in conditional probability notation: *"The probability a random case where neither parent has a college degree if it is known that the teenager didn't attend college right after high school"*. Notice that the condition is now based on the teenager, not the parent.

(b) Determine the probability from part (a). Table 2.13 on page 89 may be helpful.[23]

---

[23](a) $P(\text{parents not} \mid \text{teen not})$. (b) Equation (2.39) for conditional probability indicates we should first find $P(\text{parents not and teen not}) = 0.38$ and $P(\text{teen not}) = 0.44$. Then the ratio represents the conditional probability: $0.38/0.44 = 0.864$.

⊙ **Guided Practice 2.41** (a) Determine the probability that one of the parents has a college degree if it is known the teenager did not attend college.

(b) Using the answers from part (a) and Guided Practice 2.40(b), compute

$P(\text{parents degree} \mid \text{teen not}) + P(\text{parents not} \mid \text{teen not})$

(c) Provide an intuitive argument to explain why the sum in (b) is 1.[24]

⊙ **Guided Practice 2.42** The data indicate there is an association between parents having a college degree and their teenager attending college. Does this mean the parents' college degree(s) *caused* the teenager to go to college?[25]

2.2.3 Smallpox in Boston, 1721

The `smallpox` data set provides a sample of 6,224 individuals from the year 1721 who were exposed to smallpox in Boston.[26] Doctors at the time believed that inoculation, which involves exposing a person to the disease in a controlled form, could reduce the likelihood of death.

Each case represents one person with two variables: `inoculated` and `result`. The variable `inoculated` takes two levels: `yes` or `no`, indicating whether the person was inoculated or not. The variable `result` has outcomes `lived` or `died`. These data are summarized in Tables 2.15 and 2.16.

		inoculated		Total
		yes	no	
result	lived	238	5136	5374
	died	6	844	850
	Total	244	5980	6224

Table 2.15: Contingency table for the `smallpox` data set.

		inoculated		Total
		yes	no	
result	lived	0.0382	0.8252	0.8634
	died	0.0010	0.1356	0.1366
	Total	0.0392	0.9608	1.0000

Table 2.16: Table proportions for the `smallpox` data, computed by dividing each count by the table total, 6224.

⊙ **Guided Practice 2.43** Write out, in formal notation, the probability a randomly selected person who was not inoculated died from smallpox, and find this probability.[27]

---

[24](a) This probability is $\frac{P(\text{parents degree, teen not})}{P(\text{teen not})} = \frac{0.06}{0.44} = 0.136$. (b) The total equals 1. (c) Under the condition the teenager didn't attend college, the parents must either have a college degree or not. The complement still works for conditional probabilities, provided the probabilities are conditioned on the same information.

[25]No. While there is an association, the data are observational. Two potential confounding variables include `income` and `region`. Can you think of others?

[26]Fenner F. 1988. *Smallpox and Its Eradication (History of International Public Health, No. 6)*. Geneva: World Health Organization. ISBN 92-4-156110-6.

[27]$P(\text{result} = \text{died} \mid \text{inoculated} = \text{no}) = \frac{P(\text{result} = \text{died and inoculated} = \text{no})}{P(\text{inoculated} = \text{no})} = \frac{0.1356}{0.9608} = 0.1411$.

⊙ **Guided Practice 2.44** Determine the probability that an inoculated person died from smallpox. How does this result compare with the result of Guided Practice 2.43?[28]

⊙ **Guided Practice 2.45** The people of Boston self-selected whether or not to be inoculated. (a) Is this study observational or was this an experiment? (b) Can we infer any causal connection using these data? (c) What are some potential confounding variables that might influence whether someone `lived` or `died` and also affect whether that person was inoculated?[29]

2.2.4 General multiplication rule

Section 2.1.6 introduced the Multiplication Rule for independent processes. Here we provide the **General Multiplication Rule** for events that might not be independent.

General Multiplication Rule

If A and B represent two outcomes or events, then

$$P(A \text{ and } B) = P(A|B) \times P(B)$$

It is useful to think of A as the outcome of interest and B as the condition.

This General Multiplication Rule is simply a rearrangement of the definition for conditional probability in Equation (2.39) on page 91.

● **Example 2.46** Consider the `smallpox` data set. Suppose we are given only two pieces of information: 96.08% of residents were not inoculated, and 85.88% of the residents who were not inoculated ended up surviving. How could we compute the probability that a resident was not inoculated and lived?

We will compute our answer using the General Multiplication Rule and then verify it using Table 2.16. We want to determine

$$P(\text{result} = \text{lived and inoculated} = \text{no})$$

and we are given that

$$P(\text{result} = \text{lived} \mid \text{inoculated} = \text{no}) = 0.8588$$
$$P(\text{inoculated} = \text{no}) = 0.9608$$

Among the 96.08% of people who were not inoculated, 85.88% survived:

$$P(\text{result} = \text{lived and inoculated} = \text{no}) = 0.8588 \times 0.9608 = 0.8251$$

This is equivalent to the General Multiplication Rule. We can confirm this probability in Table 2.16 at the intersection of `no` and `lived` (with a small rounding error).

---

[28]$P(\text{result} = \text{died} \mid \text{inoculated} = \text{yes}) = \frac{P(\text{result} = \text{died and inoculated} = \text{yes})}{P(\text{inoculated} = \text{yes})} = \frac{0.0010}{0.0392} = 0.0255$. The death rate for individuals who were inoculated is only about 1 in 40 while the death rate is about 1 in 7 for those who were not inoculated.

[29]Brief answers: (a) Observational. (b) No, we cannot infer causation from this observational study. (c) Accessibility to the latest and best medical care. There are other valid answers for part (c).

⊙ **Guided Practice 2.47**   Use $P(\text{inoculated} = \text{yes}) = 0.0392$ and $P(\text{result} = \text{lived} \mid \text{inoculated} = \text{yes}) = 0.9754$ to determine the probability that a person was both inoculated and lived.[30]

⊙ **Guided Practice 2.48**   If 97.54% of the people who were inoculated lived, what proportion of inoculated people must have died?[31]

Sum of conditional probabilities

Let A_1, ..., A_k represent all the disjoint outcomes for a variable or process. Then if B is an event, possibly for another variable or process, we have:

$$P(A_1|B) + \cdots + P(A_k|B) = 1$$

The rule for complements also holds when an event and its complement are conditioned on the same information:

$$P(A|B) = 1 - P(A^c|B)$$

⊙ **Guided Practice 2.49**   Based on the probabilities computed above, does it appear that inoculation is effective at reducing the risk of death from smallpox?[32]

2.2.5   Independence considerations in conditional probability

If two events are independent, then knowing the outcome of one should provide no information about the other. We can show this is mathematically true using conditional probabilities.

⊙ **Guided Practice 2.50**   Let X and Y represent the outcomes of rolling two dice.[33]

(a) What is the probability that the first die, X, is 1?

(b) What is the probability that both X and Y are 1?

(c) Use the formula for conditional probability to compute $P(Y = 1 \mid X = 1)$.

(d) What is $P(Y = 1)$? Is this different from the answer from part (c)? Explain.

---

[30]The answer is 0.0382, which can be verified using Table 2.16.

[31]There were only two possible outcomes: lived or died. This means that 100% - 97.45% = 2.55% of the people who were inoculated died.

[32]The samples are large relative to the difference in death rates for the "inoculated" and "not inoculated" groups, so it seems there is an association between inoculated and outcome. However, as noted in the solution to Guided Practice 2.45, this is an observational study and we cannot be sure if there is a causal connection. (Further research has shown that inoculation is effective at reducing death rates.)

[33]Brief solutions: (a) 1/6. (b) 1/36. (c) $\frac{P(Y=1 \text{ and } X=1)}{P(X=1)} = \frac{1/36}{1/6} = 1/6$. (d) The probability is the same as in part (c): $P(Y = 1) = 1/6$. The probability that $Y = 1$ was unchanged by knowledge about X, which makes sense as X and Y are independent.

We can show in Guided Practice 2.50(c) that the conditioning information has no influence by using the Multiplication Rule for independence processes:

$$
\begin{aligned}
P(Y = 1 \mid X = 1) &= \frac{P(Y = 1 \text{ and } X = 1)}{P(X = 1)} \\
&= \frac{P(Y = 1) \times P(X = 1)}{P(X = 1)} \\
&= P(Y = 1)
\end{aligned}
$$

⊙ **Guided Practice 2.51**   Ron is watching a roulette table in a casino and notices that the last five outcomes were black. He figures that the chances of getting black six times in a row is very small (about 1/64) and puts his paycheck on red. What is wrong with his reasoning?[34]

2.2.6   Tree diagrams

Tree diagrams are a tool to organize outcomes and probabilities around the structure of the data. They are most useful when two or more processes occur in a sequence and each process is conditioned on its predecessors.

The smallpox data fit this description. We see the population as split by inoculation: yes and no. Following this split, survival rates were observed for each group. This structure is reflected in the **tree diagram** shown in Figure 2.17. The first branch for inoculation is said to be the **primary** branch while the other branches are **secondary**.

Inoculated   Result

yes, 0.0392
  lived, 0.9754   0.0392*0.9754 = 0.03824
  died, 0.0246   0.0392*0.0246 = 0.00096

no, 0.9608
  lived, 0.8589   0.9608*0.8589 = 0.82523
  died, 0.1411   0.9608*0.1411 = 0.13557

Figure 2.17: A tree diagram of the smallpox data set.

Tree diagrams are annotated with marginal and conditional probabilities, as shown in Figure 2.17. This tree diagram splits the smallpox data by inoculation into the yes and no groups with respective marginal probabilities 0.0392 and 0.9608. The secondary branches

---

[34]He has forgotten that the next roulette spin is independent of the previous spins. Casinos do employ this practice; they post the last several outcomes of many betting games to trick unsuspecting gamblers into believing the odds are in their favor. This is called the **gambler's fallacy**.

are conditioned on the first, so we assign conditional probabilities to these branches. For example, the top branch in Figure 2.17 is the probability that `result = lived` conditioned on the information that `inoculated = yes`. We may (and usually do) construct joint probabilities at the end of each branch in our tree by multiplying the numbers we come across as we move from left to right. These joint probabilities are computed using the General Multiplication Rule:

$$P(\text{inoculated} = \text{yes and result} = \text{lived})$$
$$= P(\text{inoculated} = \text{yes}) \times P(\text{result} = \text{lived}|\text{inoculated} = \text{yes})$$
$$= 0.0392 \times 0.9754 = 0.0382$$

● **Example 2.52** Consider the midterm and final for a statistics class. Suppose 13% of students earned an A on the midterm. Of those students who earned an A on the midterm, 47% received an A on the final, and 11% of the students who earned lower than an A on the midterm received an A on the final. You randomly pick up a final exam and notice the student received an A. What is the probability that this student earned an A on the midterm?

The end-goal is to find $P(\text{midterm} = \text{A}|\text{final} = \text{A})$. To calculate this conditional probability, we need the following probabilities:

$$P(\text{midterm} = \text{A and final} = \text{A}) \quad \text{and} \quad P(\text{final} = \text{A})$$

However, this information is not provided, and it is not obvious how to calculate these probabilities. Since we aren't sure how to proceed, it is useful to organize the information into a tree diagram, as shown in Figure 2.18. When constructing a tree diagram, variables provided with marginal probabilities are often used to create the tree's primary branches; in this case, the marginal probabilities are provided for midterm grades. The final grades, which correspond to the conditional probabilities provided, will be shown on the secondary branches.

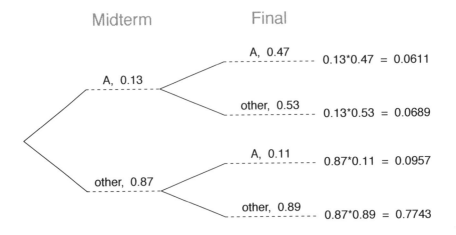

Figure 2.18: A tree diagram describing the `midterm` and `final` variables.

With the tree diagram constructed, we may compute the required probabilities:

$$P(\texttt{midterm} = \texttt{A} \text{ and } \texttt{final} = \texttt{A}) = 0.0611$$
$$P(\underline{\texttt{final} = \texttt{A}})$$
$$= P(\texttt{midterm} = \texttt{other} \text{ and } \underline{\texttt{final} = \texttt{A}}) + P(\texttt{midterm} = \texttt{A} \text{ and } \underline{\texttt{final} = \texttt{A}})$$
$$= 0.0957 + 0.0611 = 0.1568$$

The marginal probability, $P(\texttt{final} = \texttt{A})$, was calculated by adding up all the joint probabilities on the right side of the tree that correspond to $\texttt{final} = \texttt{A}$. We may now finally take the ratio of the two probabilities:

$$P(\texttt{midterm} = \texttt{A}|\texttt{final} = \texttt{A}) = \frac{P(\texttt{midterm} = \texttt{A} \text{ and } \texttt{final} = \texttt{A})}{P(\texttt{final} = \texttt{A})}$$
$$= \frac{0.0611}{0.1568} = 0.3897$$

The probability the student also earned an A on the midterm is about 0.39.

⊙ **Guided Practice 2.53** After an introductory statistics course, 78% of students can successfully construct tree diagrams. Of those who can construct tree diagrams, 97% passed, while only 57% of those students who could not construct tree diagrams passed. (a) Organize this information into a tree diagram. (b) What is the probability that a randomly selected student passed? (c) Compute the probability a student is able to construct a tree diagram if it is known that she passed.[35]

2.2.7 Bayes' Theorem

In many instances, we are given a conditional probability of the form

$$P(\text{statement about variable 1} \mid \text{statement about variable 2})$$

but we would really like to know the inverted conditional probability:

$$P(\text{statement about variable 2} \mid \text{statement about variable 1})$$

Tree diagrams can be used to find the second conditional probability when given the first. However, sometimes it is not possible to draw the scenario in a tree diagram. In these cases, we can apply a very useful and general formula: Bayes' Theorem.

We first take a critical look at an example of inverting conditional probabilities where we still apply a tree diagram.

---

[35](a) The tree diagram is shown to the right. (b) Identify which two joint probabilities represent students who passed, and add them: $P(\text{passed}) = 0.7566 + 0.1254 = 0.8820$. (c) $P(\text{construct tree diagram} \mid \text{passed}) = \frac{0.7566}{0.8820} = 0.8578$.

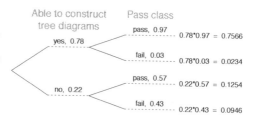

● **Example 2.54** In Canada, about 0.35% of women over 40 will develop breast cancer in any given year. A common screening test for cancer is the mammogram, but this test is not perfect. In about 11% of patients with breast cancer, the test gives a **false negative**: it indicates a woman does not have breast cancer when she does have breast cancer. Similarly, the test gives a **false positive** in 7% of patients who do not have breast cancer: it indicates these patients have breast cancer when they actually do not.[36] If we tested a random woman over 40 for breast cancer using a mammogram and the test came back positive – that is, the test suggested the patient has cancer – what is the probability that the patient actually has breast cancer?

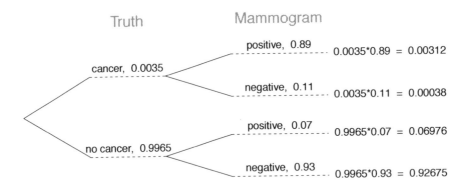

Figure 2.19: Tree diagram for Example 2.54, computing the probability a random patient who tests positive on a mammogram actually has breast cancer.

Notice that we are given sufficient information to quickly compute the probability of testing positive if a woman has breast cancer ($1.00 - 0.11 = 0.89$). However, we seek the inverted probability of cancer given a positive test result. (Watch out for the non-intuitive medical language: a *positive* test result suggests the possible presence of cancer in a mammogram screening.) This inverted probability may be broken into two pieces:

$$P(\text{has BC} \mid \text{mammogram}^+) = \frac{P(\text{has BC and mammogram}^+)}{P(\text{mammogram}^+)}$$

where "has BC" is an abbreviation for the patient actually having breast cancer and "mammogram$^+$" means the mammogram screening was positive. A tree diagram is useful for identifying each probability and is shown in Figure 2.19. The probability the patient has breast cancer and the mammogram is positive is

$$P(\text{has BC and mammogram}^+) = P(\text{mammogram}^+ \mid \text{has BC})P(\text{has BC})$$
$$= 0.89 \times 0.0035 = 0.00312$$

---

[36]The probabilities reported here were obtained using studies reported at www.breastcancer.org and www.ncbi.nlm.nih.gov/pmc/articles/PMC1173421.

The probability of a positive test result is the sum of the two corresponding scenarios:

$$P(\underline{\text{mammogram}^+}) = P(\underline{\text{mammogram}^+ \text{ and has BC}}) + P(\underline{\text{mammogram}^+ \text{ and no BC}})$$
$$= P(\text{has BC})P(\text{mammogram}^+ \mid \text{has BC})$$
$$+ P(\text{no BC})P(\text{mammogram}^+ \mid \text{no BC})$$
$$= 0.0035 \times 0.89 + 0.9965 \times 0.07 = 0.07288$$

Then if the mammogram screening is positive for a patient, the probability the patient has breast cancer is

$$P(\text{has BC} \mid \text{mammogram}^+) = \frac{P(\text{has BC and mammogram}^+)}{P(\text{mammogram}^+)}$$
$$= \frac{0.00312}{0.07288} \approx 0.0428$$

That is, even if a patient has a positive mammogram screening, there is still only a 4% chance that she has breast cancer.

Example 2.54 highlights why doctors often run more tests regardless of a first positive test result. When a medical condition is rare, a single positive test isn't generally definitive.

Consider again the last equation of Example 2.54. Using the tree diagram, we can see that the numerator (the top of the fraction) is equal to the following product:

$$P(\text{has BC and mammogram}^+) = P(\text{mammogram}^+ \mid \text{has BC})P(\text{has BC})$$

The denominator – the probability the screening was positive – is equal to the sum of probabilities for each positive screening scenario:

$$P(\underline{\text{mammogram}^+}) = P(\underline{\text{mammogram}^+ \text{ and no BC}}) + P(\underline{\text{mammogram}^+ \text{ and has BC}})$$

In the example, each of the probabilities on the right side was broken down into a product of a conditional probability and marginal probability using the tree diagram.

$$P(\text{mammogram}^+) = P(\text{mammogram}^+ \text{ and no BC}) + P(\text{mammogram}^+ \text{ and has BC})$$
$$= P(\text{mammogram}^+ \mid \text{no BC})P(\text{no BC})$$
$$+ P(\text{mammogram}^+ \mid \text{has BC})P(\text{has BC})$$

We can see an application of Bayes' Theorem by substituting the resulting probability expressions into the numerator and denominator of the original conditional probability.

$$P(\text{has BC} \mid \text{mammogram}^+)$$
$$= \frac{P(\text{mammogram}^+ \mid \text{has BC})P(\text{has BC})}{P(\text{mammogram}^+ \mid \text{no BC})P(\text{no BC}) + P(\text{mammogram}^+ \mid \text{has BC})P(\text{has BC})}$$

Bayes' Theorem: inverting probabilities

Consider the following conditional probability for variable 1 and variable 2:

$$P(\text{outcome } A_1 \text{ of variable 1} \mid \text{outcome } B \text{ of variable 2})$$

Bayes' Theorem states that this conditional probability can be identified as the following fraction:

$$\frac{P(B|A_1)P(A_1)}{P(B|A_1)P(A_1) + P(B|A_2)P(A_2) + \cdots + P(B|A_k)P(A_k)} \qquad (2.55)$$

where A_2, A_3, ..., and A_k represent all other possible outcomes of the first variable.

Bayes' Theorem is just a generalization of what we have done using tree diagrams. The numerator identifies the probability of getting both A_1 and B. The denominator is the marginal probability of getting B. This bottom component of the fraction appears long and complicated since we have to add up probabilities from all of the different ways to get B. We always completed this step when using tree diagrams. However, we usually did it in a separate step so it didn't seem as complex.

To apply Bayes' Theorem correctly, there are two preparatory steps:

(1) First identify the marginal probabilities of each possible outcome of the first variable: $P(A_1)$, $P(A_2)$, ..., $P(A_k)$.

(2) Then identify the probability of the outcome B, conditioned on each possible scenario for the first variable: $P(B|A_1)$, $P(B|A_2)$, ..., $P(B|A_k)$.

Once each of these probabilities are identified, they can be applied directly within the formula.

TIP: Only use Bayes' Theorem when tree diagrams are difficult

Drawing a tree diagram makes it easier to understand how two variables are connected. Use Bayes' Theorem only when there are so many scenarios that drawing a tree diagram would be complex.

⊙ **Guided Practice 2.56** Jose visits campus every Thursday evening. However, some days the parking garage is full, often due to college events. There are academic events on 35% of evenings, sporting events on 20% of evenings, and no events on 45% of evenings. When there is an academic event, the garage fills up about 25% of the time, and it fills up 70% of evenings with sporting events. On evenings when there are no events, it only fills up about 5% of the time. If Jose comes to campus and finds the garage full, what is the probability that there is a sporting event? Use a tree diagram to solve this problem.[37]

● **Example 2.57** Here we solve the same problem presented in Guided Practice 2.56, except this time we use Bayes' Theorem.

The outcome of interest is whether there is a sporting event (call this A_1), and the condition is that the lot is full (B). Let A_2 represent an academic event and A_3 represent there being no event on campus. Then the given probabilities can be written as

$$P(A_1) = 0.2 \qquad P(A_2) = 0.35 \qquad P(A_3) = 0.45$$
$$P(B|A_1) = 0.7 \qquad P(B|A_2) = 0.25 \qquad P(B|A_3) = 0.05$$

Bayes' Theorem can be used to compute the probability of a sporting event (A_1) under the condition that the parking lot is full (B):

$$\begin{aligned}
P(A_1|B) &= \frac{P(B|A_1)P(A_1)}{P(B|A_1)P(A_1) + P(B|A_2)P(A_2) + P(B|A_3)P(A_3)} \\
&= \frac{(0.7)(0.2)}{(0.7)(0.2) + (0.25)(0.35) + (0.05)(0.45)} \\
&= 0.56
\end{aligned}$$

Based on the information that the garage is full, there is a 56% probability that a sporting event is being held on campus that evening.

⊙ **Guided Practice 2.58** Use the information in the previous exercise and example to verify the probability that there is an academic event conditioned on the parking lot being full is 0.35.[38]

---

[37]The tree diagram, with three primary branches, is shown to the right. Next, we identify two probabilities from the tree diagram. (1) The probability that there is a sporting event and the garage is full: 0.14. (2) The probability the garage is full: $0.0875 + 0.14 + 0.0225 = 0.25$. Then the solution is the ratio of these probabilities: $\frac{0.14}{0.25} = 0.56$. If the garage is full, there is a 56% probability that there is a sporting event.

[38]Short answer:

$$\begin{aligned}
P(A_2|B) &= \frac{P(B|A_2)P(A_2)}{P(B|A_1)P(A_1) + P(B|A_2)P(A_2) + P(B|A_3)P(A_3)} \\
&= \frac{(0.25)(0.35)}{(0.7)(0.2) + (0.25)(0.35) + (0.05)(0.45)} \\
&= 0.35
\end{aligned}$$

⊙ **Guided Practice 2.59**   In Guided Practice 2.56 and 2.58, you found that if the parking lot is full, the probability there is a sporting event is 0.56 and the probability there is an academic event is 0.35. Using this information, compute $P(\text{no event} \mid \text{the lot is full})$.[39]

The last several exercises offered a way to update our belief about whether there is a sporting event, academic event, or no event going on at the school based on the information that the parking lot was full. This strategy of *updating beliefs* using Bayes' Theorem is actually the foundation of an entire section of statistics called **Bayesian statistics**. While Bayesian statistics is very important and useful, we will not have time to cover much more of it in this book.

2.3   Sampling from a small population (special topic)

● **Example 2.60**   Professors sometimes select a student at random to answer a question. If each student has an equal chance of being selected and there are 15 people in your class, what is the chance that she will pick you for the next question?

If there are 15 people to ask and none are skipping class, then the probability is 1/15, or about 0.067.

● **Example 2.61**   If the professor asks 3 questions, what is the probability that you will not be selected? Assume that she will not pick the same person twice in a given lecture.

For the first question, she will pick someone else with probability 14/15. When she asks the second question, she only has 14 people who have not yet been asked. Thus, if you were not picked on the first question, the probability you are again not picked is 13/14. Similarly, the probability you are again not picked on the third question is 12/13, and the probability of not being picked for any of the three questions is

$$P(\text{not picked in 3 questions})$$
$$= P(\texttt{Q1} = \texttt{not\_picked}, \texttt{Q2} = \texttt{not\_picked}, \texttt{Q3} = \texttt{not\_picked}.)$$
$$= \frac{14}{15} \times \frac{13}{14} \times \frac{12}{13} = \frac{12}{15} = 0.80$$

⊙ **Guided Practice 2.62**   What rule permitted us to multiply the probabilities in Example 2.61?[40]

---

[39]Each probability is conditioned on the same information that the garage is full, so the complement may be used: $1.00 - 0.56 - 0.35 = 0.09$.

[40]The three probabilities we computed were actually one marginal probability, $P(\texttt{Q1}=\texttt{not\_picked})$, and two conditional probabilities:

$$P(\texttt{Q2} = \texttt{not\_picked} \mid \texttt{Q1} = \texttt{not\_picked})$$
$$P(\texttt{Q3} = \texttt{not\_picked} \mid \texttt{Q1} = \texttt{not\_picked}, \texttt{Q2} = \texttt{not\_picked})$$

Using the General Multiplication Rule, the product of these three probabilities is the probability of not being picked in 3 questions.

● **Example 2.63** Suppose the professor randomly picks without regard to who she already selected, i.e. students can be picked more than once. What is the probability that you will not be picked for any of the three questions?

Each pick is independent, and the probability of not being picked for any individual question is 14/15. Thus, we can use the Multiplication Rule for independent processes.

$$P(\text{not picked in 3 questions})$$
$$= P(\texttt{Q1 = not\_picked, Q2 = not\_picked, Q3 = not\_picked.})$$
$$= \frac{14}{15} \times \frac{14}{15} \times \frac{14}{15} = 0.813$$

You have a slightly higher chance of not being picked compared to when she picked a new person for each question. However, you now may be picked more than once.

⊙ **Guided Practice 2.64** Under the setup of Example 2.63, what is the probability of being picked to answer all three questions?[41]

If we sample from a small population **without replacement**, we no longer have independence between our observations. In Example 2.61, the probability of not being picked for the second question was conditioned on the event that you were not picked for the first question. In Example 2.63, the professor sampled her students **with replacement**: she repeatedly sampled the entire class without regard to who she already picked.

⊙ **Guided Practice 2.65** Your department is holding a raffle. They sell 30 tickets and offer seven prizes. (a) They place the tickets in a hat and draw one for each prize. The tickets are sampled without replacement, i.e. the selected tickets are not placed back in the hat. What is the probability of winning a prize if you buy one ticket? (b) What if the tickets are sampled with replacement?[42]

⊙ **Guided Practice 2.66** Compare your answers in Guided Practice 2.65. How much influence does the sampling method have on your chances of winning a prize?[43]

Had we repeated Guided Practice 2.65 with 300 tickets instead of 30, we would have found something interesting: the results would be nearly identical. The probability would be 0.0233 without replacement and 0.0231 with replacement. When the sample size is only a small fraction of the population (under 10%), observations are nearly independent even when sampling without replacement.

---

[41]$P(\text{being picked to answer all three questions}) = \left(\frac{1}{15}\right)^3 = 0.00030$.

[42](a) First determine the probability of not winning. The tickets are sampled without replacement, which means the probability you do not win on the first draw is 29/30, 28/29 for the second, ..., and 23/24 for the seventh. The probability you win no prize is the product of these separate probabilities: 23/30. That is, the probability of winning a prize is $1 - 23/30 = 7/30 = 0.233$. (b) When the tickets are sampled with replacement, there are seven independent draws. Again we first find the probability of not winning a prize: $(29/30)^7 = 0.789$. Thus, the probability of winning (at least) one prize when drawing with replacement is 0.211.

[43]There is about a 10% larger chance of winning a prize when using sampling without replacement. However, at most one prize may be won under this sampling procedure.

2.4   Random variables (special topic)

● **Example 2.67** Two books are assigned for a statistics class: a textbook and its corresponding study guide. The university bookstore determined 20% of enrolled students do not buy either book, 55% buy the textbook only, and 25% buy both books, and these percentages are relatively constant from one term to another. If there are 100 students enrolled, how many books should the bookstore expect to sell to this class?

Around 20 students will not buy either book (0 books total), about 55 will buy one book (55 books total), and approximately 25 will buy two books (totaling 50 books for these 25 students). The bookstore should expect to sell about 105 books for this class.

⊙ **Guided Practice 2.68**   Would you be surprised if the bookstore sold slightly more or less than 105 books?[44]

● **Example 2.69**   The textbook costs $137 and the study guide $33. How much revenue should the bookstore expect from this class of 100 students?

About 55 students will just buy a textbook, providing revenue of

$$\$137 \times 55 = \$7,535$$

The roughly 25 students who buy both the textbook and the study guide would pay a total of

$$(\$137 + \$33) \times 25 = \$170 \times 25 = \$4,250$$

Thus, the bookstore should expect to generate about $7,535 + $4,250 = $11,785 from these 100 students for this one class. However, there might be some *sampling variability* so the actual amount may differ by a little bit.

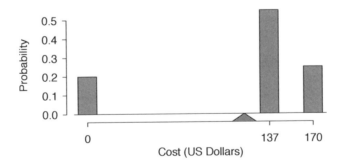

Figure 2.20: Probability distribution for the bookstore's revenue from a single student. The distribution balances on a triangle representing the average revenue per student.

---

[44]If they sell a little more or a little less, this should not be a surprise. Hopefully Chapter 1 helped make clear that there is natural variability in observed data. For example, if we would flip a coin 100 times, it will not usually come up heads exactly half the time, but it will probably be close.

● **Example 2.70** What is the average revenue per student for this course?

The expected total revenue is \$11,785, and there are 100 students. Therefore the expected revenue per student is $\$11,785/100 = \$117.85$.

2.4.1 Expectation

We call a variable or process with a numerical outcome a **random variable**, and we usually represent this random variable with a capital letter such as X, Y, or Z. The amount of money a single student will spend on her statistics books is a random variable, and we represent it by X.

Random variable

A random process or variable with a numerical outcome.

The possible outcomes of X are labeled with a corresponding lower case letter x and subscripts. For example, we write $x_1 = \$0$, $x_2 = \$137$, and $x_3 = \$170$, which occur with probabilities 0.20, 0.55, and 0.25. The distribution of X is summarized in Figure 2.20 and Table 2.21.

i	1	2	3	Total
x_i	\$0	\$137	\$170	–
$P(X = x_i)$	0.20	0.55	0.25	1.00

Table 2.21: The probability distribution for the random variable X, representing the bookstore's revenue from a single student.

We computed the average outcome of X as \$117.85 in Example 2.70. We call this average the **expected value** of X, denoted by $E(X)$. The expected value of a random variable is computed by adding each outcome weighted by its probability:

$E(X)$
Expected value of X

$$E(X) = 0 \times P(X = 0) + 137 \times P(X = 137) + 170 \times P(X = 170)$$
$$= 0 \times 0.20 + 137 \times 0.55 + 170 \times 0.25 = 117.85$$

Expected value of a Discrete Random Variable

If X takes outcomes x_1, ..., x_k with probabilities $P(X = x_1)$, ..., $P(X = x_k)$, the expected value of X is the sum of each outcome multiplied by its corresponding probability:

$$E(X) = x_1 \times P(X = x_1) + \cdots + x_k \times P(X = x_k)$$
$$= \sum_{i=1}^{k} x_i P(X = x_i) \tag{2.71}$$

The Greek letter μ may be used in place of the notation $E(X)$.

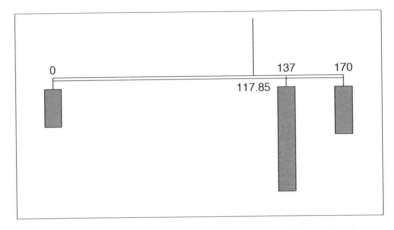

Figure 2.22: A weight system representing the probability distribution for X. The string holds the distribution at the mean to keep the system balanced.

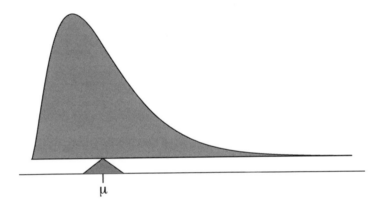

Figure 2.23: A continuous distribution can also be balanced at its mean.

The expected value for a random variable represents the average outcome. For example, $E(X) = 117.85$ represents the average amount the bookstore expects to make from a single student, which we could also write as $\mu = 117.85$.

It is also possible to compute the expected value of a continuous random variable (see Section 2.5). However, it requires a little calculus and we save it for a later class.[45]

In physics, the expectation holds the same meaning as the center of gravity. The distribution can be represented by a series of weights at each outcome, and the mean represents the balancing point. This is represented in Figures 2.20 and 2.22. The idea of a center of gravity also expands to continuous probability distributions. Figure 2.23 shows a continuous probability distribution balanced atop a wedge placed at the mean.

2.4.2 Variability in random variables

Suppose you ran the university bookstore. Besides how much revenue you expect to generate, you might also want to know the volatility (variability) in your revenue.

---

[45]$\mu = \int x f(x) dx$ where $f(x)$ represents a function for the density curve.

The variance and standard deviation can be used to describe the variability of a random variable. Section 1.6.4 introduced a method for finding the variance and standard deviation for a data set. We first computed deviations from the mean $(x_i - \mu)$, squared those deviations, and took an average to get the variance. In the case of a random variable, we again compute squared deviations. However, we take their sum weighted by their corresponding probabilities, just like we did for the expectation. This weighted sum of squared deviations equals the variance, and we calculate the standard deviation by taking the square root of the variance, just as we did in Section 1.6.4.

General variance formula

If X takes outcomes x_1, ..., x_k with probabilities $P(X = x_1)$, ..., $P(X = x_k)$ and expected value $\mu = E(X)$, then the variance of X, denoted by $Var(X)$ or the symbol σ^2, is

$$\sigma^2 = (x_1 - \mu)^2 \times P(X = x_1) + \cdots$$
$$\cdots + (x_k - \mu)^2 \times P(X = x_k)$$
$$= \sum_{j=1}^{k} (x_j - \mu)^2 P(X = x_j) \qquad (2.72)$$

The standard deviation of X, labeled σ, is the square root of the variance.

$Var(X)$
Variance of X

● **Example 2.73** Compute the expected value, variance, and standard deviation of X, the revenue of a single statistics student for the bookstore.

It is useful to construct a table that holds computations for each outcome separately, then add up the results.

i	1	2	3	Total
x_i	$0	$137	$170	
$P(X = x_i)$	0.20	0.55	0.25	
$x_i \times P(X = x_i)$	0	75.35	42.50	117.85

Thus, the expected value is $\mu = 117.85$, which we computed earlier. The variance can be constructed by extending this table:

i	1	2	3	Total
x_i	$0	$137	$170	
$P(X = x_i)$	0.20	0.55	0.25	
$x_i \times P(X = x_i)$	0	75.35	42.50	117.85
$x_i - \mu$	-117.85	19.15	52.15	
$(x_i - \mu)^2$	13888.62	366.72	2719.62	
$(x_i - \mu)^2 \times P(X = x_i)$	2777.7	201.7	679.9	3659.3

The variance of X is $\sigma^2 = 3659.3$, which means the standard deviation is $\sigma = \sqrt{3659.3} = \$60.49$.

⊙ **Guided Practice 2.74**   The bookstore also offers a chemistry textbook for $159 and a book supplement for $41. From past experience, they know about 25% of chemistry students just buy the textbook while 60% buy both the textbook and supplement.[46]

(a) What proportion of students don't buy either book? Assume no students buy the supplement without the textbook.

(b) Let Y represent the revenue from a single student. Write out the probability distribution of Y, i.e. a table for each outcome and its associated probability.

(c) Compute the expected revenue from a single chemistry student.

(d) Find the standard deviation to describe the variability associated with the revenue from a single student.

2.4.3   Linear combinations of random variables

So far, we have thought of each variable as being a complete story in and of itself. Sometimes it is more appropriate to use a combination of variables. For instance, the amount of time a person spends commuting to work each week can be broken down into several daily commutes. Similarly, the total gain or loss in a stock portfolio is the sum of the gains and losses in its components.

● **Example 2.75**   John travels to work five days a week. We will use X_1 to represent his travel time on Monday, X_2 to represent his travel time on Tuesday, and so on. Write an equation using X_1, ..., X_5 that represents his travel time for the week, denoted by W.

His total weekly travel time is the sum of the five daily values:

$$W = X_1 + X_2 + X_3 + X_4 + X_5$$

Breaking the weekly travel time W into pieces provides a framework for understanding each source of randomness and is useful for modeling W.

● **Example 2.76**   It takes John an average of 18 minutes each day to commute to work. What would you expect his average commute time to be for the week?

We were told that the average (i.e. expected value) of the commute time is 18 minutes per day: $E(X_i) = 18$. To get the expected time for the sum of the five days, we can

---

[46](a) 100% - 25% - 60% = 15% of students do not buy any books for the class. Part (b) is represented by the first two lines in the table below. The expectation for part (c) is given as the total on the line $y_i \times P(Y = y_i)$. The result of part (d) is the square-root of the variance listed on in the total on the last line: $\sigma = \sqrt{Var(Y)} = \$69.28$.

i (scenario)	1 (noBook)	2 (textbook)	3 (both)	Total
y_i	0.00	159.00	200.00	
$P(Y = y_i)$	0.15	0.25	0.60	
$y_i \times P(Y = y_i)$	0.00	39.75	120.00	$E(Y) = 159.75$
$y_i - E(Y)$	-159.75	-0.75	40.25	
$(y_i - E(Y))^2$	25520.06	0.56	1620.06	
$(y_i - E(Y))^2 \times P(Y)$	3828.0	0.1	972.0	$Var(Y) \approx 4800$

add up the expected time for each individual day:

$$E(W) = E(X_1 + X_2 + X_3 + X_4 + X_5)$$
$$= E(X_1) + E(X_2) + E(X_3) + E(X_4) + E(X_5)$$
$$= 18 + 18 + 18 + 18 + 18 = 90 \text{ minutes}$$

The expectation of the total time is equal to the sum of the expected individual times. More generally, the expectation of a sum of random variables is always the sum of the expectation for each random variable.

⊙ **Guided Practice 2.77** Elena is selling a TV at a cash auction and also intends to buy a toaster oven in the auction. If X represents the profit for selling the TV and Y represents the cost of the toaster oven, write an equation that represents the net change in Elena's cash.[47]

⊙ **Guided Practice 2.78** Based on past auctions, Elena figures she should expect to make about $175 on the TV and pay about $23 for the toaster oven. In total, how much should she expect to make or spend?[48]

⊙ **Guided Practice 2.79** Would you be surprised if John's weekly commute wasn't exactly 90 minutes or if Elena didn't make exactly $152? Explain.[49]

Two important concepts concerning combinations of random variables have so far been introduced. First, a final value can sometimes be described as the sum of its parts in an equation. Second, intuition suggests that putting the individual average values into this equation gives the average value we would expect in total. This second point needs clarification – it is guaranteed to be true in what are called *linear combinations of random variables*.

A **linear combination** of two random variables X and Y is a fancy phrase to describe a combination

$$aX + bY$$

where a and b are some fixed and known numbers. For John's commute time, there were five random variables – one for each work day – and each random variable could be written as having a fixed coefficient of 1:

$$1X_1 + 1X_2 + 1X_3 + 1X_4 + 1X_5$$

For Elena's net gain or loss, the X random variable had a coefficient of +1 and the Y random variable had a coefficient of -1.

When considering the average of a linear combination of random variables, it is safe to plug in the mean of each random variable and then compute the final result. For a few examples of nonlinear combinations of random variables – cases where we cannot simply plug in the means – see the footnote.[50]

---

[47]She will make X dollars on the TV but spend Y dollars on the toaster oven: $X - Y$.

[48]$E(X - Y) = E(X) - E(Y) = 175 - 23 = \$152$. She should expect to make about $152.

[49]No, since there is probably some variability. For example, the traffic will vary from one day to next, and auction prices will vary depending on the quality of the merchandise and the interest of the attendees.

[50]If X and Y are random variables, consider the following combinations: X^{1+Y}, $X \times Y$, X/Y. In such cases, plugging in the average value for each random variable and computing the result will not generally lead to an accurate average value for the end result.

Linear combinations of random variables and the average result

If X and Y are random variables, then a linear combination of the random variables is given by

$$aX + bY \qquad (2.80)$$

where a and b are some fixed numbers. To compute the average value of a linear combination of random variables, plug in the average of each individual random variable and compute the result:

$$a \times E(X) + b \times E(Y)$$

Recall that the expected value is the same as the mean, e.g. $E(X) = \mu_X$.

● **Example 2.81** Leonard has invested $6000 in Google Inc. (stock ticker: GOOG) and $2000 in Exxon Mobil Corp. (XOM). If X represents the change in Google's stock next month and Y represents the change in Exxon Mobil stock next month, write an equation that describes how much money will be made or lost in Leonard's stocks for the month.

For simplicity, we will suppose X and Y are not in percents but are in decimal form (e.g. if Google's stock increases 1%, then $X = 0.01$; or if it loses 1%, then $X = -0.01$). Then we can write an equation for Leonard's gain as

$$\$6000 \times X + \$2000 \times Y$$

If we plug in the change in the stock value for X and Y, this equation gives the change in value of Leonard's stock portfolio for the month. A positive value represents a gain, and a negative value represents a loss.

⊙ **Guided Practice 2.82** Suppose Google and Exxon Mobil stocks have recently been rising 2.1% and 0.4% per month, respectively. Compute the expected change in Leonard's stock portfolio for next month.[51]

⊙ **Guided Practice 2.83** You should have found that Leonard expects a positive gain in Guided Practice 2.82. However, would you be surprised if he actually had a loss this month?[52]

2.4.4 Variability in linear combinations of random variables

Quantifying the average outcome from a linear combination of random variables is helpful, but it is also important to have some sense of the uncertainty associated with the total outcome of that combination of random variables. The expected net gain or loss of Leonard's stock portfolio was considered in Guided Practice 2.82. However, there was no quantitative discussion of the volatility of this portfolio. For instance, while the average monthly gain might be about $134 according to the data, that gain is not guaranteed. Figure 2.24 shows the monthly changes in a portfolio like Leonard's during the 36 months from 2009 to 2011. The gains and losses vary widely, and quantifying these fluctuations is important when investing in stocks.

---

[51]$E(\$6000 \times X + \$2000 \times Y) = \$6000 \times 0.021 + \$2000 \times 0.004 = \$134$.
[52]No. While stocks tend to rise over time, they are often volatile in the short term.

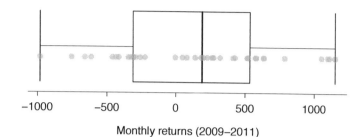

Monthly returns (2009–2011)

Figure 2.24: The change in a portfolio like Leonard's for the 36 months from 2009 to 2011, where $6000 is in Google's stock and $2000 is in Exxon Mobil's.

	Mean ($\bar{x}$)	Standard deviation (s)	Variance (s^2)
GOOG	0.0210	0.0846	0.0072
XOM	0.0038	0.0519	0.0027

Table 2.25: The mean, standard deviation, and variance of the GOOG and XOM stocks. These statistics were estimated from historical stock data, so notation used for sample statistics has been used.

Just as we have done in many previous cases, we use the variance and standard deviation to describe the uncertainty associated with Leonard's monthly returns. To do so, the variances of each stock's monthly return will be useful, and these are shown in Table 2.25. The stocks' returns are nearly independent.

Here we use an equation from probability theory to describe the uncertainty of Leonard's monthly returns; we leave the proof of this method to a dedicated probability course. The variance of a linear combination of random variables can be computed by plugging in the variances of the individual random variables and squaring the coefficients of the random variables:

$$Var(aX + bY) = a^2 \times Var(X) + b^2 \times Var(Y)$$

It is important to note that this equality assumes the random variables are independent; if independence doesn't hold, then more advanced methods are necessary. This equation can be used to compute the variance of Leonard's monthly return:

$$
\begin{aligned}
Var(6000 \times X + 2000 \times Y) &= 6000^2 \times Var(X) + 2000^2 \times Var(Y) \\
&= 36,000,000 \times 0.0072 + 4,000,000 \times 0.0027 \\
&= 270,000
\end{aligned}
$$

The standard deviation is computed as the square root of the variance: $\sqrt{270,000} = \$520$. While an average monthly return of $134 on an $8000 investment is nothing to scoff at, the monthly returns are so volatile that Leonard should not expect this income to be very stable.

Variability of linear combinations of random variables

The variance of a linear combination of random variables may be computed by squaring the constants, substituting in the variances for the random variables, and computing the result:

$$Var(aX + bY) = a^2 \times Var(X) + b^2 \times Var(Y)$$

This equation is valid as long as the random variables are independent of each other. The standard deviation of the linear combination may be found by taking the square root of the variance.

● **Example 2.84** Suppose John's daily commute has a standard deviation of 4 minutes. What is the uncertainty in his total commute time for the week?

The expression for John's commute time was

$$X_1 + X_2 + X_3 + X_4 + X_5$$

Each coefficient is 1, and the variance of each day's time is $4^2 = 16$. Thus, the variance of the total weekly commute time is

$$\text{variance} = 1^2 \times 16 + 1^2 \times 16 + 1^2 \times 16 + 1^2 \times 16 + 1^2 \times 16 = 5 \times 16 = 80$$

$$\text{standard deviation} = \sqrt{\text{variance}} = \sqrt{80} = 8.94$$

The standard deviation for John's weekly work commute time is about 9 minutes.

⊙ **Guided Practice 2.85** The computation in Example 2.84 relied on an important assumption: the commute time for each day is independent of the time on other days of that week. Do you think this is valid? Explain.[53]

⊙ **Guided Practice 2.86** Consider Elena's two auctions from Guided Practice 2.77 on page 109. Suppose these auctions are approximately independent and the variability in auction prices associated with the TV and toaster oven can be described using standard deviations of $25 and $8. Compute the standard deviation of Elena's net gain.[54]

Consider again Guided Practice 2.86. The negative coefficient for Y in the linear combination was eliminated when we squared the coefficients. This generally holds true: negatives in a linear combination will have no impact on the variability computed for a linear combination, but they do impact the expected value computations.

---

[53]One concern is whether traffic patterns tend to have a weekly cycle (e.g. Fridays may be worse than other days). If that is the case, and John drives, then the assumption is probably not reasonable. However, if John walks to work, then his commute is probably not affected by any weekly traffic cycle.

[54]The equation for Elena can be written as

$$(1) \times X + (-1) \times Y$$

The variances of X and Y are 625 and 64. We square the coefficients and plug in the variances:

$$(1)^2 \times Var(X) + (-1)^2 \times Var(Y) = 1 \times 625 + 1 \times 64 = 689$$

The variance of the linear combination is 689, and the standard deviation is the square root of 689: about $26.25.

2.5 Continuous distributions (special topic)

● **Example 2.87** Figure 2.26 shows a few different hollow histograms of the variable height for 3 million US adults from the mid-90's.[55] How does changing the number of bins allow you to make different interpretations of the data?

Adding more bins provides greater detail. This sample is extremely large, which is why much smaller bins still work well. Usually we do not use so many bins with smaller sample sizes since small counts per bin mean the bin heights are very volatile.

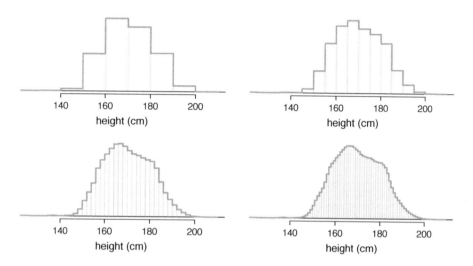

Figure 2.26: Four hollow histograms of US adults heights with varying bin widths.

● **Example 2.88** What proportion of the sample is between 180 cm and 185 cm tall (about 5'11" to 6'1")?

We can add up the heights of the bins in the range 180 cm and 185 and divide by the sample size. For instance, this can be done with the two shaded bins shown in Figure 2.27. The two bins in this region have counts of 195,307 and 156,239 people, resulting in the following estimate of the probability:

$$\frac{195307 + 156239}{3,000,000} = 0.1172$$

This fraction is the same as the proportion of the histogram's area that falls in the range 180 to 185 cm.

---

[55]This sample can be considered a simple random sample from the US population. It relies on the USDA Food Commodity Intake Database.

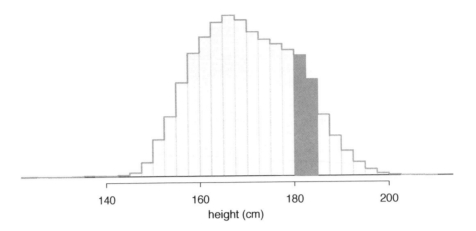

Figure 2.27: A histogram with bin sizes of 2.5 cm. The shaded region represents individuals with heights between 180 and 185 cm.

2.5.1   From histograms to continuous distributions

Examine the transition from a boxy hollow histogram in the top-left of Figure 2.26 to the much smoother plot in the lower-right. In this last plot, the bins are so slim that the hollow histogram is starting to resemble a smooth curve. This suggests the population height as a *continuous* numerical variable might best be explained by a curve that represents the outline of extremely slim bins.

This smooth curve represents a **probability density function** (also called a **density** or **distribution**), and such a curve is shown in Figure 2.28 overlaid on a histogram of the sample. A density has a special property: the total area under the density's curve is 1.

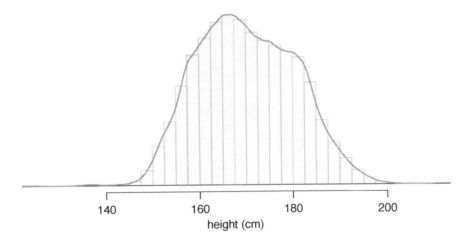

Figure 2.28: The continuous probability distribution of heights for US adults.

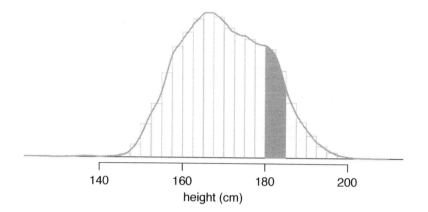

Figure 2.29: Density for heights in the US adult population with the area between 180 and 185 cm shaded. Compare this plot with Figure 2.27.

2.5.2   Probabilities from continuous distributions

We computed the proportion of individuals with heights 180 to 185 cm in Example 2.88 as a fraction:

$$\frac{\text{number of people between 180 and 185}}{\text{total sample size}}$$

We found the number of people with heights between 180 and 185 cm by determining the fraction of the histogram's area in this region. Similarly, we can use the area in the shaded region under the curve to find a probability (with the help of a computer):

$$P(\texttt{height between 180 and 185}) = \text{area between 180 and 185} = 0.1157$$

The probability that a randomly selected person is between 180 and 185 cm is 0.1157. This is very close to the estimate from Example 2.88: 0.1172.

⊙ **Guided Practice 2.89**   Three US adults are randomly selected. The probability a single adult is between 180 and 185 cm is 0.1157.[56]

    (a) What is the probability that all three are between 180 and 185 cm tall?

    (b) What is the probability that none are between 180 and 185 cm?

● **Example 2.90**   What is the probability that a randomly selected person is **exactly** 180 cm? Assume you can measure perfectly.

This probability is zero. A person might be close to 180 cm, but not exactly 180 cm tall. This also makes sense with the definition of probability as area; there is no area captured between 180 cm and 180 cm.

⊙ **Guided Practice 2.91**   Suppose a person's height is rounded to the nearest centimeter. Is there a chance that a random person's **measured** height will be 180 cm?[57]

---

[56]Brief answers: (a) $0.1157 \times 0.1157 \times 0.1157 = 0.0015$. (b) $(1 - 0.1157)^3 = 0.692$

[57]This has positive probability. Anyone between 179.5 cm and 180.5 cm will have a *measured* height of 180 cm. This is probably a more realistic scenario to encounter in practice versus Example 2.90.

2.6 Exercises

2.6.1 Defining probability

2.1 True or false. Determine if the statements below are true or false, and explain your reasoning.

(a) If a fair coin is tossed many times and the last eight tosses are all heads, then the chance that the next toss will be heads is somewhat less than 50%.

(b) Drawing a face card (jack, queen, or king) and drawing a red card from a full deck of playing cards are mutually exclusive events.

(c) Drawing a face card and drawing an ace from a full deck of playing cards are mutually exclusive events.

2.2 Roulette wheel. The game of roulette involves spinning a wheel with 38 slots: 18 red, 18 black, and 2 green. A ball is spun onto the wheel and will eventually land in a slot, where each slot has an equal chance of capturing the ball.

(a) You watch a roulette wheel spin 3 consecutive times and the ball lands on a red slot each time. What is the probability that the ball will land on a red slot on the next spin?

(b) You watch a roulette wheel spin 300 consecutive times and the ball lands on a red slot each time. What is the probability that the ball will land on a red slot on the next spin?

(c) Are you equally confident of your answers to parts (a) and (b)? Why or why not?

Photo by Håkan Dahlström
(http://flic.kr/p/93fEzp)
CC BY 2.0 license

2.3 Four games, one winner. Below are four versions of the same game. Your archnemisis gets to pick the version of the game, and then you get to choose how many times to flip a coin: 10 times or 100 times. Identify how many coin flips you should choose for each version of the game. It costs $1 to play each game. Explain your reasoning.

(a) If the proportion of heads is larger than 0.60, you win $1.

(b) If the proportion of heads is larger than 0.40, you win $1.

(c) If the proportion of heads is between 0.40 and 0.60, you win $1.

(d) If the proportion of heads is smaller than 0.30, you win $1.

2.4 Backgammon. Backgammon is a board game for two players in which the playing pieces are moved according to the roll of two dice. Players win by removing all of their pieces from the board, so it is usually good to roll high numbers. You are playing backgammon with a friend and you roll two 6s in your first roll and two 6s in your second roll. Your friend rolls two 3s in his first roll and again in his second row. Your friend claims that you are cheating, because rolling double 6s twice in a row is very unlikely. Using probability, show that your rolls were just as likely as his.

2.5 Coin flips. If you flip a fair coin 10 times, what is the probability of

(a) getting all tails?

(b) getting all heads?

(c) getting at least one tails?

2.6 Dice rolls. If you roll a pair of fair dice, what is the probability of

(a) getting a sum of 1?

(b) getting a sum of 5?

(c) getting a sum of 12?

2.7 Swing voters. A 2012 Pew Research survey asked 2,373 randomly sampled registered voters their political affiliation (Republican, Democrat, or Independent) and whether or not they identify as swing voters. 35% of respondents identified as Independent, 23% identified as swing voters, and 11% identified as both.[58]

(a) Are being Independent and being a swing voter disjoint, i.e. mutually exclusive?

(b) Draw a Venn diagram summarizing the variables and their associated probabilities.

(c) What percent of voters are Independent but not swing voters?

(d) What percent of voters are Independent or swing voters?

(e) What percent of voters are neither Independent nor swing voters?

(f) Is the event that someone is a swing voter independent of the event that someone is a political Independent?

2.8 Poverty and language. The American Community Survey is an ongoing survey that provides data every year to give communities the current information they need to plan investments and services. The 2010 American Community Survey estimates that 14.6% of Americans live below the poverty line, 20.7% speak a language other than English (foreign language) at home, and 4.2% fall into both categories.[59]

(a) Are living below the poverty line and speaking a foreign language at home disjoint?

(b) Draw a Venn diagram summarizing the variables and their associated probabilities.

(c) What percent of Americans live below the poverty line and only speak English at home?

(d) What percent of Americans live below the poverty line or speak a foreign language at home?

(e) What percent of Americans live above the poverty line and only speak English at home?

(f) Is the event that someone lives below the poverty line independent of the event that the person speaks a foreign language at home?

2.9 Disjoint vs. independent. In parts (a) and (b), identify whether the events are disjoint, independent, or neither (events cannot be both disjoint and independent).

(a) You and a randomly selected student from your class both earn A's in this course.

(b) You and your class study partner both earn A's in this course.

(c) If two events can occur at the same time, must they be dependent?

2.10 Guessing on an exam. In a multiple choice exam, there are 5 questions and 4 choices for each question (a, b, c, d). Nancy has not studied for the exam at all and decides to randomly guess the answers. What is the probability that:

(a) the first question she gets right is the 5th question?

(b) she gets all of the questions right?

(c) she gets at least one question right?

---

[58]Pew Research Center, With Voters Focused on Economy, Obama Lead Narrows, data collected between April 4-15, 2012.

[59]U.S. Census Bureau, 2010 American Community Survey 1-Year Estimates, Characteristics of People by Language Spoken at Home.

2.11 Educational attainment of couples. The table below shows the distribution of education level attained by US residents by gender based on data collected during the 2010 American Community Survey.[60]

		Gender	
		Male	Female
	Less than 9th grade	0.07	0.13
	9th to 12th grade, no diploma	0.10	0.09
Highest	HS graduate (or equivalent)	0.30	0.20
education	Some college, no degree	0.22	0.24
attained	Associate's degree	0.06	0.08
	Bachelor's degree	0.16	0.17
	Graduate or professional degree	0.09	0.09
	Total	1.00	1.00

(a) What is the probability that a randomly chosen man has at least a Bachelor's degree?

(b) What is the probability that a randomly chosen woman has at least a Bachelor's degree?

(c) What is the probability that a man and a woman getting married both have at least a Bachelor's degree? Note any assumptions you must make to answer this question.

(d) If you made an assumption in part (c), do you think it was reasonable? If you didn't make an assumption, double check your earlier answer and then return to this part.

2.12 School absences. Data collected at elementary schools in DeKalb County, GA suggest that each year roughly 25% of students miss exactly one day of school, 15% miss 2 days, and 28% miss 3 or more days due to sickness.[61]

(a) What is the probability that a student chosen at random doesn't miss any days of school due to sickness this year?

(b) What is the probability that a student chosen at random misses no more than one day?

(c) What is the probability that a student chosen at random misses at least one day?

(d) If a parent has two kids at a DeKalb County elementary school, what is the probability that neither kid will miss any school? Note any assumption you must make to answer this question.

(e) If a parent has two kids at a DeKalb County elementary school, what is the probability that both kids will miss some school, i.e. at least one day? Note any assumption you make.

(f) If you made an assumption in part (d) or (e), do you think it was reasonable? If you didn't make any assumptions, double check your earlier answers.

2.13 Grade distributions. Each row in the table below is a proposed grade distribution for a class. Identify each as a valid or invalid probability distribution, and explain your reasoning.

	Grades				
	A	B	C	D	F
(a)	0.3	0.3	0.3	0.2	0.1
(b)	0	0	1	0	0
(c)	0.3	0.3	0.3	0	0
(d)	0.3	0.5	0.2	0.1	-0.1
(e)	0.2	0.4	0.2	0.1	0.1
(f)	0	-0.1	1.1	0	0

---

[60]U.S. Census Bureau, 2010 American Community Survey 1-Year Estimates, Educational Attainment.
[61]S.S. Mizan et al. "Absence, Extended Absence, and Repeat Tardiness Related to Asthma Status among Elementary School Children". In: *Journal of Asthma* 48.3 (2011), pp. 228–234.

2.14 Health coverage, frequencies. The Behavioral Risk Factor Surveillance System (BRFSS) is an annual telephone survey designed to identify risk factors in the adult population and report emerging health trends. The following table summarizes two variables for the respondents: health status and health coverage, which describes whether each respondent had health insurance.[62]

		Excellent	Very good	Good	Fair	Poor	Total
Health	No	459	727	854	385	99	2,524
Coverage	Yes	4,198	6,245	4,821	1,634	578	17,476
	Total	4,657	6,972	5,675	2,019	677	20,000

(Health Status header spans Excellent, Very good, Good, Fair, Poor columns)

(a) If we draw one individual at random, what is the probability that the respondent has excellent health and doesn't have health coverage?

(b) If we draw one individual at random, what is the probability that the respondent has excellent health or doesn't have health coverage?

2.6.2 Conditional probability

2.15 Joint and conditional probabilities. $P(A) = 0.3$, $P(B) = 0.7$

(a) Can you compute $P(A \text{ and } B)$ if you only know $P(A)$ and $P(B)$?

(b) Assuming that events A and B arise from independent random processes,

    i. what is $P(A \text{ and } B)$?

    ii. what is $P(A \text{ or } B)$?

    iii. what is $P(A|B)$?

(c) If we are given that $P(A \text{ and } B) = 0.1$, are the random variables giving rise to events A and B independent?

(d) If we are given that $P(A \text{ and } B) = 0.1$, what is $P(A|B)$?

2.16 PB & J. Suppose 80% of people like peanut butter, 89% like jelly, and 78% like both. Given that a randomly sampled person likes peanut butter, what's the probability that he also likes jelly?

2.17 Global warming. A 2010 Pew Research poll asked 1,306 Americans "From what you've read and heard, is there solid evidence that the average temperature on earth has been getting warmer over the past few decades, or not?". The table below shows the distribution of responses by party and ideology, where the counts have been replaced with relative frequencies.[63]

		Earth is warming	Not warming	Don't Know Refuse	Total
	Conservative Republican	0.11	0.20	0.02	0.33
Party and	Mod/Lib Republican	0.06	0.06	0.01	0.13
Ideology	Mod/Cons Democrat	0.25	0.07	0.02	0.34
	Liberal Democrat	0.18	0.01	0.01	0.20
	Total	0.60	0.34	0.06	1.00

(Response header spans Earth is warming, Not warming, Don't Know Refuse columns)

(a) Are believing that the earth is warming and being a liberal Democrat mutually exclusive?

(b) What is the probability that a randomly chosen respondent believes the earth is warming or is a liberal Democrat? **(See the next page for parts (c)-(f).)**

---

[62]Office of Surveillance, Epidemiology, and Laboratory Services Behavioral Risk Factor Surveillance System, BRFSS 2010 Survey Data.

[63]Pew Research Center, Majority of Republicans No Longer See Evidence of Global Warming, data collected on October 27, 2010.

(c) What is the probability that a randomly chosen respondent believes the earth is warming given that he is a liberal Democrat?

(d) What is the probability that a randomly chosen respondent believes the earth is warming given that he is a conservative Republican?

(e) Does it appear that whether or not a respondent believes the earth is warming is independent of their party and ideology? Explain your reasoning.

(f) What is the probability that a randomly chosen respondent is a moderate/liberal Republican given that he does not believe that the earth is warming?

2.18 Health coverage, relative frequencies. The Behavioral Risk Factor Surveillance System (BRFSS) is an annual telephone survey designed to identify risk factors in the adult population and report emerging health trends. The following table displays the distribution of health status of respondents to this survey (excellent, very good, good, fair, poor) and whether or not they have health insurance.

		Excellent	Very good	Good	Fair	Poor	Total
Health	No	0.0230	0.0364	0.0427	0.0192	0.0050	0.1262
Coverage	Yes	0.2099	0.3123	0.2410	0.0817	0.0289	0.8738
	Total	0.2329	0.3486	0.2838	0.1009	0.0338	1.0000

(a) Are being in excellent health and having health coverage mutually exclusive?

(b) What is the probability that a randomly chosen individual has excellent health?

(c) What is the probability that a randomly chosen individual has excellent health given that he has health coverage?

(d) What is the probability that a randomly chosen individual has excellent health given that he doesn't have health coverage?

(e) Do having excellent health and having health coverage appear to be independent?

2.19 Burger preferences. A 2010 SurveyUSA poll asked 500 Los Angeles residents, "What is the best hamburger place in Southern California? Five Guys Burgers? In-N-Out Burger? Fat Burger? Tommy's Hamburgers? Umami Burger? Or somewhere else?" The distribution of responses by gender is shown below.[64]

		Male	Female	Total
	Five Guys Burgers	5	6	11
	In-N-Out Burger	162	181	343
Best	Fat Burger	10	12	22
hamburger	Tommy's Hamburgers	27	27	54
place	Umami Burger	5	1	6
	Other	26	20	46
	Not Sure	13	5	18
	Total	248	252	500

(a) Are being female and liking Five Guys Burgers mutually exclusive?

(b) What is the probability that a randomly chosen male likes In-N-Out the best?

(c) What is the probability that a randomly chosen female likes In-N-Out the best?

(d) What is the probability that a man and a woman who are dating both like In-N-Out the best? Note any assumption you make and evaluate whether you think that assumption is reasonable.

(e) What is the probability that a randomly chosen person likes Umami best or that person is female?

[64]SurveyUSA, Results of SurveyUSA News Poll #17718, data collected on December 2, 2010.

2.20 Assortative mating. Assortative mating is a nonrandom mating pattern where individuals with similar genotypes and/or phenotypes mate with one another more frequently than what would be expected under a random mating pattern. Researchers studying this topic collected data on eye colors of 204 Scandinavian men and their female partners. The table below summarizes the results. For simplicity, we only include heterosexual relationships in this exercise.[65]

		Partner (female)			
		Blue	Brown	Green	Total
Self (male)	Blue	78	23	13	114
	Brown	19	23	12	54
	Green	11	9	16	36
	Total	108	55	41	204

(a) What is the probability that a randomly chosen male respondent or his partner has blue eyes?

(b) What is the probability that a randomly chosen male respondent with blue eyes has a partner with blue eyes?

(c) What is the probability that a randomly chosen male respondent with brown eyes has a partner with blue eyes? What about the probability of a randomly chosen male respondent with green eyes having a partner with blue eyes?

(d) Does it appear that the eye colors of male respondents and their partners are independent? Explain your reasoning.

2.21 Drawing box plots. After an introductory statistics course, 80% of students can successfully construct box plots. Of those who can construct box plots, 86% passed, while only 65% of those students who could not construct box plots passed.

(a) Construct a tree diagram of this scenario.

(b) Calculate the probability that a student is able to construct a box plot if it is known that he passed.

2.22 Predisposition for thrombosis. A genetic test is used to determine if people have a predisposition for *thrombosis*, which is the formation of a blood clot inside a blood vessel that obstructs the flow of blood through the circulatory system. It is believed that 3% of people actually have this predisposition. The genetic test is 99% accurate if a person actually has the predisposition, meaning that the probability of a positive test result when a person actually has the predisposition is 0.99. The test is 98% accurate if a person does not have the predisposition. What is the probability that a randomly selected person who tests positive for the predisposition by the test actually has the predisposition?

2.23 HIV in Swaziland. Swaziland has the highest HIV prevalence in the world: 25.9% of this country's population is infected with HIV.[66] The ELISA test is one of the first and most accurate tests for HIV. For those who carry HIV, the ELISA test is 99.7% accurate. For those who do not carry HIV, the test is 92.6% accurate. If an individual from Swaziland has tested positive, what is the probability that he carries HIV?

2.24 Exit poll. Edison Research gathered exit poll results from several sources for the Wisconsin recall election of Scott Walker. They found that 53% of the respondents voted in favor of Scott Walker. Additionally, they estimated that of those who did vote in favor for Scott Walker, 37% had a college degree, while 44% of those who voted against Scott Walker had a college degree. Suppose we randomly sampled a person who participated in the exit poll and found that he had a college degree. What is the probability that he voted in favor of Scott Walker?[67]

[65]B. Laeng et al. "Why do blue-eyed men prefer women with the same eye color?" In: *Behavioral Ecology and Sociobiology* 61.3 (2007), pp. 371–384.

[66]Source: CIA Factbook, Country Comparison: HIV/AIDS - Adult Prevalence Rate.

[67]New York Times, Wisconsin recall exit polls.

2.25 It's never lupus. Lupus is a medical phenomenon where antibodies that are supposed to attack foreign cells to prevent infections instead see plasma proteins as foreign bodies, leading to a high risk of blood clotting. It is believed that 2% of the population suffer from this disease. The test is 98% accurate if a person actually has the disease. The test is 74% accurate if a person does not have the disease. There is a line from the Fox television show *House* that is often used after a patient tests positive for lupus: "It's never lupus." Do you think there is truth to this statement? Use appropriate probabilities to support your answer.

2.26 Twins. About 30% of human twins are identical, and the rest are fraternal. Identical twins are necessarily the same sex – half are males and the other half are females. One-quarter of fraternal twins are both male, one-quarter both female, and one-half are mixes: one male, one female. You have just become a parent of twins and are told they are both girls. Given this information, what is the probability that they are identical?

2.6.3 Sampling from a small population

2.27 Marbles in an urn. Imagine you have an urn containing 5 red, 3 blue, and 2 orange marbles in it.

(a) What is the probability that the first marble you draw is blue?

(b) Suppose you drew a blue marble in the first draw. If drawing with replacement, what is the probability of drawing a blue marble in the second draw?

(c) Suppose you instead drew an orange marble in the first draw. If drawing with replacement, what is the probability of drawing a blue marble in the second draw?

(d) If drawing with replacement, what is the probability of drawing two blue marbles in a row?

(e) When drawing with replacement, are the draws independent? Explain.

2.28 Socks in a drawer. In your sock drawer you have 4 blue, 5 gray, and 3 black socks. Half asleep one morning you grab 2 socks at random and put them on. Find the probability you end up wearing

(a) 2 blue socks

(b) no gray socks

(c) at least 1 black sock

(d) a green sock

(e) matching socks

2.29 Chips in a bag. Imagine you have a bag containing 5 red, 3 blue, and 2 orange chips.

(a) Suppose you draw a chip and it is blue. If drawing without replacement, what is the probability the next is also blue?

(b) Suppose you draw a chip and it is orange, and then you draw a second chip without replacement. What is the probability this second chip is blue?

(c) If drawing without replacement, what is the probability of drawing two blue chips in a row?

(d) When drawing without replacement, are the draws independent? Explain.

2.30 Books on a bookshelf. The table below shows the distribution of books on a bookcase based on whether they are nonfiction or fiction and hardcover or paperback.

		Format		Total
		Hardcover	Paperback	
Type	Fiction	13	59	72
	Nonfiction	15	8	23
	Total	28	67	95

(a) Find the probability of drawing a hardcover book first then a paperback fiction book second when drawing without replacement.

(b) Determine the probability of drawing a fiction book first and then a hardcover book second, when drawing without replacement.

(c) Calculate the probability of the scenario in part (b), except this time complete the calculations under the scenario where the first book is placed back on the bookcase before randomly drawing the second book.

(d) The final answers to parts (b) and (c) are very similar. Explain why this is the case.

2.31 Student outfits. In a classroom with 24 students, 7 students are wearing jeans, 4 are wearing shorts, 8 are wearing skirts, and the rest are wearing leggings. If we randomly select 3 students without replacement, what is the probability that one of the selected students is wearing leggings and the other two are wearing jeans? Note that these are mutually exclusive clothing options.

2.32 The birthday problem. Suppose we pick three people at random. For each of the following questions, ignore the special case where someone might be born on February 29th, and assume that births are evenly distributed throughout the year.

(a) What is the probability that the first two people share a birthday?

(b) What is the probability that at least two people share a birthday?

2.6.4 Random variables

2.33 College smokers. At a university, 13% of students smoke.

(a) Calculate the expected number of smokers in a random sample of 100 students from this university.

(b) The university gym opens at 9 am on Saturday mornings. One Saturday morning at 8:55 am there are 27 students outside the gym waiting for it to open. Should you use the same approach from part (a) to calculate the expected number of smokers among these 27 students?

2.34 Ace of clubs wins. Consider the following card game with a well-shuffled deck of cards. If you draw a red card, you win nothing. If you get a spade, you win $5. For any club, you win $10 plus an extra $20 for the ace of clubs.

(a) Create a probability model for the amount you win at this game. Also, find the expected winnings for a single game and the standard deviation of the winnings.

(b) What is the maximum amount you would be willing to pay to play this game? Explain your reasoning.

2.35 Hearts win. In a new card game, you start with a well-shuffled full deck and draw 3 cards without replacement. If you draw 3 hearts, you win $50. If you draw 3 black cards, you win $25. For any other draws, you win nothing.

(a) Create a probability model for the amount you win at this game, and find the expected winnings. Also compute the standard deviation of this distribution.

(b) If the game costs $5 to play, what would be the expected value and standard deviation of the net profit (or loss)? *(Hint: profit = winnings − cost; X − 5)*

(c) If the game costs $5 to play, should you play this game? Explain.

2.36 Is it worth it? Andy is always looking for ways to make money fast. Lately, he has been trying to make money by gambling. Here is the game he is considering playing: The game costs $2 to play. He draws a card from a deck. If he gets a number card (2-10), he wins nothing. For any face card (jack, queen or king), he wins $3. For any ace, he wins $5, and he wins an *extra* $20 if he draws the ace of clubs.

(a) Create a probability model and find Andy's expected profit per game.

(b) Would you recommend this game to Andy as a good way to make money? Explain.

2.37 Portfolio return. A portfolio's value increases by 18% during a financial boom and by 9% during normal times. It decreases by 12% during a recession. What is the expected return on this portfolio if each scenario is equally likely?

2.38 Baggage fees. An airline charges the following baggage fees: $25 for the first bag and $35 for the second. Suppose 54% of passengers have no checked luggage, 34% have one piece of checked luggage and 12% have two pieces. We suppose a negligible portion of people check more than two bags.

(a) Build a probability model, compute the average revenue per passenger, and compute the corresponding standard deviation.

(b) About how much revenue should the airline expect for a flight of 120 passengers? With what standard deviation? Note any assumptions you make and if you think they are justified.

2.39 American roulette. The game of American roulette involves spinning a wheel with 38 slots: 18 red, 18 black, and 2 green. A ball is spun onto the wheel and will eventually land in a slot, where each slot has an equal chance of capturing the ball. Gamblers can place bets on red or black. If the ball lands on their color, they double their money. If it lands on another color, they lose their money. Suppose you bet $1 on red. What's the expected value and standard deviation of your winnings?

2.40 European roulette. The game of European roulette involves spinning a wheel with 37 slots: 18 red, 18 black, and 1 green. A ball is spun onto the wheel and will eventually land in a slot, where each slot has an equal chance of capturing the ball. Gamblers can place bets on red or black. If the ball lands on their color, they double their money. If it lands on another color, they lose their money.

(a) Suppose you play roulette and bet $3 on a single round. What is the expected value and standard deviation of your total winnings?

(b) Suppose you bet $1 in three different rounds. What is the expected value and standard deviation of your total winnings?

(c) How do your answers to parts (a) and (b) compare? What does this say about the riskiness of the two games?

2.41 Cost of breakfast. Sally gets a cup of coffee and a muffin every day for breakfast from one of the many coffee shops in her neighborhood. She picks a coffee shop each morning at random and independently of previous days. The average price of a cup of coffee is $1.40 with a standard deviation of 30¢($0.30), the average price of a muffin is $2.50 with a standard deviation of 15¢, and the two prices are independent of each other.

(a) What is the mean and standard deviation of the amount she spends on breakfast daily?

(b) What is the mean and standard deviation of the amount she spends on breakfast weekly (7 days)?

2.42 Scooping ice cream. Ice cream usually comes in 1. 5 quart boxes (48 fluid ounces), and ice cream scoops hold about 2 ounces. However, there is some variability in the amount of ice cream in a box as well as the amount of ice cream scooped out. We represent the amount of ice cream in the box as X and the amount scooped out as Y. Suppose these random variables have the following means, standard deviations, and variances:

	mean	SD	variance
X	48	1	1
Y	2	0.25	0.0625

(a) An entire box of ice cream, plus 3 scoops from a second box is served at a party. How much ice cream do you expect to have been served at this party? What is the standard deviation of the amount of ice cream served?

(b) How much ice cream would you expect to be left in the box after scooping out one scoop of ice cream? That is, find the expected value of $X - Y$. What is the standard deviation of the amount left in the box?

(c) Using the context of this exercise, explain why we add variances when we subtract one random variable from another.

2.6.5 Continuous distributions

2.43 Cat weights. The histogram shown below represents the weights (in kg) of 47 female and 97 male cats.[68]

(a) What fraction of these cats weigh less than 2.5 kg?

(b) What fraction of these cats weigh between 2.5 and 2.75 kg?

(c) What fraction of these cats weigh between 2.75 and 3.5 kg?

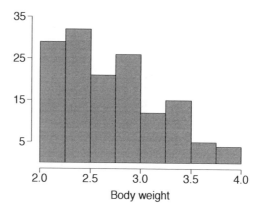

---

[68]W. N. Venables and B. D. Ripley. *Modern Applied Statistics with S.* Fourth Edition. www.stats.ox.ac.uk/pub/MASS4. New York: Springer, 2002.

2.44 Income and gender. The relative frequency table below displays the distribution of annual total personal income (in 2009 inflation-adjusted dollars) for a representative sample of 96,420,486 Americans. These data come from the American Community Survey for 2005-2009. This sample is comprised of 59% males and 41% females.[69]

(a) Describe the distribution of total personal income.

(b) What is the probability that a randomly chosen US resident makes less than $50,000 per year?

(c) What is the probability that a randomly chosen US resident makes less than $50,000 per year and is female? Note any assumptions you make.

(d) The same data source indicates that 71.8% of females make less than $50,000 per year. Use this value to determine whether or not the assumption you made in part (c) is valid.

Income	Total
$1 to $9,999 or loss	2.2%
$10,000 to $14,999	4.7%
$15,000 to $24,999	15.8%
$25,000 to $34,999	18.3%
$35,000 to $49,999	21.2%
$50,000 to $64,999	13.9%
$65,000 to $74,999	5.8%
$75,000 to $99,999	8.4%
$100,000 or more	9.7%

---

[69]U.S. Census Bureau, 2005-2009 American Community Survey.

Chapter 3

Distributions of random variables

3.1 Normal distribution

Among all the distributions we see in practice, one is overwhelmingly the most common. The symmetric, unimodal, bell curve is ubiquitous throughout statistics. Indeed it is so common, that people often know it as the **normal curve** or **normal distribution**,[1] shown in Figure 3.1. Variables such as SAT scores and heights of US adult males closely follow the normal distribution.

> **Normal distribution facts**
>
> Many variables are nearly normal, but none are exactly normal. Thus the normal distribution, while not perfect for any single problem, is very useful for a variety of problems. We will use it in data exploration and to solve important problems in statistics.

---

[1]It is also introduced as the Gaussian distribution after Frederic Gauss, the first person to formalize its mathematical expression.

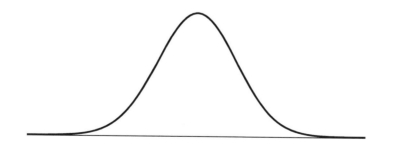

Figure 3.1: A normal curve.

3.1.1   Normal distribution model

The normal distribution model always describes a symmetric, unimodal, bell-shaped curve. However, these curves can look different depending on the details of the model. Specifically, the normal distribution model can be adjusted using two parameters: mean and standard deviation. As you can probably guess, changing the mean shifts the bell curve to the left or right, while changing the standard deviation stretches or constricts the curve. Figure 3.2 shows the normal distribution with mean 0 and standard deviation 1 in the left panel and the normal distributions with mean 19 and standard deviation 4 in the right panel. Figure 3.3 shows these distributions on the same axis.

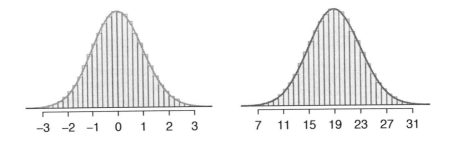

Figure 3.2: Both curves represent the normal distribution, however, they differ in their center and spread. The normal distribution with mean 0 and standard deviation 1 is called the **standard normal distribution**.

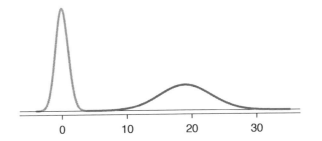

Figure 3.3: The normal models shown in Figure 3.2 but plotted together and on the same scale.

$N(\mu, \sigma)$

Normal dist. with mean μ & st. dev. σ

If a normal distribution has mean μ and standard deviation σ, we may write the distribution as $N(\mu, \sigma)$. The two distributions in Figure 3.3 can be written as

$$N(\mu = 0, \sigma = 1) \quad \text{and} \quad N(\mu = 19, \sigma = 4)$$

Because the mean and standard deviation describe a normal distribution exactly, they are called the distribution's **parameters**.

⊙ **Guided Practice 3.1**   Write down the short-hand for a normal distribution with[2]

(a) mean 5 and standard deviation 3,

(b) mean -100 and standard deviation 10, and

(c) mean 2 and standard deviation 9.

---

[2](a) $N(\mu = 5, \sigma = 3)$. (b) $N(\mu = -100, \sigma = 10)$. (c) $N(\mu = 2, \sigma = 9)$.

	SAT	ACT
Mean	1500	21
SD	300	5

Table 3.4: Mean and standard deviation for the SAT and ACT.

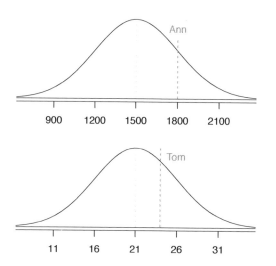

Figure 3.5: Ann's and Tom's scores shown with the distributions of SAT and ACT scores.

3.1.2 Standardizing with Z-scores

● **Example 3.2** Table 3.4 shows the mean and standard deviation for total scores on the SAT and ACT. The distribution of SAT and ACT scores are both nearly normal. Suppose Ann scored 1800 on her SAT and Tom scored 24 on his ACT. Who performed better?

We use the standard deviation as a guide. Ann is 1 standard deviation above average on the SAT: $1500 + 300 = 1800$. Tom is 0.6 standard deviations above the mean on the ACT: $21 + 0.6 \times 5 = 24$. In Figure 3.5, we can see that Ann tends to do better with respect to everyone else than Tom did, so her score was better.

Example 3.2 used a standardization technique called a Z-score, a method most commonly employed for nearly normal observations but that may be used with any distribution. The **Z-score** of an observation is defined as the number of standard deviations it falls above or below the mean. If the observation is one standard deviation above the mean, its Z-score is 1. If it is 1.5 standard deviations *below* the mean, then its Z-score is -1.5. If x is an observation from a distribution $N(\mu, \sigma)$, we define the Z-score mathematically as

Z

Z-score, the standardized observation

$$Z = \frac{x - \mu}{\sigma}$$

Using $\mu_{SAT} = 1500$, $\sigma_{SAT} = 300$, and $x_{Ann} = 1800$, we find Ann's Z-score:

$$Z_{Ann} = \frac{x_{Ann} - \mu_{SAT}}{\sigma_{SAT}} = \frac{1800 - 1500}{300} = 1$$

The Z-score

The Z-score of an observation is the number of standard deviations it falls above or below the mean. We compute the Z-score for an observation x that follows a distribution with mean μ and standard deviation σ using

$$Z = \frac{x - \mu}{\sigma}$$

⊙ **Guided Practice 3.3**   Use Tom's ACT score, 24, along with the ACT mean and standard deviation to compute his Z-score.[3]

Observations above the mean always have positive Z-scores while those below the mean have negative Z-scores. If an observation is equal to the mean (e.g. SAT score of 1500), then the Z-score is 0.

⊙ **Guided Practice 3.4**   Let X represent a random variable from $N(\mu = 3, \sigma = 2)$, and suppose we observe $x = 5.19$. (a) Find the Z-score of x. (b) Use the Z-score to determine how many standard deviations above or below the mean x falls.[4]

⊙ **Guided Practice 3.5**   Head lengths of brushtail possums follow a nearly normal distribution with mean 92.6 mm and standard deviation 3.6 mm. Compute the Z-scores for possums with head lengths of 95.4 mm and 85.8 mm.[5]

We can use Z-scores to roughly identify which observations are more unusual than others. One observation x_1 is said to be more unusual than another observation x_2 if the absolute value of its Z-score is larger than the absolute value of the other observation's Z-score: $|Z_1| > |Z_2|$. This technique is especially insightful when a distribution is symmetric.

⊙ **Guided Practice 3.6**   Which of the observations in Guided Practice 3.5 is more unusual?[6]

3.1.3   Normal probability table

● **Example 3.7**   Ann from Example 3.2 earned a score of 1800 on her SAT with a corresponding $Z = 1$. She would like to know what percentile she falls in among all SAT test-takers.

Ann's **percentile** is the percentage of people who earned a lower SAT score than Ann. We shade the area representing those individuals in Figure 3.6. The total area under the normal curve is always equal to 1, and the proportion of people who scored below Ann on the SAT is equal to the *area* shaded in Figure 3.6: 0.8413. In other words, Ann is in the 84th percentile of SAT takers.

---

[3]$Z_{Tom} = \frac{x_{Tom} - \mu_{ACT}}{\sigma_{ACT}} = \frac{24-21}{5} = 0.6$

[4](a) Its Z-score is given by $Z = \frac{x-\mu}{\sigma} = \frac{5.19-3}{2} = 2.19/2 = 1.095$. (b) The observation x is 1.095 standard deviations *above* the mean. We know it must be above the mean since Z is positive.

[5]For $x_1 = 95.4$ mm: $Z_1 = \frac{x_1-\mu}{\sigma} = \frac{95.4-92.6}{3.6} = 0.78$. For $x_2 = 85.8$ mm: $Z_2 = \frac{85.8-92.6}{3.6} = -1.89$.

[6]Because the *absolute value* of Z-score for the second observation is larger than that of the first, the second observation has a more unusual head length.

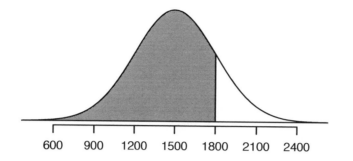

Figure 3.6: The normal model for SAT scores, shading the area of those individuals who scored below Ann.

Figure 3.7: The area to the left of Z represents the percentile of the observation.

We can use the normal model to find percentiles. A **normal probability table**, which lists Z-scores and corresponding percentiles, can be used to identify a percentile based on the Z-score (and vice versa). Statistical software can also be used.

A normal probability table is given in Appendix B.1 on page 427 and abbreviated in Table 3.8. We use this table to identify the percentile corresponding to any particular Z-score. For instance, the percentile of $Z = 0.43$ is shown in row 0.4 and column 0.03 in Table 3.8: 0.6664, or the 66.64^{th} percentile. Generally, we round Z to two decimals, identify the proper row in the normal probability table up through the first decimal, and then determine the column representing the second decimal value. The intersection of this row and column is the percentile of the observation.

We can also find the Z-score associated with a percentile. For example, to identify Z for the 80th percentile, we look for the value closest to 0.8000 in the middle portion of the table: 0.7995. We determine the Z-score for the 80th percentile by combining the row and column Z values: 0.84.

⊙ **Guided Practice 3.8**   Determine the proportion of SAT test takers who scored better than Ann on the SAT.[7]

---

[7]If 84% had lower scores than Ann, the proportion of people who had better scores must be 16%. (Generally ties are ignored when the normal model, or any other continuous distribution, is used.)

Z	Second decimal place of Z									
	0.00	0.01	0.02	*0.03*	**0.04**	0.05	0.06	0.07	0.08	0.09
0.0	0.5000	0.5040	0.5080	0.5120	0.5160	0.5199	0.5239	0.5279	0.5319	0.5359
0.1	0.5398	0.5438	0.5478	0.5517	0.5557	0.5596	0.5636	0.5675	0.5714	0.5753
0.2	0.5793	0.5832	0.5871	0.5910	0.5948	0.5987	0.6026	0.6064	0.6103	0.6141
0.3	0.6179	0.6217	0.6255	0.6293	0.6331	0.6368	0.6406	0.6443	0.6480	0.6517
0.4	0.6554	0.6591	0.6628	*0.6664*	0.6700	0.6736	0.6772	0.6808	0.6844	0.6879
0.5	0.6915	0.6950	0.6985	0.7019	0.7054	0.7088	0.7123	0.7157	0.7190	0.7224
0.6	0.7257	0.7291	0.7324	0.7357	0.7389	0.7422	0.7454	0.7486	0.7517	0.7549
0.7	0.7580	0.7611	0.7642	0.7673	0.7704	0.7734	0.7764	0.7794	0.7823	0.7852
0.8	0.7881	0.7910	0.7939	0.7967	**0.7995**	0.8023	0.8051	0.8078	0.8106	0.8133
0.9	0.8159	0.8186	0.8212	0.8238	0.8264	0.8289	0.8315	0.8340	0.8365	0.8389
1.0	0.8413	0.8438	0.8461	0.8485	0.8508	0.8531	0.8554	0.8577	0.8599	0.8621
1.1	0.8643	0.8665	0.8686	0.8708	0.8729	0.8749	0.8770	0.8790	0.8810	0.8830
⋮	⋮	⋮	⋮	⋮	⋮	⋮	⋮	⋮	⋮	⋮

Table 3.8: A section of the normal probability table. The percentile for a normal random variable with $Z = 0.43$ has been *highlighted*, and the percentile closest to 0.8000 has also been **highlighted**.

3.1.4   Normal probability examples

Cumulative SAT scores are approximated well by a normal model, $N(\mu = 1500, \sigma = 300)$.

● **Example 3.9** Shannon is a randomly selected SAT taker, and nothing is known about Shannon's SAT aptitude. What is the probability Shannon scores at least 1630 on her SATs?

First, always draw and label a picture of the normal distribution. (Drawings need not be exact to be useful.) We are interested in the chance she scores above 1630, so we shade this upper tail:

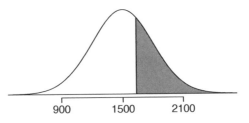

The picture shows the mean and the values at 2 standard deviations above and below the mean. The simplest way to find the shaded area under the curve makes use of the Z-score of the cutoff value. With $\mu = 1500$, $\sigma = 300$, and the cutoff value $x = 1630$, the Z-score is computed as

$$Z = \frac{x - \mu}{\sigma} = \frac{1630 - 1500}{300} = \frac{130}{300} = 0.43$$

We look up the percentile of $Z = 0.43$ in the normal probability table shown in Table 3.8 or in Appendix B.1 on page 427, which yields 0.6664. However, the percentile describes those who had a Z-score *lower* than 0.43. To find the area *above* $Z = 0.43$, we compute one minus the area of the lower tail:

$$1.0000 \quad - \quad 0.6664 \quad = \quad 0.3336$$

The probability Shannon scores at least 1630 on the SAT is 0.3336.

TIP: always draw a picture first, and find the Z-score second

For any normal probability situation, *always always always* draw and label the normal curve and shade the area of interest first. The picture will provide an estimate of the probability.

After drawing a figure to represent the situation, identify the Z-score for the observation of interest.

⊙ **Guided Practice 3.10**   If the probability of Shannon scoring at least 1630 is 0.3336, then what is the probability she scores less than 1630? Draw the normal curve representing this exercise, shading the lower region instead of the upper one.[8]

● **Example 3.11**   Edward earned a 1400 on his SAT. What is his percentile?

First, a picture is needed. Edward's percentile is the proportion of people who do not get as high as a 1400. These are the scores to the left of 1400.

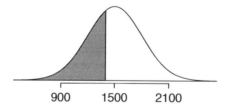

Identifying the mean $\mu = 1500$, the standard deviation $\sigma = 300$, and the cutoff for the tail area $x = 1400$ makes it easy to compute the Z-score:

$$Z = \frac{x - \mu}{\sigma} = \frac{1400 - 1500}{300} = -0.33$$

Using the normal probability table, identify the row of -0.3 and column of 0.03, which corresponds to the probability 0.3707. Edward is at the 37th percentile.

⊙ **Guided Practice 3.12**   Use the results of Example 3.11 to compute the proportion of SAT takers who did better than Edward. Also draw a new picture.[9]

---

[8]We found the probability in Example 3.9: 0.6664. A picture for this exercise is represented by the shaded area below "0.6664" in Example 3.9.

[9]If Edward did better than 37% of SAT takers, then about 63% must have done better than him.

> **TIP: areas to the right**
> The normal probability table in most books gives the area to the left. If you would like the area to the right, first find the area to the left and then subtract this amount from one.

⊙ **Guided Practice 3.13**   Stuart earned an SAT score of 2100. Draw a picture for each part. (a) What is his percentile? (b) What percent of SAT takers did better than Stuart?[10]

Based on a sample of 100 men,[11] the heights of male adults between the ages 20 and 62 in the US is nearly normal with mean 70.0" and standard deviation 3.3".

⊙ **Guided Practice 3.14**   Mike is 5'7" and Jim is 6'4". (a) What is Mike's height percentile? (b) What is Jim's height percentile? Also draw one picture for each part.[12]

The last several problems have focused on finding the probability or percentile for a particular observation. What if you would like to know the observation corresponding to a particular percentile?

● **Example 3.15**   Erik's height is at the 40th percentile. How tall is he?

As always, first draw the picture.

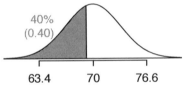

In this case, the lower tail probability is known (0.40), which can be shaded on the diagram. We want to find the observation that corresponds to this value. As a first step in this direction, we determine the Z-score associated with the 40th percentile.

Because the percentile is below 50%, we know Z will be negative. Looking in the negative part of the normal probability table, we search for the probability *inside* the table closest to 0.4000. We find that 0.4000 falls in row -0.2 and between columns 0.05 and 0.06. Since it falls closer to 0.05, we take this one: $Z = -0.25$.

Knowing $Z_{Erik} = -0.25$ and the population parameters $\mu = 70$ and $\sigma = 3.3$ inches, the Z-score formula can be set up to determine Erik's unknown height, labeled x_{Erik}:

$$-0.25 = Z_{Erik} = \frac{x_{Erik} - \mu}{\sigma} = \frac{x_{Erik} - 70}{3.3}$$

Solving for x_{Erik} yields the height 69.18 inches. That is, Erik is about 5'9" (this is notation for 5-feet, 9-inches).

---

[10]Numerical answers: (a) 0.9772. (b) 0.0228.

[11]This sample was taken from the USDA Food Commodity Intake Database.

[12]First put the heights into inches: 67 and 76 inches. Figures are shown below. (a) $Z_{Mike} = \frac{67-70}{3.3} = -0.91 \rightarrow 0.1814$. (b) $Z_{Jim} = \frac{76-70}{3.3} = 1.82 \rightarrow 0.9656$.

● **Example 3.16**   What is the adult male height at the 82nd percentile?

Again, we draw the figure first.

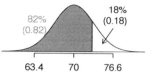

Next, we want to find the Z-score at the 82nd percentile, which will be a positive value. Looking in the Z-table, we find Z falls in row 0.9 and the nearest column is 0.02, i.e. $Z = 0.92$. Finally, the height x is found using the Z-score formula with the known mean μ, standard deviation σ, and Z-score $Z = 0.92$:

$$0.92 = Z = \frac{x - \mu}{\sigma} = \frac{x - 70}{3.3}$$

This yields 73.04 inches or about 6'1" as the height at the 82nd percentile.

⊙ **Guided Practice 3.17**   (a) What is the 95th percentile for SAT scores? (b) What is the 97.5^{th} percentile of the male heights? As always with normal probability problems, first draw a picture.[13]

⊙ **Guided Practice 3.18**   (a) What is the probability that a randomly selected male adult is at least 6'2" (74 inches)? (b) What is the probability that a male adult is shorter than 5'9" (69 inches)?[14]

● **Example 3.19**   What is the probability that a random adult male is between 5'9" and 6'2"?

These heights correspond to 69 inches and 74 inches. First, draw the figure. The area of interest is no longer an upper or lower tail.

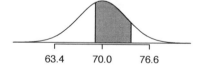

The total area under the curve is 1. If we find the area of the two tails that are not shaded (from Guided Practice 3.18, these areas are 0.3821 and 0.1131), then we can find the middle area:

$$1.0000 \ - \ 0.3821 \ - \ 0.1131 \ = \ 0.5048$$

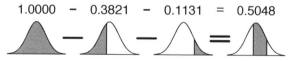

That is, the probability of being between 5'9" and 6'2" is 0.5048.

---

[13]Remember: draw a picture first, then find the Z-score. (We leave the pictures to you.) The Z-score can be found by using the percentiles and the normal probability table. (a) We look for 0.95 in the probability portion (middle part) of the normal probability table, which leads us to row 1.6 and (about) column 0.05, i.e. $Z_{95} = 1.65$. Knowing $Z_{95} = 1.65$, $\mu = 1500$, and $\sigma = 300$, we setup the Z-score formula: $1.65 = \frac{x_{95} - 1500}{300}$. We solve for x_{95}: $x_{95} = 1995$. (b) Similarly, we find $Z_{97.5} = 1.96$, again setup the Z-score formula for the heights, and calculate $x_{97.5} = 76.5$.

[14]Numerical answers: (a) 0.1131. (b) 0.3821.

⊙ **Guided Practice 3.20**   What percent of SAT takers get between 1500 and 2000?[15]

⊙ **Guided Practice 3.21**   What percent of adult males are between 5'5" and 5'7"?[16]

▶ Calculator videos

Videos covering calculations for the normal distribution using TI and Casio graphing calculators are available at openintro.org/videos.

3.1.5   68-95-99.7 rule

Here, we present a useful rule of thumb for the probability of falling within 1, 2, and 3 standard deviations of the mean in the normal distribution. This will be useful in a wide range of practical settings, especially when trying to make a quick estimate without a calculator or Z-table.

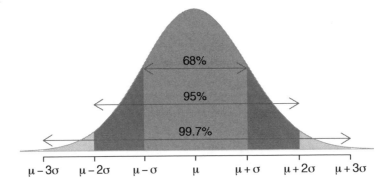

Figure 3.9: Probabilities for falling within 1, 2, and 3 standard deviations of the mean in a normal distribution.

⊙ **Guided Practice 3.22**   Use the Z-table to confirm that about 68%, 95%, and 99.7% of observations fall within 1, 2, and 3, standard deviations of the mean in the normal distribution, respectively. For instance, first find the area that falls between $Z = -1$ and $Z = 1$, which should have an area of about 0.68. Similarly there should be an area of about 0.95 between $Z = -2$ and $Z = 2$.[17]

It is possible for a normal random variable to fall 4, 5, or even more standard deviations from the mean. However, these occurrences are very rare if the data are nearly normal. The probability of being further than 4 standard deviations from the mean is about 1-in-15,000. For 5 and 6 standard deviations, it is about 1-in-2 million and 1-in-500 million, respectively.

---

[15]This is an abbreviated solution. (Be sure to draw a figure!) First find the percent who get below 1500 and the percent that get above 2000: $Z_{1500} = 0.00 \to 0.5000$ (area below), $Z_{2000} = 1.67 \to 0.0475$ (area above). Final answer: $1.0000 - 0.5000 - 0.0475 = 0.4525$.

[16]5'5" is 65 inches. 5'7" is 67 inches. Numerical solution: $1.000 - 0.0649 - 0.8183 = 0.1168$, i.e. 11.68%.

[17]First draw the pictures. To find the area between $Z = -1$ and $Z = 1$, use the normal probability table to determine the areas below $Z = -1$ and above $Z = 1$. Next verify the area between $Z = -1$ and $Z = 1$ is about 0.68. Repeat this for $Z = -2$ to $Z = 2$ and also for $Z = -3$ to $Z = 3$.

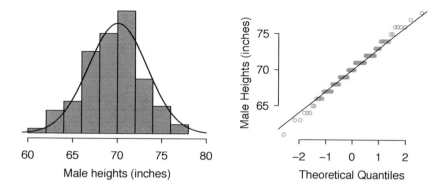

Figure 3.10: A sample of 100 male heights. The observations are rounded to the nearest whole inch, explaining why the points appear to jump in increments in the normal probability plot.

⊙ **Guided Practice 3.23** SAT scores closely follow the normal model with mean $\mu = 1500$ and standard deviation $\sigma = 300$. (a) About what percent of test takers score 900 to 2100? (b) What percent score between 1500 and 2100?[18]

3.2 Evaluating the normal approximation

Many processes can be well approximated by the normal distribution. We have already seen two good examples: SAT scores and the heights of US adult males. While using a normal model can be extremely convenient and helpful, it is important to remember normality is always an approximation. Testing the appropriateness of the normal assumption is a key step in many data analyses.

Example 3.15 suggests the distribution of heights of US males is well approximated by the normal model. We are interested in proceeding under the assumption that the data are normally distributed, but first we must check to see if this is reasonable.

There are two visual methods for checking the assumption of normality, which can be implemented and interpreted quickly. The first is a simple histogram with the best fitting normal curve overlaid on the plot, as shown in the left panel of Figure 3.10. The sample mean $\bar{x}$ and standard deviation s are used as the parameters of the best fitting normal curve. The closer this curve fits the histogram, the more reasonable the normal model assumption. Another more common method is examining a **normal probability plot**,[19] shown in the right panel of Figure 3.10. The closer the points are to a perfect straight line, the more confident we can be that the data follow the normal model.

---

[18](a) 900 and 2100 represent two standard deviations above and below the mean, which means about 95% of test takers will score between 900 and 2100. (b) Since the normal model is symmetric, then half of the test takers from part (a) ($\frac{95\%}{2} = 47.5\%$ of all test takers) will score 900 to 1500 while 47.5% score between 1500 and 2100.

[19]Also commonly called a **quantile-quantile plot**.

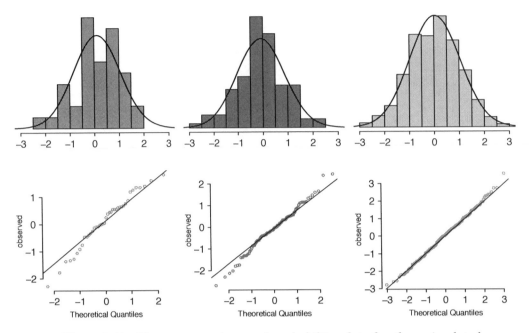

Figure 3.11: Histograms and normal probability plots for three simulated normal data sets; $n = 40$ (left), $n = 100$ (middle), $n = 400$ (right).

● **Example 3.24** Three data sets of 40, 100, and 400 samples were simulated from a normal distribution, and the histograms and normal probability plots of the data sets are shown in Figure 3.11. These will provide a benchmark for what to look for in plots of real data.

The left panels show the histogram (top) and normal probability plot (bottom) for the simulated data set with 40 observations. The data set is too small to really see clear structure in the histogram. The normal probability plot also reflects this, where there are some deviations from the line. We should expect deviations of this amount for such a small data set.

The middle panels show diagnostic plots for the data set with 100 simulated observations. The histogram shows more normality and the normal probability plot shows a better fit. While there are a few observations that deviate noticeably from the line, they are not particularly extreme.

The data set with 400 observations has a histogram that greatly resembles the normal distribution, while the normal probability plot is nearly a perfect straight line. Again in the normal probability plot there is one observation (the largest) that deviates slightly from the line. If that observation had deviated 3 times further from the line, it would be of greater importance in a real data set. Apparent outliers can occur in normally distributed data but they are rare.

Notice the histograms look more normal as the sample size increases, and the normal probability plot becomes straighter and more stable.

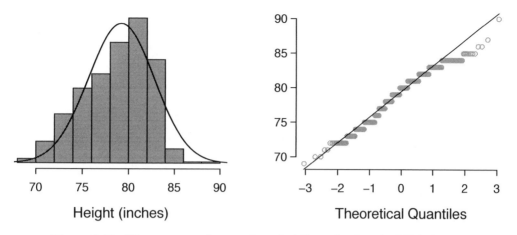

Figure 3.12: Histogram and normal probability plot for the NBA heights from the 2008-9 season.

● **Example 3.25** Are NBA player heights normally distributed? Consider all 435 NBA players from the 2008-9 season presented in Figure 3.12.[20]

We first create a histogram and normal probability plot of the NBA player heights. The histogram in the left panel is slightly left skewed, which contrasts with the symmetric normal distribution. The points in the normal probability plot do not appear to closely follow a straight line but show what appears to be a "wave". We can compare these characteristics to the sample of 400 normally distributed observations in Example 3.24 and see that they represent much stronger deviations from the normal model. NBA player heights do not appear to come from a normal distribution.

● **Example 3.26** Can we approximate poker winnings by a normal distribution? We consider the poker winnings of an individual over 50 days. A histogram and normal probability plot of these data are shown in Figure 3.13.

The data are very strongly right skewed in the histogram, which corresponds to the very strong deviations on the upper right component of the normal probability plot. If we compare these results to the sample of 40 normal observations in Example 3.24, it is apparent that these data show very strong deviations from the normal model.

⊙ **Guided Practice 3.27** Determine which data sets represented in Figure 3.14 plausibly come from a nearly normal distribution. Are you confident in all of your conclusions? There are 100 (top left), 50 (top right), 500 (bottom left), and 15 points (bottom right) in the four plots.[21]

---

[20]These data were collected from www.nba.com.

[21]Answers may vary a little. The top-left plot shows some deviations in the smallest values in the data set; specifically, the left tail of the data set has some outliers we should be wary of. The top-right and bottom-left plots do not show any obvious or extreme deviations from the lines for their respective sample sizes, so a normal model would be reasonable for these data sets. The bottom-right plot has a consistent curvature that suggests it is not from the normal distribution. If we examine just the vertical coordinates of these observations, we see that there is a lot of data between -20 and 0, and then about five observations scattered between 0 and 70. This describes a distribution that has a strong right skew.

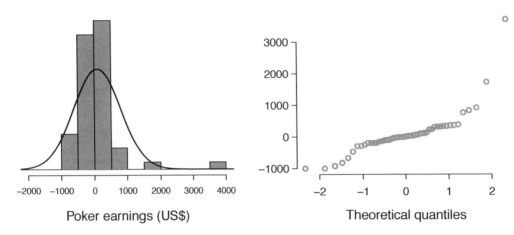

Figure 3.13: A histogram of poker data with the best fitting normal plot and a normal probability plot.

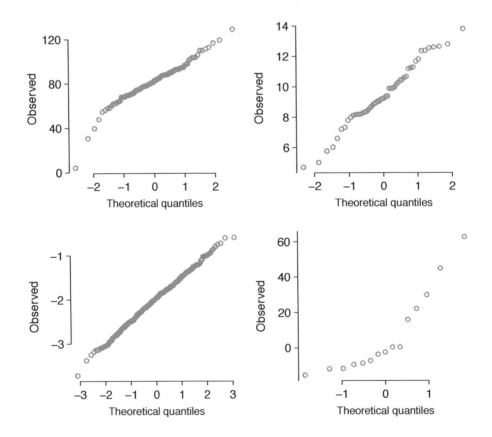

Figure 3.14: Four normal probability plots for Guided Practice 3.27.

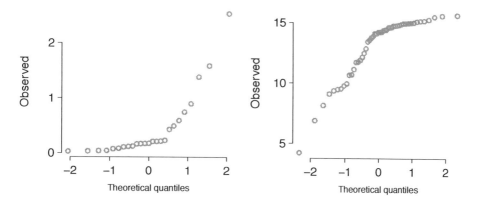

Figure 3.15: Normal probability plots for Guided Practice 3.28.

⊙ **Guided Practice 3.28** Figure 3.15 shows normal probability plots for two distributions that are skewed. One distribution is skewed to the low end (left skewed) and the other to the high end (right skewed). Which is which?[22]

3.3 Geometric distribution (special topic)

How long should we expect to flip a coin until it turns up `heads`? Or how many times should we expect to roll a die until we get a 1? These questions can be answered using the geometric distribution. We first formalize each trial – such as a single coin flip or die toss – using the Bernoulli distribution, and then we combine these with our tools from probability (Chapter 2) to construct the geometric distribution.

3.3.1 Bernoulli distribution

Stanley Milgram began a series of experiments in 1963 to estimate what proportion of people would willingly obey an authority and give severe shocks to a stranger. Milgram found that about 65% of people would obey the authority and give such shocks. Over the years, additional research suggested this number is approximately consistent across communities and time.[23]

Each person in Milgram's experiment can be thought of as a **trial**. We label a person a **success** if she refuses to administer the worst shock. A person is labeled a **failure** if she administers the worst shock. Because only 35% of individuals refused to administer the most severe shock, we denote the **probability of a success** with $p = 0.35$. The probability of a failure is sometimes denoted with $q = 1 - p$.

Thus, `success` or `failure` is recorded for each person in the study. When an individual trial only has two possible outcomes, it is called a **Bernoulli random variable**.

---

[22]Examine where the points fall along the vertical axis. In the first plot, most points are near the low end with fewer observations scattered along the high end; this describes a distribution that is skewed to the high end. The second plot shows the opposite features, and this distribution is skewed to the low end.

[23]Find further information on Milgram's experiment at
www.cnr.berkeley.edu/ucce50/ag-labor/7article/article35.htm.

> **Bernoulli random variable, descriptive**
>
> A Bernoulli random variable has exactly two possible outcomes. We typically label one of these outcomes a "success" and the other outcome a "failure". We may also denote a success by 1 and a failure by 0.

> **TIP: "success" need not be something positive**
>
> We chose to label a person who refuses to administer the worst shock a "success" and all others as "failures". However, we could just as easily have reversed these labels. The mathematical framework we will build does not depend on which outcome is labeled a success and which a failure, as long as we are consistent.

Bernoulli random variables are often denoted as 1 for a success and 0 for a failure. In addition to being convenient in entering data, it is also mathematically handy. Suppose we observe ten trials:

$$0\ 1\ 1\ 1\ 1\ 0\ 1\ 0\ 1\ 0\ 0$$

Then the **sample proportion**, $\hat{p}$, is the sample mean of these observations:

$$\hat{p} = \frac{\#\text{ of successes}}{\#\text{ of trials}} = \frac{0+1+1+1+1+0+1+1+0+0}{10} = 0.6$$

This mathematical inquiry of Bernoulli random variables can be extended even further. Because 0 and 1 are numerical outcomes, we can define the mean and standard deviation of a Bernoulli random variable.[24]

> **Bernoulli random variable, mathematical**
>
> If X is a random variable that takes value 1 with probability of success p and 0 with probability $1 - p$, then X is a Bernoulli random variable with mean and standard deviation
>
> $$\mu = p \qquad\qquad \sigma = \sqrt{p(1-p)}$$

In general, it is useful to think about a Bernoulli random variable as a random process with only two outcomes: a success or failure. Then we build our mathematical framework using the numerical labels 1 and 0 for successes and failures, respectively.

---

[24]If p is the true probability of a success, then the mean of a Bernoulli random variable X is given by

$$\mu = E[X] = P(X = 0) \times 0 + P(X = 1) \times 1$$
$$= (1 - p) \times 0 + p \times 1 = 0 + p = p$$

Similarly, the variance of X can be computed:

$$\sigma^2 = P(X = 0)(0 - p)^2 + P(X = 1)(1 - p)^2$$
$$= (1 - p)p^2 + p(1 - p)^2 = p(1 - p)$$

The standard deviation is $\sigma = \sqrt{p(1-p)}$.

3.3.2 Geometric distribution

● **Example 3.29** Dr. Smith wants to repeat Milgram's experiments but she only wants to sample people until she finds someone who will not inflict the worst shock.[25] If the probability a person will *not* give the most severe shock is still 0.35 and the subjects are independent, what are the chances that she will stop the study after the first person? The second person? The third? What about if it takes her $n - 1$ individuals who will administer the worst shock before finding her first success, i.e. the first success is on the nth person? (If the first success is the fifth person, then we say $n = 5$.)

The probability of stopping after the first person is just the chance the first person will not administer the worst shock: $1 - 0.65 = 0.35$. The probability it will be the second person is

$$P(\text{second person is the first to not administer the worst shock})$$
$$= P(\text{the first will, the second won't}) = (0.65)(0.35) = 0.228$$

Likewise, the probability it will be the third person is $(0.65)(0.65)(0.35) = 0.148$.

If the first success is on the nth person, then there are $n - 1$ failures and finally 1 success, which corresponds to the probability $(0.65)^{n-1}(0.35)$. This is the same as $(1 - 0.35)^{n-1}(0.35)$.

Example 3.29 illustrates what is called the geometric distribution, which describes the waiting time until a success for **independent and identically distributed (iid)** Bernoulli random variables. In this case, the *independence* aspect just means the individuals in the example don't affect each other, and *identical* means they each have the same probability of success.

The geometric distribution from Example 3.29 is shown in Figure 3.16. In general, the probabilities for a geometric distribution decrease **exponentially** fast.

While this text will not derive the formulas for the mean (expected) number of trials needed to find the first success or the standard deviation or variance of this distribution, we present general formulas for each.

Geometric Distribution

If the probability of a success in one trial is p and the probability of a failure is $1 - p$, then the probability of finding the first success in the nth trial is given by

$$(1 - p)^{n-1}p \tag{3.30}$$

The mean (i.e. expected value), variance, and standard deviation of this wait time are given by

$$\mu = \frac{1}{p} \qquad\qquad \sigma^2 = \frac{1 - p}{p^2} \qquad\qquad \sigma = \sqrt{\frac{1 - p}{p^2}} \tag{3.31}$$

It is no accident that we use the symbol μ for both the mean and expected value. The mean and the expected value are one and the same.

---

[25]This is hypothetical since, in reality, this sort of study probably would not be permitted any longer under current ethical standards.

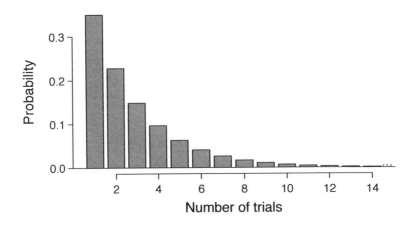

Figure 3.16: The geometric distribution when the probability of success is $p = 0.35$.

The left side of Equation (3.31) says that, on average, it takes $1/p$ trials to get a success. This mathematical result is consistent with what we would expect intuitively. If the probability of a success is high (e.g. 0.8), then we don't usually wait very long for a success: $1/0.8 = 1.25$ trials on average. If the probability of a success is low (e.g. 0.1), then we would expect to view many trials before we see a success: $1/0.1 = 10$ trials.

⊙ **Guided Practice 3.32** The probability that an individual would refuse to administer the worst shock is said to be about 0.35. If we were to examine individuals until we found one that did not administer the shock, how many people should we expect to check? The first expression in Equation (3.31) may be useful.[26]

● **Example 3.33** What is the chance that Dr. Smith will find the first success within the first 4 people?

This is the chance it is the first ($n = 1$), second ($n = 2$), third ($n = 3$), or fourth ($n = 4$) person as the first success, which are four disjoint outcomes. Because the individuals in the sample are randomly sampled from a large population, they are independent. We compute the probability of each case and add the separate results:

$$P(n = 1, 2, 3, \text{ or } 4)$$
$$= P(n = 1) + P(n = 2) + P(n = 3) + P(n = 4)$$
$$= (0.65)^{1-1}(0.35) + (0.65)^{2-1}(0.35) + (0.65)^{3-1}(0.35) + (0.65)^{4-1}(0.35)$$
$$= 0.82$$

There is an 82% chance that she will end the study within 4 people.

⊙ **Guided Practice 3.34** Determine a more clever way to solve Example 3.33. Show that you get the same result.[27]

---

[26]We would expect to see about $1/0.35 = 2.86$ individuals to find the first success.

[27]First find the probability of the complement: $P(\text{no success in first 4 trials}) = 0.65^4 = 0.18$. Next, compute one minus this probability: $1 - P(\text{no success in 4 trials}) = 1 - 0.18 = 0.82$.

● **Example 3.35** Suppose in one region it was found that the proportion of people who would administer the worst shock was "only" 55%. If people were randomly selected from this region, what is the expected number of people who must be checked before one was found that would be deemed a success? What is the standard deviation of this waiting time?

A success is when someone will **not** inflict the worst shock, which has probability $p = 1 - 0.55 = 0.45$ for this region. The expected number of people to be checked is $1/p = 1/0.45 = 2.22$ and the standard deviation is $\sqrt{(1-p)/p^2} = 1.65$.

⊙ **Guided Practice 3.36** Using the results from Example 3.35, $\mu = 2.22$ and $\sigma = 1.65$, would it be appropriate to use the normal model to find what proportion of experiments would end in 3 or fewer trials?[28]

The independence assumption is crucial to the geometric distribution's accurate description of a scenario. Mathematically, we can see that to construct the probability of the success on the nth trial, we had to use the Multiplication Rule for Independent Processes. It is no simple task to generalize the geometric model for dependent trials.

3.4 Binomial distribution (special topic)

● **Example 3.37** Suppose we randomly selected four individuals to participate in the "shock" study. What is the chance exactly one of them will be a success? Let's call the four people Allen (A), Brittany (B), Caroline (C), and Damian (D) for convenience. Also, suppose 35% of people are successes as in the previous version of this example.

Let's consider a scenario where one person refuses:

$$P(A = \texttt{refuse}, \ B = \texttt{shock}, \ C = \texttt{shock}, \ D = \texttt{shock})$$
$$= P(A = \texttt{refuse}) \ P(B = \texttt{shock}) \ P(C = \texttt{shock}) \ P(D = \texttt{shock})$$
$$= (0.35)(0.65)(0.65)(0.65) = (0.35)^1(0.65)^3 = 0.096$$

But there are three other scenarios: Brittany, Caroline, or Damian could have been the one to refuse. In each of these cases, the probability is again $(0.35)^1(0.65)^3$. These four scenarios exhaust all the possible ways that exactly one of these four people could refuse to administer the most severe shock, so the total probability is $4 \times (0.35)^1(0.65)^3 = 0.38$.

⊙ **Guided Practice 3.38** Verify that the scenario where Brittany is the only one to refuse to give the most severe shock has probability $(0.35)^1(0.65)^3$. [29]

---

[28]No. The geometric distribution is always right skewed and can never be well-approximated by the normal model.

[29]$P(A = \texttt{shock}, \ B = \texttt{refuse}, \ C = \texttt{shock}, \ D = \texttt{shock}) = (0.65)(0.35)(0.65)(0.65) = (0.35)^1(0.65)^3$.

3.4.1 The binomial distribution

The scenario outlined in Example 3.37 is a special case of what is called the binomial distribution. The **binomial distribution** describes the probability of having exactly k successes in n independent Bernoulli trials with probability of a success p (in Example 3.37, $n = 4$, $k = 1$, $p = 0.35$). We would like to determine the probabilities associated with the binomial distribution more generally, i.e. we want a formula where we can use n, k, and p to obtain the probability. To do this, we reexamine each part of the example.

There were four individuals who could have been the one to refuse, and each of these four scenarios had the same probability. Thus, we could identify the final probability as

$$[\# \text{ of scenarios}] \times P(\text{single scenario}) \tag{3.39}$$

The first component of this equation is the number of ways to arrange the $k = 1$ successes among the $n = 4$ trials. The second component is the probability of any of the four (equally probable) scenarios.

Consider $P(\text{single scenario})$ under the general case of k successes and $n - k$ failures in the n trials. In any such scenario, we apply the Multiplication Rule for independent events:

$$p^k (1 - p)^{n-k}$$

This is our general formula for $P(\text{single scenario})$.

Secondly, we introduce a general formula for the number of ways to choose k successes in n trials, i.e. arrange k successes and $n - k$ failures:

$$\binom{n}{k} = \frac{n!}{k!(n-k)!}$$

The quantity $\binom{n}{k}$ is read **n choose k**.[30] The exclamation point notation (e.g. $k!$) denotes a **factorial** expression.

$$0! = 1$$
$$1! = 1$$
$$2! = 2 \times 1 = 2$$
$$3! = 3 \times 2 \times 1 = 6$$
$$4! = 4 \times 3 \times 2 \times 1 = 24$$
$$\vdots$$
$$n! = n \times (n-1) \times ... \times 3 \times 2 \times 1$$

Using the formula, we can compute the number of ways to choose $k = 1$ successes in $n = 4$ trials:

$$\binom{4}{1} = \frac{4!}{1!(4-1)!} = \frac{4!}{1!3!} = \frac{4 \times 3 \times 2 \times 1}{(1)(3 \times 2 \times 1)} = 4$$

This result is exactly what we found by carefully thinking of each possible scenario in Example 3.37.

Substituting n choose k for the number of scenarios and $p^k (1 - p)^{n-k}$ for the single scenario probability in Equation (3.39) yields the general binomial formula.

---

[30]Other notation for n choose k includes $_nC_k$, C_n^k, and $C(n, k)$.

Binomial distribution

Suppose the probability of a single trial being a success is p. Then the probability of observing exactly k successes in n independent trials is given by

$$\binom{n}{k} p^k (1-p)^{n-k} = \frac{n!}{k!(n-k)!} p^k (1-p)^{n-k} \tag{3.40}$$

Additionally, the mean, variance, and standard deviation of the number of observed successes are

$$\mu = np \qquad \sigma^2 = np(1-p) \qquad \sigma = \sqrt{np(1-p)} \tag{3.41}$$

TIP: Is it binomial? Four conditions to check.
(1) The trials are independent.
(2) The number of trials, n, is fixed.
(3) Each trial outcome can be classified as a *success* or *failure*.
(4) The probability of a success, p, is the same for each trial.

● **Example 3.42** What is the probability that 3 of 8 randomly selected students will refuse to administer the worst shock, i.e. 5 of 8 will?

We would like to apply the binomial model, so we check our conditions. The number of trials is fixed ($n = 8$) (condition 2) and each trial outcome can be classified as a success or failure (condition 3). Because the sample is random, the trials are independent (condition 1) and the probability of a success is the same for each trial (condition 4).

In the outcome of interest, there are $k = 3$ successes in $n = 8$ trials, and the probability of a success is $p = 0.35$. So the probability that 3 of 8 will refuse is given by

$$\binom{8}{3}(0.35)^3(1-0.35)^{8-3} = \frac{8!}{3!(8-3)!}(0.35)^3(1-0.35)^{8-3}$$
$$= \frac{8!}{3!5!}(0.35)^3(0.65)^5$$

Dealing with the factorial part:

$$\frac{8!}{3!5!} = \frac{8 \times 7 \times 6 \times 5 \times 4 \times 3 \times 2 \times 1}{(3 \times 2 \times 1)(5 \times 4 \times 3 \times 2 \times 1)} = \frac{8 \times 7 \times 6}{3 \times 2 \times 1} = 56$$

Using $(0.35)^3(0.65)^5 \approx 0.005$, the final probability is about $56 * 0.005 = 0.28$.

TIP: computing binomial probabilities
The first step in using the binomial model is to check that the model is appropriate. The second step is to identify n, p, and k. The final step is to apply the formulas and interpret the results.

TIP: computing n choose k

In general, it is useful to do some cancelation in the factorials immediately. Alternatively, many computer programs and calculators have built in functions to compute n choose k, factorials, and even entire binomial probabilities.

⊙ **Guided Practice 3.43**   If you ran a study and randomly sampled 40 students, how many would you expect to refuse to administer the worst shock? What is the standard deviation of the number of people who would refuse? Equation (3.41) may be useful.[31]

⊙ **Guided Practice 3.44**   The probability that a random smoker will develop a severe lung condition in his or her lifetime is about 0.3. If you have 4 friends who smoke, are the conditions for the binomial model satisfied?[32]

⊙ **Guided Practice 3.45**   Suppose these four friends do not know each other and we can treat them as if they were a random sample from the population. Is the binomial model appropriate? What is the probability that (a) none of them will develop a severe lung condition? (b) One will develop a severe lung condition? (c) That no more than one will develop a severe lung condition?[33]

⊙ **Guided Practice 3.46**   What is the probability that at least 2 of your 4 smoking friends will develop a severe lung condition in their lifetimes?[34]

⊙ **Guided Practice 3.47**   Suppose you have 7 friends who are smokers and they can be treated as a random sample of smokers. (a) How many would you expect to develop a severe lung condition, i.e. what is the mean? (b) What is the probability that at most 2 of your 7 friends will develop a severe lung condition.[35]

Next we consider the first term in the binomial probability, n choose k under some special scenarios.

---

[31]We are asked to determine the expected number (the mean) and the standard deviation, both of which can be directly computed from the formulas in Equation (3.41): $\mu = np = 40 \times 0.35 = 14$ and $\sigma = \sqrt{np(1-p)} = \sqrt{40 \times 0.35 \times 0.65} = 3.02$. Because very roughly 95% of observations fall within 2 standard deviations of the mean (see Section 1.6.4), we would probably observe at least 8 but less than 20 individuals in our sample who would refuse to administer the shock.

[32]One possible answer: if the friends know each other, then the independence assumption is probably not satisfied. For example, acquaintances may have similar smoking habits.

[33]To check if the binomial model is appropriate, we must verify the conditions. (i) Since we are supposing we can treat the friends as a random sample, they are independent. (ii) We have a fixed number of trials ($n = 4$). (iii) Each outcome is a success or failure. (iv) The probability of a success is the same for each trials since the individuals are like a random sample ($p = 0.3$ if we say a "success" is someone getting a lung condition, a morbid choice). Compute parts (a) and (b) from the binomial formula in Equation (3.40): $P(0) = \binom{4}{0}(0.3)^0(0.7)^4 = 1 \times 1 \times 0.7^4 = 0.2401$, $P(1) = \binom{4}{1}(0.3)^1(0.7)^3 = 0.4116$. Note: $0! = 1$, as shown on page 146. Part (c) can be computed as the sum of parts (a) and (b): $P(0) + P(1) = 0.2401 + 0.4116 = 0.6517$. That is, there is about a 65% chance that no more than one of your four smoking friends will develop a severe lung condition.

[34]The complement (no more than one will develop a severe lung condition) as computed in Guided Practice 3.45 as 0.6517, so we compute one minus this value: 0.3483.

[35](a) $\mu = 0.3 \times 7 = 2.1$. (b) $P(0, 1, \text{ or } 2 \text{ develop severe lung condition}) = P(k = 0) + P(k = 1) + P(k = 2) = 0.6471$.

⊙ **Guided Practice 3.48**   Why is it true that $\binom{n}{0} = 1$ and $\binom{n}{n} = 1$ for any number n?[36]

⊙ **Guided Practice 3.49**   How many ways can you arrange one success and $n - 1$ failures in n trials? How many ways can you arrange $n - 1$ successes and one failure in n trials?[37]

⧉ Calculator videos

Videos covering calculations for the binomial coefficient and binomial formula using TI and Casio graphing calculators are available at openintro.org/videos.

3.4.2   Normal approximation to the binomial distribution

The binomial formula is cumbersome when the sample size (n) is large, particularly when we consider a range of observations. In some cases we may use the normal distribution as an easier and faster way to estimate binomial probabilities.

● **Example 3.50**   Approximately 20% of the US population smokes cigarettes. A local government believed their community had a lower smoker rate and commissioned a survey of 400 randomly selected individuals. The survey found that only 59 of the 400 participants smoke cigarettes. If the true proportion of smokers in the community was really 20%, what is the probability of observing 59 or fewer smokers in a sample of 400 people?

We leave the usual verification that the four conditions for the binomial model are valid as an exercise.

The question posed is equivalent to asking, what is the probability of observing $k = 0$, 1, ..., 58, or 59 smokers in a sample of $n = 400$ when $p = 0.20$? We can compute these 60 different probabilities and add them together to find the answer:

$$P(k = 0 \text{ or } k = 1 \text{ or } \cdots \text{ or } k = 59)$$
$$= P(k = 0) + P(k = 1) + \cdots + P(k = 59)$$
$$= 0.0041$$

If the true proportion of smokers in the community is $p = 0.20$, then the probability of observing 59 or fewer smokers in a sample of $n = 400$ is less than 0.0041.

The computations in Example 3.50 are tedious and long. In general, we should avoid such work if an alternative method exists that is faster, easier, and still accurate. Recall that calculating probabilities of a range of values is much easier in the normal model. We

---

[36] Frame these expressions into words. How many different ways are there to arrange 0 successes and n failures in n trials? (1 way.) How many different ways are there to arrange n successes and 0 failures in n trials? (1 way.)

[37] One success and $n - 1$ failures: there are exactly n unique places we can put the success, so there are n ways to arrange one success and $n - 1$ failures. A similar argument is used for the second question. Mathematically, we show these results by verifying the following two equations:

$$\binom{n}{1} = n, \qquad \binom{n}{n-1} = n$$

might wonder, is it reasonable to use the normal model in place of the binomial distribution? Surprisingly, yes, if certain conditions are met.

⊙ **Guided Practice 3.51**   Here we consider the binomial model when the probability of a success is $p = 0.10$. Figure 3.17 shows four hollow histograms for simulated samples from the binomial distribution using four different sample sizes: $n = 10, 30, 100, 300$. What happens to the shape of the distributions as the sample size increases? What distribution does the last hollow histogram resemble?[38]

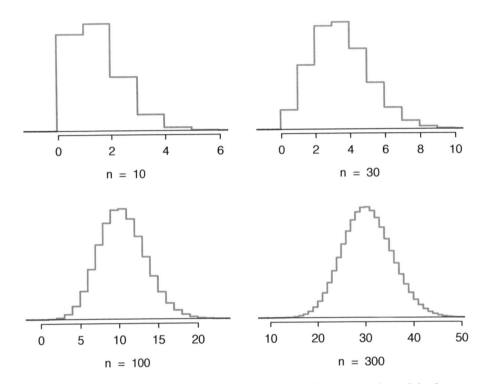

n = 10   n = 30

n = 100   n = 300

Figure 3.17: Hollow histograms of samples from the binomial model when $p = 0.10$. The sample sizes for the four plots are $n = 10$, 30, 100, and 300, respectively.

Normal approximation of the binomial distribution

The binomial distribution with probability of success p is nearly normal when the sample size n is sufficiently large that np and $n(1 - p)$ are both at least 10. The approximate normal distribution has parameters corresponding to the mean and standard deviation of the binomial distribution:

$$\mu = np \qquad\qquad \sigma = \sqrt{np(1 - p)}$$

---

[38]The distribution is transformed from a blocky and skewed distribution into one that rather resembles the normal distribution in last hollow histogram

The normal approximation may be used when computing the range of many possible successes. For instance, we may apply the normal distribution to the setting of Example 3.50.

● **Example 3.52** How can we use the normal approximation to estimate the probability of observing 59 or fewer smokers in a sample of 400, if the true proportion of smokers is $p = 0.20$?

Showing that the binomial model is reasonable was a suggested exercise in Example 3.50. We also verify that both np and $n(1 - p)$ are at least 10:

$$np = 400 \times 0.20 = 80 \qquad\qquad n(1 - p) = 400 \times 0.8 = 320$$

With these conditions checked, we may use the normal approximation in place of the binomial distribution using the mean and standard deviation from the binomial model:

$$\mu = np = 80 \qquad\qquad \sigma = \sqrt{np(1 - p)} = 8$$

We want to find the probability of observing fewer than 59 smokers using this model.

⊙ **Guided Practice 3.53** Use the normal model $N(\mu = 80, \sigma = 8)$ to estimate the probability of observing fewer than 59 smokers. Your answer should be approximately equal to the solution of Example 3.50: 0.0041.[39]

3.4.3 The normal approximation breaks down on small intervals

---

Caution: The normal approximation may fail on small intervals

The normal approximation to the binomial distribution tends to perform poorly when estimating the probability of a small range of counts, even when the conditions are met.

Suppose we wanted to compute the probability of observing 69, 70, or 71 smokers in 400 when $p = 0.20$. With such a large sample, we might be tempted to apply the normal approximation and use the range 69 to 71. However, we would find that the binomial solution and the normal approximation notably differ:

$$\text{Binomial: } 0.0703 \qquad\qquad \text{Normal: } 0.0476$$

We can identify the cause of this discrepancy using Figure 3.18, which shows the areas representing the binomial probability (outlined) and normal approximation (shaded). Notice that the width of the area under the normal distribution is 0.5 units too slim on both sides of the interval.

TIP: Improving the accuracy of the normal approximation to the binomial distribution

The normal approximation to the binomial distribution for intervals of values is usually improved if cutoff values are modified slightly. The cutoff values for the lower end of a shaded region should be reduced by 0.5, and the cutoff value for the upper end should be increased by 0.5.

---

[39]Compute the Z-score first: $Z = \frac{59 - 80}{8} = -2.63$. The corresponding left tail area is 0.0043.

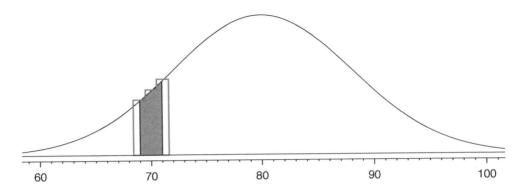

Figure 3.18: A normal curve with the area between 69 and 71 shaded. The outlined area represents the exact binomial probability.

The tip to add extra area when applying the normal approximation is most often useful when examining a range of observations. While it is possible to apply it when computing a tail area, the benefit of the modification usually disappears since the total interval is typically quite wide.

3.5   More discrete distributions (special topic)

3.5.1   Negative binomial distribution

The geometric distribution describes the probability of observing the first success on the nth trial. The **negative binomial distribution** is more general: it describes the probability of observing the kth success on the nth trial.

⬤ **Example 3.54** Each day a high school football coach tells his star kicker, Brian, that he can go home after he successfully kicks four 35 yard field goals. Suppose we say each kick has a probability p of being successful. If p is small – e.g. close to 0.1 – would we expect Brian to need many attempts before he successfully kicks his fourth field goal?

We are waiting for the fourth success ($k = 4$). If the probability of a success (p) is small, then the number of attempts (n) will probably be large. This means that Brian is more likely to need many attempts before he gets $k = 4$ successes. To put this another way, the probability of n being small is low.

To identify a negative binomial case, we check 4 conditions. The first three are common to the binomial distribution.[40]

TIP: Is it negative binomial? Four conditions to check.
(1) The trials are independent.
(2) Each trial outcome can be classified as a success or failure.
(3) The probability of a success (p) is the same for each trial.
(4) The last trial must be a success.

---

[40]See a similar guide for the binomial distribution on page 147.

⊙ **Guided Practice 3.55** Suppose Brian is very diligent in his attempts and he makes each 35 yard field goal with probability $p = 0.8$. Take a guess at how many attempts he would need before making his fourth kick.[41]

⬤ **Example 3.56** In yesterday's practice, it took Brian only 6 tries to get his fourth field goal. Write out each of the possible sequence of kicks.

Because it took Brian six tries to get the fourth success, we know the last kick must have been a success. That leaves three successful kicks and two unsuccessful kicks (we label these as failures) that make up the first five attempts. There are ten possible sequences of these first five kicks, which are shown in Table 3.19. If Brian achieved his fourth success ($k = 4$) on his sixth attempt ($n = 6$), then his order of successes and failures must be one of these ten possible sequences.

Kick Attempt

	1	2	3	4	5	6
1	F	F	S^1	S^2	S^3	S^4
2	F	S^1	F	S^2	S^3	S^4
3	F	S^1	S^2	F	S^3	S^4
4	F	S^1	S^2	S^3	F	S^4
5	S^1	F	F	S^2	S^3	S^4
6	S^1	F	S^2	F	S^3	S^4
7	S^1	F	S^2	S^3	F	S^4
8	S^1	S^2	F	F	S^3	S^4
9	S^1	S^2	F	S^3	F	S^4
10	S^1	S^2	S^3	F	F	S^4

Table 3.19: The ten possible sequences when the fourth successful kick is on the sixth attempt.

⊙ **Guided Practice 3.57** Each sequence in Table 3.19 has exactly two failures and four successes with the last attempt always being a success. If the probability of a success is $p = 0.8$, find the probability of the first sequence.[42]

If the probability Brian kicks a 35 yard field goal is $p = 0.8$, what is the probability it takes Brian exactly six tries to get his fourth successful kick? We can write this as

P(it takes Brian six tries to make four field goals)

    $= P$(Brian makes three of his first five field goals, and he makes the sixth one)

    $= P(1^{st}$ sequence OR 2nd sequence OR ... OR 10th sequence)

---

[41]One possible answer: since he is likely to make each field goal attempt, it will take him at least 4 attempts but probably not more than 6 or 7.
[42]The first sequence: $0.2 \times 0.2 \times 0.8 \times 0.8 \times 0.8 \times 0.8 = 0.0164$.

where the sequences are from Table 3.19. We can break down this last probability into the sum of ten disjoint possibilities:

$$P(1^{st} \text{ sequence OR } 2^{nd} \text{ sequence OR ... OR } 10^{th} \text{ sequence})$$
$$= P(1^{st} \text{ sequence}) + P(2^{nd} \text{ sequence}) + \cdots + P(10^{th} \text{ sequence})$$

The probability of the first sequence was identified in Guided Practice 3.57 as 0.0164, and each of the other sequences have the same probability. Since each of the ten sequence has the same probability, the total probability is ten times that of any individual sequence.

The way to compute this negative binomial probability is similar to how the binomial problems were solved in Section 3.4. The probability is broken into two pieces:

$$P(\text{it takes Brian six tries to make four field goals})$$
$$= [\text{Number of possible sequences}] \times P(\text{Single sequence})$$

Each part is examined separately, then we multiply to get the final result.

We first identify the probability of a single sequence. One particular case is to first observe all the failures ($n - k$ of them) followed by the k successes:

$$P(\text{Single sequence})$$
$$= P(n - k \text{ failures and then } k \text{ successes})$$
$$= (1 - p)^{n-k} p^k$$

We must also identify the number of sequences for the general case. Above, ten sequences were identified where the fourth success came on the sixth attempt. These sequences were identified by fixing the last observation as a success and looking for all the ways to arrange the other observations. In other words, how many ways could we arrange $k - 1$ successes in $n - 1$ trials? This can be found using the n choose k coefficient but for $n - 1$ and $k - 1$ instead:

$$\binom{n-1}{k-1} = \frac{(n-1)!}{(k-1)!\,((n-1)-(k-1))!} = \frac{(n-1)!}{(k-1)!\,(n-k)!}$$

This is the number of different ways we can order $k - 1$ successes and $n - k$ failures in $n - 1$ trials. If the factorial notation (the exclamation point) is unfamiliar, see page 146.

Negative binomial distribution

The negative binomial distribution describes the probability of observing the kth success on the nth trial:

$$P(\text{the } k^{th} \text{ success on the } n^{th} \text{ trial}) = \binom{n-1}{k-1} p^k (1 - p)^{n-k} \qquad (3.58)$$

where p is the probability an individual trial is a success. All trials are assumed to be independent.

● **Example 3.59** Show using Equation (3.58) that the probability Brian kicks his fourth successful field goal on the sixth attempt is 0.164.

The probability of a single success is $p = 0.8$, the number of successes is $k = 4$, and the number of necessary attempts under this scenario is $n = 6$.

$$\binom{n-1}{k-1} p^k (1-p)^{n-k} = \frac{5!}{3!2!}(0.8)^4(0.2)^2 = 10 \times 0.0164 = 0.164$$

⊙ **Guided Practice 3.60** The negative binomial distribution requires that each kick attempt by Brian is independent. Do you think it is reasonable to suggest that each of Brian's kick attempts are independent?[43]

⊙ **Guided Practice 3.61** Assume Brian's kick attempts are independent. What is the probability that Brian will kick his fourth field goal within 5 attempts?[44]

TIP: Binomial versus negative binomial

In the binomial case, we typically have a fixed number of trials and instead consider the number of successes. In the negative binomial case, we examine how many trials it takes to observe a fixed number of successes and require that the last observation be a success.

⊙ **Guided Practice 3.62** On 70% of days, a hospital admits at least one heart attack patient. On 30% of the days, no heart attack patients are admitted. Identify each case below as a binomial or negative binomial case, and compute the probability.[45]

(a) What is the probability the hospital will admit a heart attack patient on exactly three days this week?

(b) What is the probability the second day with a heart attack patient will be the fourth day of the week?

(c) What is the probability the fifth day of next month will be the first day with a heart attack patient?

---

[43]Answers may vary. We cannot conclusively say they are or are not independent. However, many statistical reviews of athletic performance suggests such attempts are very nearly independent.

[44]If his fourth field goal ($k = 4$) is within five attempts, it either took him four or five tries ($n = 4$ or $n = 5$). We have $p = 0.8$ from earlier. Use Equation (3.58) to compute the probability of $n = 4$ tries and $n = 5$ tries, then add those probabilities together:

$$P(n = 4 \text{ OR } n = 5) = P(n = 4) + P(n = 5)$$

$$= \binom{4-1}{4-1}0.8^4 + \binom{5-1}{4-1}(0.8)^4(1-0.8) = 1 \times 0.41 + 4 \times 0.082 = 0.41 + 0.33 = 0.74$$

[45]In each part, $p = 0.7$. (a) The number of days is fixed, so this is binomial. The parameters are $k = 3$ and $n = 7$: 0.097. (b) The last "success" (admitting a heart attack patient) is fixed to the last day, so we should apply the negative binomial distribution. The parameters are $k = 2$, $n = 4$: 0.132. (c) This problem is negative binomial with $k = 1$ and $n = 5$: 0.006. Note that the negative binomial case when $k = 1$ is the same as using the geometric distribution.

3.5.2   Poisson distribution

● **Example 3.63**   There are about 8 million individuals in New York City. How many individuals might we expect to be hospitalized for acute myocardial infarction (AMI), i.e. a heart attack, each day? According to historical records, the average number is about 4.4 individuals. However, we would also like to know the approximate distribution of counts. What would a histogram of the number of AMI occurrences each day look like if we recorded the daily counts over an entire year?

A histogram of the number of occurrences of AMI on 365 days for NYC is shown in Figure 3.20.[46] The sample mean (4.38) is similar to the historical average of 4.4. The sample standard deviation is about 2, and the histogram indicates that about 70% of the data fall between 2.4 and 6.4. The distribution's shape is unimodal and skewed to the right.

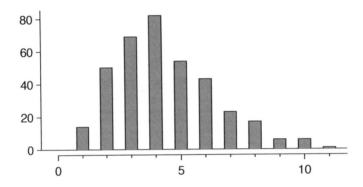

Figure 3.20: A histogram of the number of occurrences of AMI on 365 separate days in NYC.

The **Poisson distribution** is often useful for estimating the number of events in a large population over a unit of time. For instance, consider each of the following events:

- having a heart attack,

- getting married, and

- getting struck by lightning.

The Poisson distribution helps us describe the number of such events that will occur in a short unit of time for a fixed population if the individuals within the population are independent.

The histogram in Figure 3.20 approximates a Poisson distribution with rate equal to 4.4. The **rate** for a Poisson distribution is the average number of occurrences in a mostly-fixed population per unit of time. In Example 3.63, the time unit is a day, the population is all New York City residents, and the historical rate is 4.4. The parameter in the Poisson distribution is the rate – or how many events we expect to observe – and it is typically denoted by λ (the Greek letter *lambda*) or μ. Using the rate, we can describe the probability of observing exactly k events in a single unit of time.

λ
Rate for the
Poisson dist.

---

[46]These data are simulated. In practice, we should check for an association between successive days.

Poisson distribution

Suppose we are watching for events and the number of observed events follows a Poisson distribution with rate λ. Then

$$P(\text{observe } k \text{ events}) = \frac{\lambda^k e^{-\lambda}}{k!}$$

where k may take a value 0, 1, 2, and so on, and $k!$ represents k-factorial, as described on page 146. The letter $e \approx 2.718$ is the base of the natural logarithm. The mean and standard deviation of this distribution are λ and $\sqrt{\lambda}$, respectively.

We will leave a rigorous set of conditions for the Poisson distribution to a later course. However, we offer a few simple guidelines that can be used for an initial evaluation of whether the Poisson model would be appropriate.

TIP: Is it Poisson?

A random variable may follow a Poisson distribution if we are looking for the number of events, the population that generates such events is large, and the events occur independently of each other.

Even when events are not really independent – for instance, Saturdays and Sundays are especially popular for weddings – a Poisson model may sometimes still be reasonable if we allow it to have a different rate for different times. In the wedding example, the rate would be modeled as higher on weekends than on weekdays. The idea of modeling rates for a Poisson distribution against a second variable such as dayOfTheWeek forms the foundation of some more advanced methods that fall in the realm of **generalized linear models**. In Chapters 7 and 8, we will discuss a foundation of linear models.

3.6 Exercises

3.6.1 Normal distribution

3.1 Area under the curve, Part I. What percent of a standard normal distribution $N(\mu = 0, \sigma = 1)$ is found in each region? Be sure to draw a graph.

(a) $Z < -1.35$ (b) $Z > 1.48$ (c) $-0.4 < Z < 1.5$ (d) $|Z| > 2$

3.2 Area under the curve, Part II. What percent of a standard normal distribution $N(\mu = 0, \sigma = 1)$ is found in each region? Be sure to draw a graph.

(a) $Z > -1.13$ (b) $Z < 0.18$ (c) $Z > 8$ (d) $|Z| < 0.5$

3.3 GRE scores, Part I. Sophia who took the Graduate Record Examination (GRE) scored 160 on the Verbal Reasoning section and 157 on the Quantitative Reasoning section. The mean score for Verbal Reasoning section for all test takers was 151 with a standard deviation of 7, and the mean score for the Quantitative Reasoning was 153 with a standard deviation of 7.67. Suppose that both distributions are nearly normal.

(a) Write down the short-hand for these two normal distributions.

(b) What is Sophia's Z-score on the Verbal Reasoning section? On the Quantitative Reasoning section? Draw a standard normal distribution curve and mark these two Z-scores.

(c) What do these Z-scores tell you?

(d) Relative to others, which section did she do better on?

(e) Find her percentile scores for the two exams.

(f) What percent of the test takers did better than her on the Verbal Reasoning section? On the Quantitative Reasoning section?

(g) Explain why simply comparing raw scores from the two sections could lead to an incorrect conclusion as to which section a student did better on.

(h) If the distributions of the scores on these exams are not nearly normal, would your answers to parts (b) - (f) change? Explain your reasoning.

3.4 Triathlon times, Part I. In triathlons, it is common for racers to be placed into age and gender groups. Friends Leo and Mary both completed the Hermosa Beach Triathlon, where Leo competed in the *Men, Ages 30 - 34* group while Mary competed in the *Women, Ages 25 - 29* group. Leo completed the race in 1:22:28 (4948 seconds), while Mary completed the race in 1:31:53 (5513 seconds). Obviously Leo finished faster, but they are curious about how they did within their respective groups. Can you help them? Here is some information on the performance of their groups:

- The finishing times of the *Men, Ages 30 - 34* group has a mean of 4313 seconds with a standard deviation of 583 seconds.
- The finishing times of the *Women, Ages 25 - 29* group has a mean of 5261 seconds with a standard deviation of 807 seconds.
- The distributions of finishing times for both groups are approximately Normal.

Remember: a better performance corresponds to a faster finish.

(a) Write down the short-hand for these two normal distributions.

(b) What are the Z-scores for Leo's and Mary's finishing times? What do these Z-scores tell you?

(c) Did Leo or Mary rank better in their respective groups? Explain your reasoning.

(d) What percent of the triathletes did Leo finish faster than in his group?

(e) What percent of the triathletes did Mary finish faster than in her group?

(f) If the distributions of finishing times are not nearly normal, would your answers to parts (b) - (e) change? Explain your reasoning.

3.5 GRE scores, Part II. In Exercise 3.3 we saw two distributions for GRE scores: $N(\mu = 151, \sigma = 7)$ for the verbal part of the exam and $N(\mu = 153, \sigma = 7.67)$ for the quantitative part. Use this information to compute each of the following:

(a) The score of a student who scored in the 80th percentile on the Quantitative Reasoning section.

(b) The score of a student who scored worse than 70% of the test takers in the Verbal Reasoning section.

3.6 Triathlon times, Part II. In Exercise 3.4 we saw two distributions for triathlon times: $N(\mu = 4313, \sigma = 583)$ for *Men, Ages 30 - 34* and $N(\mu = 5261, \sigma = 807)$ for the *Women, Ages 25 - 29* group. Times are listed in seconds. Use this information to compute each of the following:

(a) The cutoff time for the fastest 5% of athletes in the men's group, i.e. those who took the shortest 5% of time to finish.

(b) The cutoff time for the slowest 10% of athletes in the women's group.

3.7 LA weather, Part I. The average daily high temperature in June in LA is 77°F with a standard deviation of 5°F. Suppose that the temperatures in June closely follow a normal distribution.

(a) What is the probability of observing an 83°F temperature or higher in LA during a randomly chosen day in June?

(b) How cold are the coldest 10% of the days during June in LA?

3.8 CAPM. The Capital Asset Pricing Model (CAPM) is a financial model that assumes returns on a portfolio are normally distributed. Suppose a portfolio has an average annual return of 14.7% (i.e. an average gain of 14.7%) with a standard deviation of 33%. A return of 0% means the value of the portfolio doesn't change, a negative return means that the portfolio loses money, and a positive return means that the portfolio gains money.

(a) What percent of years does this portfolio lose money, i.e. have a return less than 0%?

(b) What is the cutoff for the highest 15% of annual returns with this portfolio?

3.9 LA weather, Part II. Exercise 3.7 states that average daily high temperature in June in LA is 77°F with a standard deviation of 5°F, and it can be assumed that they to follow a normal distribution. We use the following equation to convert °F (Fahrenheit) to °C (Celsius):

$$C = (F - 32) \times \frac{5}{9}.$$

(a) Write the probability model for the distribution of temperature in °C in June in LA.

(b) What is the probability of observing a 28°C (which roughly corresponds to 83°F) temperature or higher in June in LA? Calculate using the °C model from part (a).

(c) Did you get the same answer or different answers in part (b) of this question and part (a) of Exercise 3.7? Are you surprised? Explain.

(d) Estimate the IQR of the temperatures (in °C) in June in LA.

3.10 Heights of 10 year olds. Heights of 10 year olds, regardless of gender, closely follow a normal distribution with mean 55 inches and standard deviation 6 inches.

(a) What is the probability that a randomly chosen 10 year old is shorter than 48 inches?

(b) What is the probability that a randomly chosen 10 year old is between 60 and 65 inches?

(c) If the tallest 10% of the class is considered "very tall", what is the height cutoff for "very tall"?

(d) The height requirement for *Batman the Ride* at Six Flags Magic Mountain is 54 inches. What percent of 10 year olds cannot go on this ride?

3.11 Auto insurance premiums. Suppose a newspaper article states that the distribution of auto insurance premiums for residents of California is approximately normal with a mean of $1,650. The article also states that 25% of California residents pay more than $1,800.

(a) What is the Z-score that corresponds to the top 25% (or the 75th percentile) of the standard normal distribution?

(b) What is the mean insurance cost? What is the cutoff for the 75th percentile?

(c) Identify the standard deviation of insurance premiums in LA.

3.12 Speeding on the I-5, Part I. The distribution of passenger vehicle speeds traveling on the Interstate 5 Freeway (I-5) in California is nearly normal with a mean of 72.6 miles/hour and a standard deviation of 4.78 miles/hour.[47]

(a) What percent of passenger vehicles travel slower than 80 miles/hour?

(b) What percent of passenger vehicles travel between 60 and 80 miles/hour?

(c) How fast do the fastest 5% of passenger vehicles travel?

(d) The speed limit on this stretch of the I-5 is 70 miles/hour. Approximate what percentage of the passenger vehicles travel above the speed limit on this stretch of the I-5.

3.13 Overweight baggage. Suppose weights of the checked baggage of airline passengers follow a nearly normal distribution with mean 45 pounds and standard deviation 3.2 pounds. Most airlines charge a fee for baggage that weigh in excess of 50 pounds. Determine what percent of airline passengers incur this fee.

3.14 Find the SD. Find the standard deviation of the distribution in the following situations.

(a) MENSA is an organization whose members have IQs in the top 2% of the population. IQs are normally distributed with mean 100, and the minimum IQ score required for admission to MENSA is 132.

(b) Cholesterol levels for women aged 20 to 34 follow an approximately normal distribution with mean 185 milligrams per deciliter (mg/dl). Women with cholesterol levels above 220 mg/dl are considered to have high cholesterol and about 18.5% of women fall into this category.

3.15 Buying books on Ebay. The textbook you need to buy for your chemistry class is expensive at the college bookstore, so you consider buying it on Ebay instead. A look at past auctions suggest that the prices of that chemistry textbook have an approximately normal distribution with mean $89 and standard deviation $15.

(a) What is the probability that a randomly selected auction for this book closes at more than $100?

(b) Ebay allows you to set your maximum bid price so that if someone outbids you on an auction you can automatically outbid them, up to the maximum bid price you set. If you are only bidding on one auction, what are the advantages and disadvantages of setting a bid price too high or too low? What if you are bidding on multiple auctions?

(c) If you watched 10 auctions, roughly what percentile might you use for a maximum bid cutoff to be somewhat sure that you will win one of these ten auctions? Is it possible to find a cutoff point that will ensure that you win an auction?

(d) If you are willing to track up to ten auctions closely, about what price might you use as your maximum bid price if you want to be somewhat sure that you will buy one of these ten books?

3.16 SAT scores. SAT scores (out of 2400) are distributed normally with a mean of 1500 and a standard deviation of 300. Suppose a school council awards a certificate of excellence to all students who score at least 1900 on the SAT, and suppose we pick one of the recognized students at random. What is the probability this student's score will be at least 2100? (The material covered in Section 2.2 would be useful for this question.)

[47]S. Johnson and D. Murray. "Empirical Analysis of Truck and Automobile Speeds on Rural Interstates: Impact of Posted Speed Limits". In: *Transportation Research Board 89th Annual Meeting*. 2010.

3.6.2 Evaluating the normal approximation

3.17 Scores on stats final. Below are final exam scores of 20 Introductory Statistics students.

$$\overset{1}{57}, \overset{2}{66}, \overset{3}{69}, \overset{4}{71}, \overset{5}{72}, \overset{6}{73}, \overset{7}{74}, \overset{8}{77}, \overset{9}{78}, \overset{10}{78}, \overset{11}{79}, \overset{12}{79}, \overset{13}{81}, \overset{14}{81}, \overset{15}{82}, \overset{16}{83}, \overset{17}{83}, \overset{18}{88}, \overset{19}{89}, \overset{20}{94}$$

(a) The mean score is 77.7 points. with a standard deviation of 8.44 points. Use this information to determine if the scores approximately follow the 68-95-99.7% Rule.

(b) Do these data appear to follow a normal distribution? Explain your reasoning using the graphs provided below.

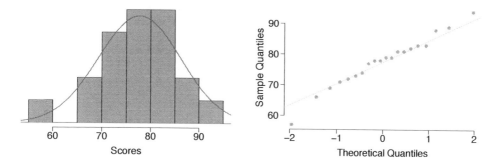

3.18 Heights of female college students. Below are heights of 25 female college students.

$$\overset{1}{54}, \overset{2}{55}, \overset{3}{56}, \overset{4}{56}, \overset{5}{57}, \overset{6}{58}, \overset{7}{58}, \overset{8}{59}, \overset{9}{60}, \overset{10}{60}, \overset{11}{60}, \overset{12}{61}, \overset{13}{61}, \overset{14}{62}, \overset{15}{62}, \overset{16}{63}, \overset{17}{63}, \overset{18}{63}, \overset{19}{64}, \overset{20}{65}, \overset{21}{65}, \overset{22}{67}, \overset{23}{67}, \overset{24}{69}, \overset{25}{73}$$

(a) The mean height is 61.52 inches with a standard deviation of 4.58 inches. Use this information to determine if the heights approximately follow the 68-95-99.7% Rule.

(b) Do these data appear to follow a normal distribution? Explain your reasoning using the graphs provided below.

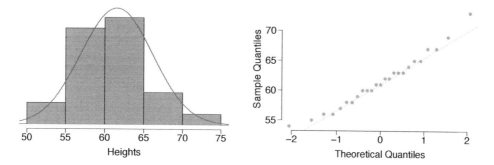

3.6.3   Geometric distribution

3.19   Is it Bernoulli? Determine if each trial can be considered an independent Bernoulli trial for the following situations.

(a) Cards dealt in a hand of poker.

(b) Outcome of each roll of a die.

3.20   With and without replacement. In the following situations assume that half of the specified population is male and the other half is female.

(a) Suppose you're sampling from a room with 10 people. What is the probability of sampling two females in a row when sampling with replacement? What is the probability when sampling without replacement?

(b) Now suppose you're sampling from a stadium with 10,000 people. What is the probability of sampling two females in a row when sampling with replacement? What is the probability when sampling without replacement?

(c) We often treat individuals who are sampled from a large population as independent. Using your findings from parts (a) and (b), explain whether or not this assumption is reasonable.

3.21   Married women. The 2010 American Community Survey estimates that 47.1% of women ages 15 years and over are married.[48]

(a) We randomly select three women between these ages. What is the probability that the third woman selected is the only one who is married?

(b) What is the probability that all three randomly selected women are married?

(c) On average, how many women would you expect to sample before selecting a married woman? What is the standard deviation?

(d) If the proportion of married women was actually 30%, how many women would you expect to sample before selecting a married woman? What is the standard deviation?

(e) Based on your answers to parts (c) and (d), how does decreasing the probability of an event affect the mean and standard deviation of the wait time until success?

3.22   Defective rate. A machine that produces a special type of transistor (a component of computers) has a 2% defective rate. The production is considered a random process where each transistor is independent of the others.

(a) What is the probability that the 10th transistor produced is the first with a defect?

(b) What is the probability that the machine produces no defective transistors in a batch of 100?

(c) On average, how many transistors would you expect to be produced before the first with a defect? What is the standard deviation?

(d) Another machine that also produces transistors has a 5% defective rate where each transistor is produced independent of the others. On average how many transistors would you expect to be produced with this machine before the first with a defect? What is the standard deviation?

(e) Based on your answers to parts (c) and (d), how does increasing the probability of an event affect the mean and standard deviation of the wait time until success?

3.23   Eye color, Part I. A husband and wife both have brown eyes but carry genes that make it possible for their children to have brown eyes (probability 0.75), blue eyes (0.125), or green eyes (0.125).

(a) What is the probability the first blue-eyed child they have is their third child? Assume that the eye colors of the children are independent of each other.

(b) On average, how many children would such a pair of parents have before having a blue-eyed child? What is the standard deviation of the number of children they would expect to have until the first blue-eyed child?

---

[48]U.S. Census Bureau, 2010 American Community Survey, Marital Status.

3.24 Speeding on the I-5, Part II. Exercise 3.12 states that the distribution of speeds of cars traveling on the Interstate 5 Freeway (I-5) in California is nearly normal with a mean of 72.6 miles/hour and a standard deviation of 4.78 miles/hour. The speed limit on this stretch of the I-5 is 70 miles/hour.

(a) A highway patrol officer is hidden on the side of the freeway. What is the probability that 5 cars pass and none are speeding? Assume that the speeds of the cars are independent of each other.

(b) On average, how many cars would the highway patrol officer expect to watch until the first car that is speeding? What is the standard deviation of the number of cars he would expect to watch?

3.6.4 Binomial distribution

3.25 Underage drinking, Part I. Data collected by the Substance Abuse and Mental Health Services Administration (SAMSHA) suggests that 69.7% of 18-20 year olds consumed alcoholic beverages in 2008.[49]

(a) Suppose a random sample of ten 18-20 year olds is taken. Is the use of the binomial distribution appropriate for calculating the probability that exactly six consumed alcoholic beverages? Explain.

(b) Calculate the probability that exactly 6 out of 10 randomly sampled 18- 20 year olds consumed an alcoholic drink.

(c) What is the probability that exactly four out of ten 18-20 year olds have *not* consumed an alcoholic beverage?

(d) What is the probability that at most 2 out of 5 randomly sampled 18-20 year olds have consumed alcoholic beverages?

(e) What is the probability that at least 1 out of 5 randomly sampled 18-20 year olds have consumed alcoholic beverages?

3.26 Chicken pox, Part I. The National Vaccine Information Center estimates that 90% of Americans have had chickenpox by the time they reach adulthood.[50]

(a) Is the use of the binomial distribution appropriate for calculating the probability that exactly 97 out of 100 randomly sampled American adults had chickenpox during childhood.

(b) Calculate the probability that exactly 97 out of 100 randomly sampled American adults had chickenpox during childhood.

(c) What is the probability that exactly 3 out of a new sample of 100 American adults have *not* had chickenpox in their childhood?

(d) What is the probability that at least 1 out of 10 randomly sampled American adults have had chickenpox?

(e) What is the probability that at most 3 out of 10 randomly sampled American adults have *not* had chickenpox?

---

[49]SAMHSA, Office of Applied Studies, National Survey on Drug Use and Health, 2007 and 2008.
[50]National Vaccine Information Center, Chickenpox, The Disease & The Vaccine Fact Sheet.

3.27 Underage drinking, Part II. We learned in Exercise 3.25 that about 70% of 18-20 year olds consumed alcoholic beverages in 2008. We now consider a random sample of fifty 18-20 year olds.

(a) How many people would you expect to have consumed alcoholic beverages? And with what standard deviation?

(b) Would you be surprised if there were 45 or more people who have consumed alcoholic beverages?

(c) What is the probability that 45 or more people in this sample have consumed alcoholic beverages? How does this probability relate to your answer to part (b)?

3.28 Chickenpox, Part II. We learned in Exercise 3.26 that about 90% of American adults had chickenpox before adulthood. We now consider a random sample of 120 American adults.

(a) How many people in this sample would you expect to have had chickenpox in their childhood? And with what standard deviation?

(b) Would you be surprised if there were 105 people who have had chickenpox in their childhood?

(c) What is the probability that 105 or fewer people in this sample have had chickenpox in their childhood? How does this probability relate to your answer to part (b)?

3.29 University admissions. Suppose a university announced that it admitted 2,500 students for the following year's freshman class. However, the university has dorm room spots for only 1,786 freshman students. If there is a 70% chance that an admitted student will decide to accept the offer and attend this university, what is the approximate probability that the university will not have enough dormitory room spots for the freshman class?

3.30 Survey response rate. Pew Research reported in 2012 that the typical response rate to their surveys is only 9%. If for a particular survey 15,000 households are contacted, what is the probability that at least 1,500 will agree to respond?[51]

3.31 Game of dreidel. A dreidel is a four-sided spinning top with the Hebrew letters *nun, gimel, hei,* and *shin,* one on each side. Each side is equally likely to come up in a single spin of the dreidel. Suppose you spin a dreidel three times. Calculate the probability of getting

(a) at least one *nun*?

(b) exactly 2 *nuns*?

(c) exactly 1 *hei*?

(d) at most 2 *gimels*?

Photo by Staccabees, cropped
(http://flic.kr/p/7gLZTf)
CC BY 2.0 license

3.32 Arachnophobia. A 2005 Gallup Poll found that 7% of teenagers (ages 13 to 17) suffer from arachnophobia and are extremely afraid of spiders. At a summer camp there are 10 teenagers sleeping in each tent. Assume that these 10 teenagers are independent of each other.[52]

(a) Calculate the probability that at least one of them suffers from arachnophobia.

(b) Calculate the probability that exactly 2 of them suffer from arachnophobia.

(c) Calculate the probability that at most 1 of them suffers from arachnophobia.

(d) If the camp counselor wants to make sure no more than 1 teenager in each tent is afraid of spiders, does it seem reasonable for him to randomly assign teenagers to tents?

---

[51]The Pew Research Center for the People and the Press, Assessing the Representativeness of Public Opinion Surveys, May 15, 2012.

[52]Gallup Poll, What Frightens America's Youth?, March 29, 2005.

3.33 Eye color, Part II. Exercise 3.23 introduces a husband and wife with brown eyes who have 0.75 probability of having children with brown eyes, 0.125 probability of having children with blue eyes, and 0.125 probability of having children with green eyes.

(a) What is the probability that their first child will have green eyes and the second will not?

(b) What is the probability that exactly one of their two children will have green eyes?

(c) If they have six children, what is the probability that exactly two will have green eyes?

(d) If they have six children, what is the probability that at least one will have green eyes?

(e) What is the probability that the first green eyed child will be the 4th child?

(f) Would it be considered unusual if only 2 out of their 6 children had brown eyes?

3.34 Sickle cell anemia. Sickle cell anemia is a genetic blood disorder where red blood cells lose their flexibility and assume an abnormal, rigid, "sickle" shape, which results in a risk of various complications. If both parents are carriers of the disease, then a child has a 25% chance of having the disease, 50% chance of being a carrier, and 25% chance of neither having the disease nor being a carrier. If two parents who are carriers of the disease have 3 children, what is the probability that

(a) two will have the disease?

(b) none will have the disease?

(c) at least one will neither have the disease nor be a carrier?

(d) the first child with the disease will the be 3rd child?

3.35 Roulette winnings. In the game of roulette, a wheel is spun and you place bets on where it will stop. One popular bet is that it will stop on a red slot; such a bet has an 18/38 chance of winning. If it stops on red, you double the money you bet. If not, you lose the money you bet. Suppose you play 3 times, each time with a $1 bet. Let Y represent the total amount won or lost. Write a probability model for Y.

3.36 Multiple choice quiz. In a multiple choice quiz there are 5 questions and 4 choices for each question (a, b, c, d). Robin has not studied for the quiz at all, and decides to randomly guess the answers. What is the probability that

(a) the first question she gets right is the 3rd question?

(b) she gets exactly 3 or exactly 4 questions right?

(c) she gets the majority of the questions right?

3.37 Exploring combinations. The formula for the number of ways to arrange n objects is $n! = n \times (n-1) \times \cdots \times 2 \times 1$. This exercise walks you through the derivation of this formula for a couple of special cases.

A small company has five employees: Anna, Ben, Carl, Damian, and Eddy. There are five parking spots in a row at the company, none of which are assigned, and each day the employees pull into a random parking spot. That is, all possible orderings of the cars in the row of spots are equally likely.

(a) On a given day, what is the probability that the employees park in alphabetical order?

(b) If the alphabetical order has an equal chance of occurring relative to all other possible orderings, how many ways must there be to arrange the five cars?

(c) Now consider a sample of 8 employees instead. How many possible ways are there to order these 8 employees' cars?

3.38 Male children. While it is often assumed that the probabilities of having a boy or a girl are the same, the actual probability of having a boy is slightly higher at 0.51. Suppose a couple plans to have 3 kids.

(a) Use the binomial model to calculate the probability that two of them will be boys.

(b) Write out all possible orderings of 3 children, 2 of whom are boys. Use these scenarios to calculate the same probability from part (a) but using the addition rule for disjoint outcomes. Confirm that your answers from parts (a) and (b) match.

(c) If we wanted to calculate the probability that a couple who plans to have 8 kids will have 3 boys, briefly describe why the approach from part (b) would be more tedious than the approach from part (a).

3.6.5   More discrete distributions

3.39 Rolling a die. Calculate the following probabilities and indicate which probability distribution model is appropriate in each case. You roll a fair die 5 times. What is the probability of rolling

(a) the first 6 on the fifth roll?

(b) exactly three 6s?

(c) the third 6 on the fifth roll?

3.40 Playing darts. Calculate the following probabilities and indicate which probability distribution model is appropriate in each case. A very good darts player can hit the bull's eye (red circle in the center of the dart board) 65% of the time. What is the probability that he

(a) hits the bullseye for the 10th time on the 15th try?

(b) hits the bullseye 10 times in 15 tries?

(c) hits the first bullseye on the third try?

3.41 Sampling at school. For a sociology class project you are asked to conduct a survey on 20 students at your school. You decide to stand outside of your dorm's cafeteria and conduct the survey on a random sample of 20 students leaving the cafeteria after dinner one evening. Your dorm is comprised of 45% males and 55% females.

(a) Which probability model is most appropriate for calculating the probability that the 4th person you survey is the 2nd female? Explain.

(b) Compute the probability from part (a).

(c) The three possible scenarios that lead to 4th person you survey being the 2nd female are

$$\{M, M, F, F\}, \{M, F, M, F\}, \{F, M, M, F\}$$

One common feature among these scenarios is that the last trial is always female. In the first three trials there are 2 males and 1 female. Use the binomial coefficient to confirm that there are 3 ways of ordering 2 males and 1 female.

(d) Use the findings presented in part (c) to explain why the formula for the coefficient for the negative binomial is $\binom{n-1}{k-1}$ while the formula for the binomial coefficient is $\binom{n}{k}$.

3.42 Serving in volleyball. A not-so-skilled volleyball player has a 15% chance of making the serve, which involves hitting the ball so it passes over the net on a trajectory such that it will land in the opposing team's court. Suppose that her serves are independent of each other.

(a) What is the probability that on the 10th try she will make her 3rd successful serve?

(b) Suppose she has made two successful serves in nine attempts. What is the probability that her 10th serve will be successful?

(c) Even though parts (a) and (b) discuss the same scenario, the probabilities you calculated should be different. Can you explain the reason for this discrepancy?

3.43 Customers at a coffee shop. A coffee shop serves an average of 75 customers per hour during the morning rush.

(a) Which distribution we have studied is most appropriate for calculating the probability of a given number of customers arriving within one hour during this time of day?

(b) What are the mean and the standard deviation of the number of customers this coffee shop serves in one hour during this time of day?

(c) Would it be considered unusually low if only 60 customers showed up to this coffee shop in one hour during this time of day?

(d) Calculate the probability that this coffee shop serves 70 customers in one hour during this time of day?

3.44 Stenographer's typos, Part I. A very skilled court stenographer makes one typographical error (typo) per hour on average.

(a) What probability distribution is most appropriate for calculating the probability of a given number of typos this stenographer makes in an hour?

(b) What are the mean and the standard deviation of the number of typos this stenographer makes?

(c) Would it be considered unusual if this stenographer made 4 typos in a given hour?

(d) Calculate the probability that this stenographer makes at most 2 typos in a given hour.

Chapter 4

Foundations for inference

Statistical inference is concerned primarily with understanding the quality of parameter estimates. For example, a classic inferential question is, "How sure are we that the estimated mean, $\bar{x}$, is near the true population mean, μ?" While the equations and details change depending on the setting, the foundations for inference are the same throughout all of statistics. We introduce these common themes in Sections 4.1-4.4 by discussing inference about the population mean, μ, and set the stage for other parameters and scenarios in Section 4.5. Understanding this chapter will make the rest of this book, and indeed the rest of statistics, seem much more familiar.

Throughout the next few sections we consider a data set called yrbss, which represents all 13,583 high school students in the Youth Risk Behavior Surveillance System (YRBSS) from 2013.[1] Part of this data set is shown in Table 4.1, and the variables are described in Table 4.2.

ID	age	gender	grade	height	weight	helmet	active	lifting
1	14	female	9			never	4	0
2	14	female	9			never	2	0
3	15	female	9	1.73	84.37	never	7	0
$\vdots$	$\vdots$	$\vdots$	$\vdots$	$\vdots$	$\vdots$	$\vdots$	$\vdots$	$\vdots$
13582	17	female	12	1.60	77.11	sometimes	5	
13583	17	female	12	1.57	52.16	did not ride	5	

Table 4.1: Five cases from the yrbss data set. Some observations are blank since there are missing data. For example, the height and weight of students 1 and 2 are missing.

We're going to consider the population of high school students who participated in the 2013 YRBSS. We took a simple random sample of this population, which is represented in Table 4.3.[2] We will use this sample, which we refer to as the yrbss_samp data set, to draw conclusions about the population of YRBSS participants. This is the practice of statistical inference in the broadest sense. Two histograms summarizing the height, weight, active, and lifting variables from yrbss_samp data set are shown in Figure 4.4.

---

[1] www.cdc.gov/healthyyouth/data/yrbs/data.htm

[2] About 10% of high schoolers for each variable chose not to answer the question, we used multiple regression (see Chapter 8) to predict what those responses would have been. For simplicity, we will assume that these predicted values are exactly the truth.

age	Age of the student.
gender	Sex of the student.
grade	Grade in high school
height	Height, in meters. There are 3.28 feet in a meter.
weight	Weight, in kilograms (2.2 pounds per kilogram).
helmet	Frequency that the student wore a helmet while biking in the last 12 months.
active	Number of days physically active for 60+ minutes in the last 7 days.
lifting	Number of days of strength training (e.g. lifting weights) in the last 7 days.

Table 4.2: Variables and their descriptions for the yrbss data set.

ID	age	gender	grade	height	weight	helmet	active	lifting
5653	16	female	11	1.50	52.62	never	0	0
9437	17	male	11	1.78	74.84	rarely	7	5
2021	17	male	11	1.75	106.60	never	7	0
⋮	⋮	⋮	⋮	⋮	⋮	⋮	⋮	⋮
2325	14	male	9	1.70	55.79	never	1	0

Table 4.3: Four observations for the yrbss_samp data set, which represents a simple random sample of 100 high schoolers from the 2013 YRBSS.

4.1 Variability in estimates 📹

We would like to estimate four features of the high schoolers in YRBSS using the sample.

(1) What is the average height of the YRBSS high schoolers?

(2) What is the average weight of the YRBSS high schoolers?

(3) On average, how many days per week are YRBSS high schoolers physically active?

(4) On average, how many days per week do YRBSS high schoolers do weight training?

While we focus on the mean in this chapter, questions regarding variation are often just as important in practice. For instance, if students are either very active or almost entirely inactive (the distribution is bimodal), we might try different strategies to promote a healthy lifestyle among students than if all high schoolers were already somewhat active.

4.1.1 Point estimates

We want to estimate the **population mean** based on the sample. The most intuitive way to go about doing this is to simply take the **sample mean**. That is, to estimate the average height of all YRBSS students, take the average height for the sample:

$$\bar{x}_{height} = \frac{1.50 + 1.78 + \cdots + 1.70}{100} = 1.697$$

The sample mean $\bar{x} = 1.697$ meters (5 feet, 6.8 inches) is called a **point estimate** of the population mean: if we can only choose one value to estimate the population mean, this is our best guess. Suppose we take a new sample of 100 people and recompute the mean; we will probably not get the exact same answer that we got using the yrbss_samp data set. Estimates generally vary from one sample to another, and this **sampling variation** suggests our estimate may be close, but it will not be exactly equal to the parameter.

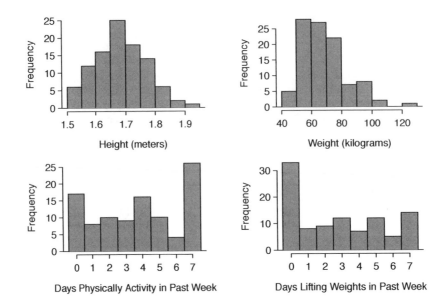

Figure 4.4: Histograms of `height`, `weight`, `activity`, and `lifting` for the sample YRBSS data. The `height` distribution is approximately symmetric, `weight` is moderately skewed to the right, `activity` is bimodal or multimodal (with unclear skew), and `lifting` is strongly right skewed.

We can also estimate the average weight of YRBSS respondents by examining the sample mean of `weight` (in kg), and average number of days physically active in a week:

$$\bar{x}_{weight} = \frac{52.6 + 74.8 + \cdots + 55.8}{100} = 68.89 \qquad \bar{x}_{active} = \frac{0 + 7 + \cdots + 1}{100} = 3.75$$

The average weight is 68.89 kilograms, which is about 151.6 pounds.

What about generating point estimates of other **population parameters**, such as the population median or population standard deviation? Once again we might estimate parameters based on sample statistics, as shown in Table 4.5. For example, the population standard deviation of `active` using the sample standard deviation, 2.56 days.

active	estimate	parameter
mean	3.75	3.90
median	4.00	4.00
st. dev.	2.556	2.564

Table 4.5: Point estimates and parameter values for the `active` variable. The parameters were obtained by computing the mean, median, and SD for all YRBSS respondents.

⊙ **Guided Practice 4.1** Suppose we want to estimate the difference in days active for men and women. If $\bar{x}_{men} = 4.3$ and $\bar{x}_{women} = 3.2$, then what would be a good point estimate for the population difference?[3]

---

[3]We could take the difference of the two sample means: $4.3 - 3.2 = 1.1$. Men are physically active about 1.1 days per week more than women on average in YRBSS.

⊙ **Guided Practice 4.2** If you had to provide a point estimate of the population IQR for the heights of participants, how might you make such an estimate using a sample?[4]

4.1.2 Point estimates are not exact

Estimates are usually not exactly equal to the truth, but they get better as more data become available. We can see this by plotting a running mean from `yrbss_samp`. A **running mean** is a sequence of means, where each mean uses one more observation in its calculation than the mean directly before it in the sequence. For example, the second mean in the sequence is the average of the first two observations and the third in the sequence is the average of the first three. The running mean for the `active` variable in the `yrbss_samp` is shown in Figure 4.6, and it approaches the true population average, 3.90 days, as more data become available.

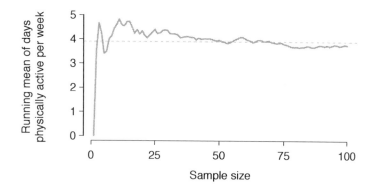

Figure 4.6: The mean computed after adding each individual to the sample. The mean tends to approach the true population average as more data become available.

Sample point estimates only approximate the population parameter, and they vary from one sample to another. If we took another simple random sample of the YRBSS students, we would find that the sample mean for the number of days active would be a little different. It will be useful to quantify how variable an estimate is from one sample to another. If this variability is small (i.e. the sample mean doesn't change much from one sample to another) then that estimate is probably very accurate. If it varies widely from one sample to another, then we should not expect our estimate to be very good.

4.1.3 Standard error of the mean

From the random sample represented in `yrbss_samp`, we guessed the average number of days a YRBSS student is physically active is 3.75 days. Suppose we take another random sample of 100 individuals and take its mean: 3.22 days. Suppose we took another (3.67 days) and another (4.10 days), and so on. If we do this many many times – which we can do only because we have all YRBSS students – we can build up a **sampling distribution** for the sample mean when the sample size is 100, shown in Figure 4.7.

---

[4]To obtain a point estimate of the height for the full set of YRBSS students, we could take the IQR of the sample.

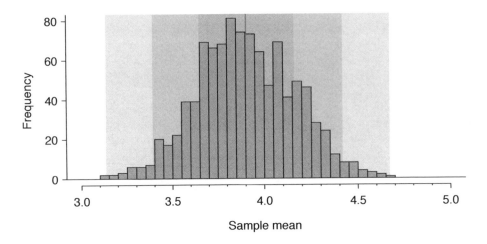

Figure 4.7: A histogram of 1000 sample means for number of days physically active per week, where the samples are of size $n = 100$.

Sampling distribution

The sampling distribution represents the distribution of the point estimates based on samples of a fixed size from a certain population. It is useful to think of a particular point estimate as being drawn from such a distribution. Understanding the concept of a sampling distribution is central to understanding statistical inference.

The sampling distribution shown in Figure 4.7 is unimodal and approximately symmetric. It is also centered exactly at the true population mean: $\mu = 3.90$. Intuitively, this makes sense. The sample means should tend to "fall around" the population mean.

We can see that the sample mean has some variability around the population mean, which can be quantified using the standard deviation of this distribution of sample means: $\sigma_{\bar{x}} = 0.26$. The standard deviation of the sample mean tells us how far the typical estimate is away from the actual population mean, 3.90 days. It also describes the typical **error** of the point estimate, and for this reason we usually call this standard deviation the **standard error (SE)** of the estimate.

SE
standard
error

Standard error of an estimate

The standard deviation associated with an estimate is called the *standard error*. It describes the typical error or uncertainty associated with the estimate.

When considering the case of the point estimate $\bar{x}$, there is one problem: there is no obvious way to estimate its standard error from a single sample. However, statistical theory provides a helpful tool to address this issue.

⊙ **Guided Practice 4.3**   (a) Would you rather use a small sample or a large sample when estimating a parameter? Why? (b) Using your reasoning from (a), would you expect a point estimate based on a small sample to have smaller or larger standard error than a point estimate based on a larger sample?[5]

In the sample of 100 students, the standard error of the sample mean is equal to the population standard deviation divided by the square root of the sample size:

$$SE_{\bar{x}} = \sigma_{\bar{x}} = \frac{\sigma_x}{\sqrt{n}} = \frac{2.6}{\sqrt{100}} = 0.26$$

where σ_x is the standard deviation of the individual observations. This is no coincidence. We can show mathematically that this equation is correct when the observations are independent using the probability tools of Section 2.4.

Computing SE for the sample mean

Given n independent observations from a population with standard deviation σ, the standard error of the sample mean is equal to

$$SE = \frac{\sigma}{\sqrt{n}} \qquad (4.4)$$

A reliable method to ensure sample observations are independent is to conduct a simple random sample consisting of less than 10% of the population.

There is one subtle issue in Equation (4.4): the population standard deviation is typically unknown. You might have already guessed how to resolve this problem: we can use the point estimate of the standard deviation from the sample. This estimate tends to be sufficiently good when the sample size is at least 30 and the population distribution is not strongly skewed. Thus, we often just use the sample standard deviation s instead of σ. When the sample size is smaller than 30, we will need to use a method to account for extra uncertainty in the standard error. If the skew condition is not met, a larger sample is needed to compensate for the extra skew. These topics are further discussed in Section 4.4.

⊙ **Guided Practice 4.5**   In the sample of 100 students, the standard deviation of student heights is $s_{height} = 0.088$ meters. In this case, we can confirm that the observations are independent by checking that the data come from a simple random sample consisting of less than 10% of the population. (a) What is the standard error of the sample mean, $\bar{x}_{height} = 1.70$ meters? (b) Would you be surprised if someone told you the average height of all YRBSS respondents was actually 1.69 meters?[6]

---

[5](a) Consider two random samples: one of size 10 and one of size 1000. Individual observations in the small sample are highly influential on the estimate while in larger samples these individual observations would more often average each other out. The larger sample would tend to provide a more accurate estimate. (b) If we think an estimate is better, we probably mean it typically has less error. Based on (a), our intuition suggests that a larger sample size corresponds to a smaller standard error.

[6](a) Use Equation (4.4) with the sample standard deviation to compute the standard error: $SE_{\bar{y}} = 0.088/\sqrt{100} = 0.0088$ meters. (b) It would not be surprising. Our sample is about 1 standard error from 1.69m. In other words, 1.69m does not seem to be implausible given that our sample was relatively close to it. (We use the standard error to identify what is close.)

⊙ **Guided Practice 4.6**   (a) Would you be more trusting of a sample that has 100 observations or 400 observations? (b) We want to show mathematically that our estimate tends to be better when the sample size is larger. If the standard deviation of the individual observations is 10, what is our estimate of the standard error when the sample size is 100? What about when it is 400? (c) Explain how your answer to part (b) mathematically justifies your intuition in part (a).[7]

4.1.4   Basic properties of point estimates

We achieved three goals in this section. First, we determined that point estimates from a sample may be used to estimate population parameters. We also determined that these point estimates are not exact: they vary from one sample to another. Lastly, we quantified the uncertainty of the sample mean using what we call the standard error, mathematically represented in Equation (4.4). While we could also quantify the standard error for other estimates – such as the median, standard deviation, or any other number of statistics – we will postpone these extensions until later chapters or courses.

4.2   Confidence intervals 📹

A point estimate provides a single plausible value for a parameter. However, a point estimate is rarely perfect; usually there is some error in the estimate. Instead of supplying just a point estimate of a parameter, a next logical step would be to provide a plausible *range of values* for the parameter.

4.2.1   Capturing the population parameter

A plausible range of values for the population parameter is called a **confidence interval**.

Using only a point estimate is like fishing in a murky lake with a spear, and using a confidence interval is like fishing with a net. We can throw a spear where we saw a fish, but we will probably miss. On the other hand, if we toss a net in that area, we have a good chance of catching the fish.

If we report a point estimate, we probably will not hit the exact population parameter. On the other hand, if we report a range of plausible values – a confidence interval – we have a good shot at capturing the parameter.

⊙ **Guided Practice 4.7**   If we want to be very certain we capture the population parameter, should we use a wider interval or a smaller interval?[8]

---

[7](a) Extra observations are usually helpful in understanding the population, so a point estimate with 400 observations seems more trustworthy. (b) The standard error when the sample size is 100 is given by $SE_{100} = 10/\sqrt{100} = 1$. For 400: $SE_{400} = 10/\sqrt{400} = 0.5$. The larger sample has a smaller standard error. (c) The standard error of the sample with 400 observations is lower than that of the sample with 100 observations. The standard error describes the typical error, and since it is lower for the larger sample, this mathematically shows the estimate from the larger sample tends to be better – though it does not guarantee that every large sample will provide a better estimate than a particular small sample.

[8]If we want to be more certain we will capture the fish, we might use a wider net. Likewise, we use a wider confidence interval if we want to be more certain that we capture the parameter.

4.2.2 An approximate 95% confidence interval

Our point estimate is the most plausible value of the parameter, so it makes sense to build the confidence interval around the point estimate. The standard error, which is a measure of the uncertainty associated with the point estimate, provides a guide for how large we should make the confidence interval.

The standard error represents the standard deviation associated with the estimate, and roughly 95% of the time the estimate will be within 2 standard errors of the parameter. If the interval spreads out 2 standard errors from the point estimate, we can be roughly 95% **confident** that we have captured the true parameter:

$$\text{point estimate} \ \pm \ 2 \times SE \tag{4.8}$$

But what does "95% confident" mean? Suppose we took many samples and built a confidence interval from each sample using Equation (4.8). Then about 95% of those intervals would contain the actual mean, μ. Figure 4.8 shows this process with 25 samples, where 24 of the resulting confidence intervals contain the average number of days per week that YRBSS students are physically active, $\mu = 3.90$ days, and one interval does not.

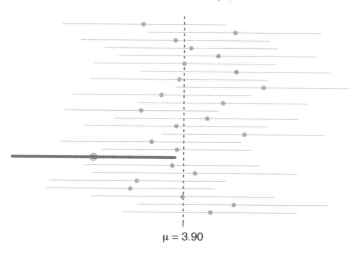

$\mu = 3.90$

Figure 4.8: Twenty-five samples of size $n = 100$ were taken from `yrbss`. For each sample, a confidence interval was created to try to capture the average number of days per week that students are physically active. Only 1 of these 25 intervals did not capture the true mean, $\mu = 3.90$ days.

⊙ **Guided Practice 4.9** In Figure 4.8, one interval does not contain 3.90 minutes. Does this imply that the mean cannot be 3.90?[9]

The rule where about 95% of observations are within 2 standard deviations of the mean is only approximately true. However, it holds very well for the normal distribution. As we will soon see, the mean tends to be normally distributed when the sample size is sufficiently large.

---

[9]Just as some observations occur more than 2 standard deviations from the mean, some point estimates will be more than 2 standard errors from the parameter. A confidence interval only provides a plausible range of values for a parameter. While we might say other values are implausible based on the data, this does not mean they are impossible.

● **Example 4.10** The sample mean of days active per week from `yrbss_samp` is 3.75 days. The standard error, as estimated using the sample standard deviation, is $SE = \frac{2.6}{\sqrt{100}} = 0.26$ days. (The population SD is unknown in most applications, so we use the sample SD here.) Calculate an approximate 95% confidence interval for the average days active per week for all YRBSS students.

We apply Equation (4.8):

$$3.75 \ \pm \ 2 \times 0.26 \quad \rightarrow \quad (3.23, 4.27)$$

Based on these data, we are about 95% confident that the average days active per week for all YRBSS students was larger than 3.23 but less than 4.27 days. Our interval extends out 2 standard errors from the point estimate, $\bar{x}_{active}$.

⊙ **Guided Practice 4.11** The sample data suggest the average YRBSS student height is $\bar{x}_{height} = 1.697$ meters with a standard error of 0.0088 meters (estimated using the sample standard deviation, 0.088 meters). What is an approximate 95% confidence interval for the average height of all of the YRBSS students?[10]

4.2.3 The sampling distribution for the mean

In Section 4.1.3, we introduced a sampling distribution for $\bar{x}$, the average days physically active per week for samples of size 100. We examined this distribution earlier in Figure 4.7. Now we'll take 100,000 samples, calculate the mean of each, and plot them in a histogram to get an especially accurate depiction of the sampling distribution. This histogram is shown in the left panel of Figure 4.9.

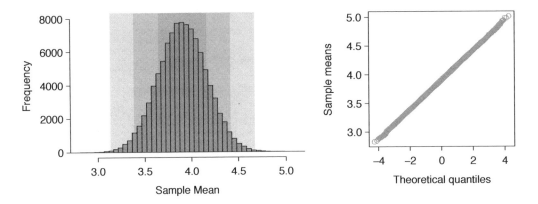

Figure 4.9: The left panel shows a histogram of the sample means for 100,000 different random samples. The right panel shows a normal probability plot of those sample means.

Does this distribution look familiar? Hopefully so! The distribution of sample means closely resembles the normal distribution (see Section 3.1). A normal probability plot of these sample means is shown in the right panel of Figure 4.9. Because all of the points

---

[10]Apply Equation (4.8): $1.697 \ \pm \ 2 \times 0.0088 \rightarrow (1.6794, 1.7146)$. We interpret this interval as follows: We are about 95% confident the average height of all YRBSS students was between 1.6794 and 1.7146 meters (5.51 to 5.62 feet).

closely fall around a straight line, we can conclude the distribution of sample means is nearly normal. This result can be explained by the Central Limit Theorem.

Central Limit Theorem, informal description

If a sample consists of at least 30 independent observations and the data are not strongly skewed, then the distribution of the sample mean is well approximated by a normal model.

We will apply this informal version of the Central Limit Theorem for now, and discuss its details further in Section 4.4.

The choice of using 2 standard errors in Equation (4.8) was based on our general guideline that roughly 95% of the time, observations are within two standard deviations of the mean. Under the normal model, we can make this more accurate by using 1.96 in place of 2.

$$\text{point estimate } \pm \ 1.96 \times SE \tag{4.12}$$

If a point estimate, such as $\bar{x}$, is associated with a normal model and standard error SE, then we use this more precise 95% confidence interval.

4.2.4 Changing the confidence level

Suppose we want to consider confidence intervals where the confidence level is somewhat higher than 95%; perhaps we would like a confidence level of 99%. Think back to the analogy about trying to catch a fish: if we want to be more sure that we will catch the fish, we should use a wider net. To create a 99% confidence level, we must also widen our 95% interval. On the other hand, if we want an interval with lower confidence, such as 90%, we could make our original 95% interval slightly slimmer.

The 95% confidence interval structure provides guidance in how to make intervals with new confidence levels. Below is a general 95% confidence interval for a point estimate that comes from a nearly normal distribution:

$$\text{point estimate } \pm \ 1.96 \times SE \tag{4.13}$$

There are three components to this interval: the point estimate, "1.96", and the standard error. The choice of $1.96 \times SE$ was based on capturing 95% of the data since the estimate is within 1.96 standard deviations of the parameter about 95% of the time. The choice of 1.96 corresponds to a 95% confidence level.

⊙ **Guided Practice 4.14** If X is a normally distributed random variable, how often will X be within 2.58 standard deviations of the mean?[11]

To create a 99% confidence interval, change 1.96 in the 95% confidence interval formula to be 2.58. Guided Practice 4.14 highlights that 99% of the time a normal random variable will be within 2.58 standard deviations of the mean. This approach – using the Z-scores in the normal model to compute confidence levels – is appropriate when $\bar{x}$ is associated with

---

[11]This is equivalent to asking how often the Z-score will be larger than -2.58 but less than 2.58. (For a picture, see Figure 4.10.) To determine this probability, look up -2.58 and 2.58 in the normal probability table (0.0049 and 0.9951). Thus, there is a $0.9951 - 0.0049 \approx 0.99$ probability that the unobserved random variable X will be within 2.58 standard deviations of μ.

a normal distribution with mean μ and standard deviation $SE_{\bar{x}}$. Thus, the formula for a 99% confidence interval is

$$\bar{x} \ \pm \ 2.58 \times SE_{\bar{x}} \tag{4.15}$$

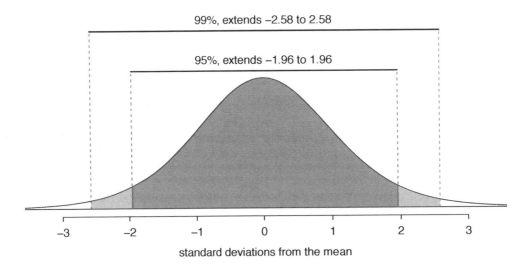

Figure 4.10: The area between $-z^{\star}$ and $z^{\star}$ increases as $|z^{\star}|$ becomes larger. If the confidence level is 99%, we choose $z^{\star}$ such that 99% of the normal curve is between $-z^{\star}$ and $z^{\star}$, which corresponds to 0.5% in the lower tail and 0.5% in the upper tail: $z^{\star} = 2.58$.

The normal approximation is crucial to the precision of these confidence intervals. Section 4.4 provides a more detailed discussion about when the normal model can safely be applied. When the normal model is not a good fit, we will use alternative distributions that better characterize the sampling distribution.

Conditions for $\bar{x}$ being nearly normal and SE being accurate

Important conditions to help ensure the sampling distribution of $\bar{x}$ is nearly normal and the estimate of SE sufficiently accurate:

- The sample observations are independent.
- The sample size is large: $n \geq 30$ is a good rule of thumb.
- The population distribution is not strongly skewed. This condition can be difficult to evaluate, so just use your best judgement.

Additionally, the larger the sample size, the more lenient we can be with the sample's skew.

How to verify sample observations are independent

If the observations are from a simple random sample and consist of fewer than 10% of the population, then they are independent.

Subjects in an experiment are considered independent if they undergo random assignment to the treatment groups.

If a sample is from a seemingly random process, e.g. the lifetimes of wrenches used in a particular manufacturing process, checking independence is more difficult. In this case, use your best judgement.

Checking for strong skew usually means checking for obvious outliers

When there are prominent outliers present, the sample should contain at least 100 observations, and in some cases, much more.

This is a first course in statistics, so you won't have perfect judgement on assessing skew. That's okay. If you're in a bind, either consult a statistician or learn about the studentized bootstrap (bootstrap-t) method.

⊙ **Guided Practice 4.16**   Create a 99% confidence interval for the average days active per week of all YRBSS students using `yrbss_samp`. The point estimate is $\bar{x}_{active} = 3.75$ and the standard error is $SE_{\bar{x}} = 0.26$.[12]

Confidence interval for any confidence level

If the point estimate follows the normal model with standard error SE, then a confidence interval for the population parameter is

$$\text{point estimate} \ \pm \ z^{\star} SE$$

where $z^{\star}$ corresponds to the confidence level selected.

Figure 4.10 provides a picture of how to identify $z^{\star}$ based on a confidence level. We select $z^{\star}$ so that the area between $-z^{\star}$ and $z^{\star}$ in the normal model corresponds to the confidence level.

Margin of error

In a confidence interval, $z^{\star} \times SE$ is called the **margin of error**.

---

[12]The observations are independent (simple random sample, < 10% of the population), the sample size is at least 30 ($n = 100$), and the distribution doesn't have a clear skew (Figure 4.4 on page 170); the normal approximation and estimate of SE should be reasonable. Apply the 99% confidence interval formula: $\bar{x}_{active} \ \pm \ 2.58 \times SE_{\bar{x}} \rightarrow (3.08, 4.42)$. We are 99% confident that the average days active per week of all YRBSS students is between 3.08 and 4.42 days.

⊙ **Guided Practice 4.17**   Use the data in Guided Practice 4.16 to create a 90% confidence interval for the average days active per week of all YRBSS students.[13]

4.2.5   Interpreting confidence intervals

A careful eye might have observed the somewhat awkward language used to describe confidence intervals. Correct interpretation:

We are XX% confident that the population parameter is between...

Incorrect language might try to describe the confidence interval as capturing the population parameter with a certain probability. This is a common error: while it might be useful to think of it as a probability, the confidence level only quantifies how plausible it is that the parameter is in the interval.

Another important consideration of confidence intervals is that they *only try to capture the population parameter*. A confidence interval says nothing about the confidence of capturing individual observations, a proportion of the observations, or about capturing point estimates. Confidence intervals only attempt to capture population parameters.

4.3   Hypothesis testing 📹

Are students lifting weights or performing other strength training exercises more or less often than they have in the past? We'll compare data from students from the 2011 YRBSS survey to our sample of 100 students from the 2013 YRBSS survey.

We'll also consider sleep behavior. A recent study found that college students average about 7 hours of sleep per night.[14] However, researchers at a rural college are interested in showing that their students sleep longer than seven hours on average. We investigate this topic in Section 4.3.4.

4.3.1   Hypothesis testing framework

Students from the 2011 YRBSS lifted weights (or performed other strength training exercises) 3.09 days per week on average. We want to determine if the yrbss_samp data set provides strong evidence that YRBSS students selected in 2013 are lifting more or less than the 2011 YRBSS students, versus the other possibility that there has been no change.[15] We simplify these three options into two competing **hypotheses**:

H_0: The average days per week that YRBSS students lifted weights was the same for 2011 and 2013.

H_A: The average days per week that YRBSS students lifted weights was *different* for 2013 than in 2011.

We call H_0 the null hypothesis and H_A the alternative hypothesis.

H_0
null hypothesis

H_A
alternative
hypothesis

[13]We first find $z^\star$ such that 90% of the distribution falls between $-z^\star$ and $z^\star$ in the standard normal model, $N(\mu = 0, \sigma = 1)$. We can look up $-z^\star$ in the normal probability table by looking for a lower tail of 5% (the other 5% is in the upper tail): $z^\star = 1.65$. The 90% confidence interval can then be computed as $\bar{x}_{active} \pm 1.65 \times SE_{\bar{x}} \rightarrow (3.32, 4.18)$. (We had already verified conditions for normality and the standard error.) That is, we are 90% confident the average days active per week is between 3.32 and 4.18 days.

[14]*Poll shows college students get least amount of sleep.* theloquitur.com/?p=1161

[15]While we could answer this question by examining the entire YRBSS data set from 2013 (yrbss), we only consider the sample data (yrbss_samp), which is more realistic since we rarely have access to population data.

Null and alternative hypotheses

The **null hypothesis** (H_0) often represents either a skeptical perspective or a claim to be tested. The **alternative hypothesis** (H_A) represents an alternative claim under consideration and is often represented by a range of possible parameter values.

The null hypothesis often represents a skeptical position or a perspective of no difference. The alternative hypothesis often represents a new perspective, such as the possibility that there has been a change.

TIP: Hypothesis testing framework

The skeptic will not reject the null hypothesis (H_0), unless the evidence in favor of the alternative hypothesis (H_A) is so strong that she rejects H_0 in favor of H_A.

The hypothesis testing framework is a very general tool, and we often use it without a second thought. If a person makes a somewhat unbelievable claim, we are initially skeptical. However, if there is sufficient evidence that supports the claim, we set aside our skepticism and reject the null hypothesis in favor of the alternative. The hallmarks of hypothesis testing are also found in the US court system.

⊙ **Guided Practice 4.18**   A US court considers two possible claims about a defendant: she is either innocent or guilty. If we set these claims up in a hypothesis framework, which would be the null hypothesis and which the alternative?[16]

Jurors examine the evidence to see whether it convincingly shows a defendant is guilty. Even if the jurors leave unconvinced of guilt beyond a reasonable doubt, this does not mean they believe the defendant is innocent. This is also the case with hypothesis testing: *even if we fail to reject the null hypothesis, we typically do not accept the null hypothesis as true.* Failing to find strong evidence for the alternative hypothesis is not equivalent to accepting the null hypothesis.

In the example with the YRBSS, the null hypothesis represents no difference in the average days per week of weight lifting in 2011 and 2013. The alternative hypothesis represents something new or more interesting: there was a difference, either an increase or a decrease. These hypotheses can be described in mathematical notation using μ_{13} as the average days of weight lifting for 2013:

H_0: $\mu_{13} = 3.09$

H_A: $\mu_{13} \neq 3.09$

where 3.09 is the average number of days per week that students from the 2011 YRBSS lifted weights. Using the mathematical notation, the hypotheses can more easily be evaluated using statistical tools. We call 3.09 the **null value** since it represents the value of the parameter if the null hypothesis is true.

---

[16]The jury considers whether the evidence is so convincing (strong) that there is no reasonable doubt regarding the person's guilt; in such a case, the jury rejects innocence (the null hypothesis) and concludes the defendant is guilty (alternative hypothesis).

4.3.2   Testing hypotheses using confidence intervals

We will use the `yrbss_samp` data set to evaluate the hypothesis test, and we start by comparing the 2013 point estimate of the number of days per week that students lifted weights: $\bar{x}_{13} = 2.78$ days. This estimate suggests that students from the 2013 YRBSS were lifting weights less than students in the 2011 YRBSS. However, to evaluate whether this provides strong evidence that there has been a change, we must consider the uncertainty associated with $\bar{x}_{13}$.

We learned in Section 4.1 that there is fluctuation from one sample to another, and it is unlikely that the sample mean will be exactly equal to the parameter; we should not expect $\bar{x}_{13}$ to exactly equal μ_{13}. Given that $\bar{x}_{13} = 2.78$, it might still be possible that the average of all students from the 2013 YRBSS survey is the same as the average from the 2011 YRBSS survey. The difference between $\bar{x}_{13}$ and 3.09 could be due to *sampling variation*, i.e. the variability associated with the point estimate when we take a random sample.

In Section 4.2, confidence intervals were introduced as a way to find a range of plausible values for the population mean.

⬤ **Example 4.19** In the sample of 100 students from the 2013 YRBSS survey, the average number of days per week that students lifted weights was 2.78 days with a standard deviation of 2.56 days (coincidentally the same as days active). Compute a 95% confidence interval for the average for all students from the 2013 YRBSS survey. You can assume the conditions for the normal model are met.

The general formula for the confidence interval based on the normal distribution is

$$\bar{x} \pm z^{\star} SE_{\bar{x}}$$

We are given $\bar{x}_{13} = 2.78$, we use $z^{\star} = 1.96$ for a 95% confidence level, and we can compute the standard error using the standard deviation divided by the square root of the sample size:

$$SE_{\bar{x}} = \frac{s_{13}}{\sqrt{n}} = \frac{2.56}{\sqrt{100}} = 0.256$$

Entering the sample mean, $z^{\star}$, and the standard error into the confidence interval formula results in (2.27, 3.29). We are 95% confident that the average number of days per week that all students from the 2013 YRBSS lifted weights was between 2.27 and 3.29 days.

Because the average of all students from the 2011 YRBSS survey is 3.09, which falls within the range of plausible values from the confidence interval, we cannot say the null hypothesis is implausible. That is, we fail to reject the null hypothesis, H_0.

TIP: Double negatives can sometimes be used in statistics

In many statistical explanations, we use double negatives. For instance, we might say that the null hypothesis is *not implausible* or we *failed to reject* the null hypothesis. Double negatives are used to communicate that while we are not rejecting a position, we are also not saying it is correct.

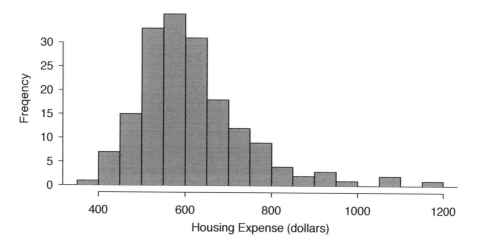

Figure 4.11: Sample distribution of student housing expense. These data are strongly skewed, which we can see by the long right tail with a few notable outliers.

⊙ **Guided Practice 4.20** Colleges frequently provide estimates of student expenses such as housing. A consultant hired by a community college claimed that the average student housing expense was $650 per month. What are the null and alternative hypotheses to test whether this claim is accurate?[17]

⊙ **Guided Practice 4.21** The community college decides to collect data to evaluate the $650 per month claim. They take a random sample of 175 students at their school and obtain the data represented in Figure 4.11. Can we apply the normal model to the sample mean?[18]

Evaluating the skew condition is challenging

Don't despair if checking the skew condition is difficult or confusing. You aren't alone – nearly all students get frustrated when checking skew. Properly assessing skew takes practice, and you won't be a pro, even at the end of this book.

But this doesn't mean you should give up. Checking skew and the other conditions is extremely important for a responsible data analysis. However, rest assured that evaluating skew isn't something you need to be a master of by the end of the book, though by that time you should be able to properly assess clear cut cases.

---

[17]H_0: The average cost is $650 per month, $\mu = \$650$.
　H_A: The average cost is different than $650 per month, $\mu \neq \$650$.
[18]Applying the normal model requires that certain conditions are met. Because the data are a simple random sample and the sample (presumably) represents no more than 10% of all students at the college, the observations are independent. The sample size is also sufficiently large ($n = 175$) and the data exhibit strong skew. While the data are strongly skewed, the sample is sufficiently large that this is acceptable, and the normal model may be applied to the sample mean.

● **Example 4.22** The sample mean for student housing is \$616.91 and the sample standard deviation is \$128.65. Construct a 95% confidence interval for the population mean and evaluate the hypotheses of Guided Practice 4.20.

The standard error associated with the mean may be estimated using the sample standard deviation divided by the square root of the sample size. Recall that $n = 175$ students were sampled.

$$SE = \frac{s}{\sqrt{n}} = \frac{128.65}{\sqrt{175}} = 9.73$$

You showed in Guided Practice 4.21 that the normal model may be applied to the sample mean. This ensures a 95% confidence interval may be accurately constructed:

$$\bar{x} \pm z^{\star}SE \quad \rightarrow \quad 616.91 \pm 1.96 \times 9.73 \quad \rightarrow \quad (597.84, 635.98)$$

Because the null value \$650 is not in the confidence interval, a true mean of \$650 is implausible and we reject the null hypothesis. The data provide statistically significant evidence that the actual average housing expense is less than \$650 per month.

4.3.3 Decision errors

Hypothesis tests are not flawless, since we can make a wrong decision in statistical hypothesis tests based on the data. For example, in the court system innocent people are sometimes wrongly convicted and the guilty sometimes walk free. However, the difference is that in statistical hypothesis tests, we have the tools necessary to quantify how often we make such errors.

There are two competing hypotheses: the null and the alternative. In a hypothesis test, we make a statement about which one might be true, but we might choose incorrectly. There are four possible scenarios, which are summarized in Table 4.12.

		Test conclusion	
		do not reject H_0	reject H_0 in favor of H_A
	H_0 true	okay	Type 1 Error
Truth	H_A true	Type 2 Error	okay

Table 4.12: Four different scenarios for hypothesis tests.

A **Type 1 Error** is rejecting the null hypothesis when H_0 is actually true. A **Type 2 Error** is failing to reject the null hypothesis when the alternative is actually true.

⊙ **Guided Practice 4.23** In a US court, the defendant is either innocent (H_0) or guilty (H_A). What does a Type 1 Error represent in this context? What does a Type 2 Error represent? Table 4.12 may be useful.[19]

⊙ **Guided Practice 4.24** How could we reduce the Type 1 Error rate in US courts? What influence would this have on the Type 2 Error rate?[20]

---

[19]If the court makes a Type 1 Error, this means the defendant is innocent (H_0 true) but wrongly convicted. A Type 2 Error means the court failed to reject H_0 (i.e. failed to convict the person) when she was in fact guilty (H_A true).

[20]To lower the Type 1 Error rate, we might raise our standard for conviction from "beyond a reasonable doubt" to "beyond a conceivable doubt" so fewer people would be wrongly convicted. However, this would also make it more difficult to convict the people who are actually guilty, so we would make more Type 2 Errors.

⊙ **Guided Practice 4.25** How could we reduce the Type 2 Error rate in US courts? What influence would this have on the Type 1 Error rate?[21]

Exercises 4.23-4.25 provide an important lesson: if we reduce how often we make one type of error, we generally make more of the other type.

Hypothesis testing is built around rejecting or failing to reject the null hypothesis. That is, we do not reject H_0 unless we have strong evidence. But what precisely does *strong evidence* mean? As a general rule of thumb, for those cases where the null hypothesis is actually true, we do not want to incorrectly reject H_0 more than 5% of the time. This corresponds to a **significance level** of 0.05. We often write the significance level using α (the Greek letter *alpha*): $\alpha = 0.05$. We discuss the appropriateness of different significance levels in Section 4.3.6.

α
significance
level of a
hypothesis test

If we use a 95% confidence interval to evaluate a hypothesis test where the null hypothesis is true, we will make an error whenever the point estimate is at least 1.96 standard errors away from the population parameter. This happens about 5% of the time (2.5% in each tail). Similarly, using a 99% confidence interval to evaluate a hypothesis is equivalent to a significance level of $\alpha = 0.01$.

A confidence interval is, in one sense, simplistic in the world of hypothesis tests. Consider the following two scenarios:

- The null value (the parameter value under the null hypothesis) is in the 95% confidence interval but just barely, so we would not reject H_0. However, we might like to somehow say, quantitatively, that it was a close decision.

- The null value is very far outside of the interval, so we reject H_0. However, we want to communicate that, not only did we reject the null hypothesis, but it wasn't even close. Such a case is depicted in Figure 4.13.

In Section 4.3.4, we introduce a tool called the *p-value* that will be helpful in these cases. The p-value method also extends to hypothesis tests where confidence intervals cannot be easily constructed or applied.

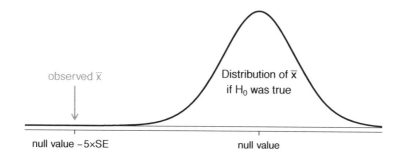

Figure 4.13: It would be helpful to quantify the strength of the evidence against the null hypothesis. In this case, the evidence is extremely strong.

---

[21]To lower the Type 2 Error rate, we want to convict more guilty people. We could lower the standards for conviction from "beyond a reasonable doubt" to "beyond a little doubt". Lowering the bar for guilt will also result in more wrongful convictions, raising the Type 1 Error rate.

4.3.4 Formal testing using p-values

The p-value is a way of quantifying the strength of the evidence against the null hypothesis and in favor of the alternative. Formally the *p-value* is a conditional probability.

---

p-value

The **p-value** is the probability of observing data at least as favorable to the alternative hypothesis as our current data set, if the null hypothesis is true. We typically use a summary statistic of the data, in this chapter the sample mean, to help compute the p-value and evaluate the hypotheses.

⊙ **Guided Practice 4.26**   A poll by the National Sleep Foundation found that college students average about 7 hours of sleep per night. Researchers at a rural school are interested in showing that students at their school sleep longer than seven hours on average, and they would like to demonstrate this using a sample of students. What would be an appropriate skeptical position for this research?[22]

We can set up the null hypothesis for this test as a skeptical perspective: the students at this school average 7 hours of sleep per night. The alternative hypothesis takes a new form reflecting the interests of the research: the students average more than 7 hours of sleep. We can write these hypotheses as

H_0: $\mu = 7$.

H_A: $\mu > 7$.

Using $\mu > 7$ as the alternative is an example of a **one-sided** hypothesis test. In this investigation, there is no apparent interest in learning whether the mean is less than 7 hours.[23] Earlier we encountered a **two-sided** hypothesis where we looked for any clear difference, greater than or less than the null value.

Always use a two-sided test unless it was made clear prior to data collection that the test should be one-sided. Switching a two-sided test to a one-sided test after observing the data is dangerous because it can inflate the Type 1 Error rate.

TIP: One-sided and two-sided tests
When you are interested in checking for an increase or a decrease, but not both, use a one-sided test. When you are interested in any difference from the null value – an increase or decrease – then the test should be two-sided.

TIP: Always write the null hypothesis as an equality
We will find it most useful if we always list the null hypothesis as an equality (e.g. $\mu = 7$) while the alternative always uses an inequality (e.g. $\mu \neq 7$, $\mu > 7$, or $\mu < 7$).

---

[22] A skeptic would have no reason to believe that sleep patterns at this school are different than the sleep patterns at another school.

[23] This is entirely based on the interests of the researchers. Had they been only interested in the opposite case – showing that their students were actually averaging fewer than seven hours of sleep but not interested in showing more than 7 hours – then our setup would have set the alternative as $\mu < 7$.

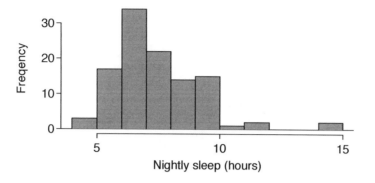

Figure 4.14: Distribution of a night of sleep for 110 college students. These data are strongly skewed.

The researchers at the rural school conducted a simple random sample of $n = 110$ students on campus. They found that these students averaged 7.42 hours of sleep and the standard deviation of the amount of sleep for the students was 1.75 hours. A histogram of the sample is shown in Figure 4.14.

Before we can use a normal model for the sample mean or compute the standard error of the sample mean, we must verify conditions. (1) Because this is a simple random sample from less than 10% of the student body, the observations are independent. (2) The sample size in the sleep study is sufficiently large since it is greater than 30. (3) The data show strong skew in Figure 4.14 and the presence of a couple of outliers. This skew and the outliers are acceptable for a sample size of $n = 110$. With these conditions verified, the normal model can be safely applied to $\bar{x}$ and we can reasonably calculate the standard error.

⊙ **Guided Practice 4.27**   In the sleep study, the sample standard deviation was 1.75 hours and the sample size is 110. Calculate the standard error of $\bar{x}$.[24]

The hypothesis test for the sleep study will be evaluated using a significance level of $\alpha = 0.05$. We want to consider the data under the scenario that the null hypothesis is true. In this case, the sample mean is from a distribution that is nearly normal and has mean 7 and standard deviation of about $SE_{\bar{x}} = 0.17$. Such a distribution is shown in Figure 4.15.

The shaded tail in Figure 4.15 represents the chance of observing such a large mean, conditional on the null hypothesis being true. That is, the shaded tail represents the p-value. We shade all means larger than our sample mean, $\bar{x} = 7.42$, because they are more favorable to the alternative hypothesis than the observed mean.

We compute the p-value by finding the tail area of this normal distribution, which we learned to do in Section 3.1. First compute the Z-score of the sample mean, $\bar{x} = 7.42$:

$$Z = \frac{\bar{x} - \text{null value}}{SE_{\bar{x}}} = \frac{7.42 - 7}{0.17} = 2.47$$

Using the normal probability table, the lower unshaded area is found to be 0.993. Thus the shaded area is $1 - 0.993 = 0.007$. *If the null hypothesis is true, the probability of observing a sample mean at least as large as 7.42 hours for a sample of 110 students is only 0.007.* That is, if the null hypothesis is true, we would not often see such a large mean.

---

[24]The standard error can be estimated from the sample standard deviation and the sample size: $SE_{\bar{x}} = \frac{s_x}{\sqrt{n}} = \frac{1.75}{\sqrt{110}} = 0.17$.

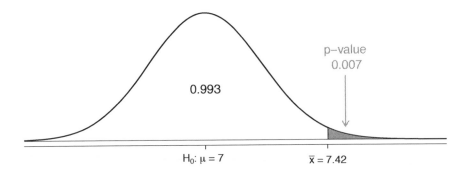

Figure 4.15: If the null hypothesis is true, then the sample mean $\bar{x}$ came from this nearly normal distribution. The right tail describes the probability of observing such a large sample mean if the null hypothesis is true.

We evaluate the hypotheses by comparing the p-value to the significance level. Because the p-value is less than the significance level (p-value $= 0.007 < 0.05 = \alpha$), we reject the null hypothesis. What we observed is so unusual with respect to the null hypothesis that it casts serious doubt on H_0 and provides strong evidence favoring H_A.

---

p-value as a tool in hypothesis testing

The smaller the p-value, the stronger the data favor H_A over H_0. A small p-value (usually < 0.05) corresponds to sufficient evidence to reject H_0 in favor of H_A.

TIP: It is useful to first draw a picture to find the p-value

It is useful to draw a picture of the distribution of $\bar{x}$ as though H_0 was true (i.e. μ equals the null value), and shade the region (or regions) of sample means that are at least as favorable to the alternative hypothesis. These shaded regions represent the p-value.

The ideas below review the process of evaluating hypothesis tests with p-values:

- The null hypothesis represents a skeptic's position or a position of no difference. We reject this position only if the evidence strongly favors H_A.

- A small p-value means that if the null hypothesis is true, there is a low probability of seeing a point estimate at least as extreme as the one we saw. We interpret this as strong evidence in favor of the alternative.

- We reject the null hypothesis if the p-value is smaller than the significance level, α, which is usually 0.05. Otherwise, we fail to reject H_0.

- We should always state the conclusion of the hypothesis test in plain language so non-statisticians can also understand the results.

The p-value is constructed in such a way that we can directly compare it to the significance level (α) to determine whether or not to reject H_0. This method ensures that the Type 1 Error rate does not exceed the significance level standard.

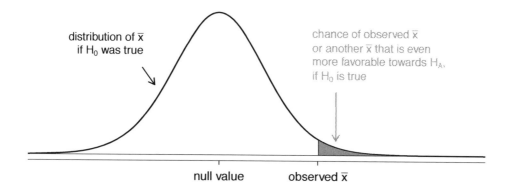

distribution of x̄
if H₀ was true

chance of observed x̄
or another x̄ that is even
more favorable towards Hₐ,
if H₀ is true

null value   observed x̄

Figure 4.16: To identify the p-value, the distribution of the sample mean is considered as if the null hypothesis was true. Then the p-value is defined and computed as the probability of the observed $\bar{x}$ or an $\bar{x}$ even more favorable to H_A under this distribution.

⊙ **Guided Practice 4.28**   If the null hypothesis is true, how often should the p-value be less than 0.05?[25]

⊙ **Guided Practice 4.29**   Suppose we had used a significance level of 0.01 in the sleep study. Would the evidence have been strong enough to reject the null hypothesis? (The p-value was 0.007.) What if the significance level was $\alpha = 0.001$? [26]

⊙ **Guided Practice 4.30**   Ebay might be interested in showing that buyers on its site tend to pay less than they would for the corresponding new item on Amazon. We'll research this topic for one particular product: a video game called *Mario Kart* for the Nintendo Wii. During early October 2009, Amazon sold this game for $46.99. Set up an appropriate (one-sided!) hypothesis test to check the claim that Ebay buyers pay less during auctions at this same time.[27]

⊙ **Guided Practice 4.31**   During early October 2009, 52 Ebay auctions were recorded for *Mario Kart*.[28] The total prices for the auctions are presented using a histogram in Figure 4.17, and we may like to apply the normal model to the sample mean. Check the three conditions required for applying the normal model: (1) independence, (2) at least 30 observations, and (3) the data are not strongly skewed.[29]

---

[25]About 5% of the time. If the null hypothesis is true, then the data only has a 5% chance of being in the 5% of data most favorable to H_A.

[26]We reject the null hypothesis whenever *p-value* < α. Thus, we would still reject the null hypothesis if $\alpha = 0.01$ but not if the significance level had been $\alpha = 0.001$.

[27]The skeptic would say the average is the same on Ebay, and we are interested in showing the average price is lower.

H_0: The average auction price on Ebay is equal to (or more than) the price on Amazon. We write only the equality in the statistical notation: $\mu_{ebay} = 46.99$.

H_A: The average price on Ebay is less than the price on Amazon, $\mu_{ebay} < 46.99$.

[28]These data were collected by OpenIntro staff.

[29](1) The independence condition is unclear. *We will make the assumption that the observations are independent, which we should report with any final results.* (2) The sample size is sufficiently large: $n = 52 \geq 30$. (3) The data distribution is not strongly skewed; it is approximately symmetric.

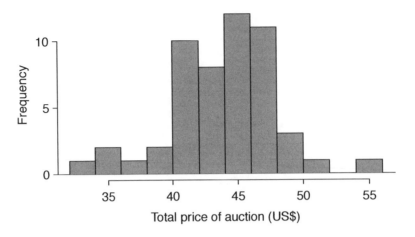

Figure 4.17: A histogram of the total auction prices for 52 Ebay auctions.

● **Example 4.32** The average sale price of the 52 Ebay auctions for *Wii Mario Kart* was $44.17 with a standard deviation of $4.15. Does this provide sufficient evidence to reject the null hypothesis in Guided Practice 4.30? Use a significance level of $\alpha = 0.01$.

The hypotheses were set up and the conditions were checked in Exercises 4.30 and 4.31. The next step is to find the standard error of the sample mean and produce a sketch to help find the p-value.

$$SE_{\bar{x}} = s/\sqrt{n} = 4.15/\sqrt{52} = 0.5755$$

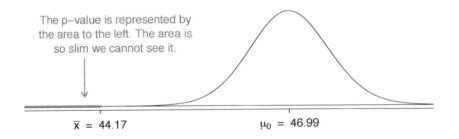

Because the alternative hypothesis says we are looking for a smaller mean, we shade the lower tail. We find this shaded area by using the Z-score and normal probability table: $Z = \frac{44.17-46.99}{0.5755} = -4.90$, which has area less than 0.0002. The area is so small we cannot really see it on the picture. This lower tail area corresponds to the p-value.

Because the p-value is so small – specifically, smaller than $\alpha = 0.01$ – this provides sufficiently strong evidence to reject the null hypothesis in favor of the alternative. The data provide statistically significant evidence that the average price on Ebay is lower than Amazon's asking price.

What's so special about 0.05?

It's common to use a threshold of 0.05 to determine whether a result is statistically significant, but why is the most common value 0.05? Maybe the standard significance level should be bigger, or maybe it should be smaller. If you're a little puzzled, that probably means you're reading with a critical eye – good job! We've made a 5-minute task to help clarify *why 0.05*:

www.openintro.org/why05

Sometimes it's also a good idea to deviate from the standard. We'll discuss when to choose a threshold different than 0.05 in Section 4.3.6.

4.3.5 Two-sided hypothesis testing with p-values

We now consider how to compute a p-value for a two-sided test. In one-sided tests, we shade the single tail in the direction of the alternative hypothesis. For example, when the alternative had the form $\mu > 7$, then the p-value was represented by the upper tail (Figure 4.16). When the alternative was $\mu < 46.99$, the p-value was the lower tail (Guided Practice 4.30). In a two-sided test, *we shade two tails* since evidence in either direction is favorable to H_A.

⊙ **Guided Practice 4.33** Earlier we talked about a research group investigating whether the students at their school slept longer than 7 hours each night. Let's consider a second group of researchers who want to evaluate whether the students at their college differ from the norm of 7 hours. Write the null and alternative hypotheses for this investigation.[30]

● **Example 4.34** The second college randomly samples 122 students and finds a mean of $\bar{x} = 6.83$ hours and a standard deviation of $s = 1.8$ hours. Does this provide strong evidence against H_0 in Guided Practice 4.33? Use a significance level of $\alpha = 0.05$.

First, we must verify assumptions. (1) A simple random sample of less than 10% of the student body means the observations are independent. (2) The sample size is 122, which is greater than 30. (3) Based on the earlier distribution and what we already know about college student sleep habits, the sample size will be acceptable.

Next we can compute the standard error ($SE_{\bar{x}} = \frac{s}{\sqrt{n}} = 0.16$) of the estimate and create a picture to represent the p-value, shown in Figure 4.18. Both tails are shaded. An estimate of 7.17 or more provides at least as strong of evidence against the null hypothesis and in favor of the alternative as the observed estimate, $\bar{x} = 6.83$.

We can calculate the tail areas by first finding the lower tail corresponding to $\bar{x}$:

$$Z = \frac{6.83 - 7.00}{0.16} = -1.06 \quad \overset{table}{\rightarrow} \quad \text{left tail} = 0.1446$$

Because the normal model is symmetric, the right tail will have the same area as the left tail. The p-value is found as the sum of the two shaded tails:

$$\text{p-value} = \text{left tail} + \text{right tail} = 2 \times (\text{left tail}) = 0.2892$$

---

[30]Because the researchers are interested in any difference, they should use a two-sided setup: $H_0 : \mu = 7$, $H_A : \mu \neq 7$.

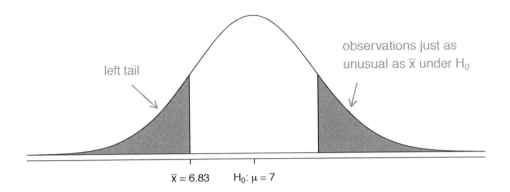

Figure 4.18: H_A is two-sided, so *both* tails must be counted for the p-value.

This p-value is relatively large (larger than $\alpha = 0.05$), so we should not reject H_0. That is, if H_0 is true, it would not be very unusual to see a sample mean this far from 7 hours simply due to sampling variation. Thus, we do not have sufficient evidence to conclude that the mean is different than 7 hours.

● **Example 4.35** It is never okay to change two-sided tests to one-sided tests after observing the data. In this example we explore the consequences of ignoring this advice. Using $\alpha = 0.05$, we show that freely switching from two-sided tests to one-sided tests will cause us to make twice as many Type 1 Errors as intended.

Suppose the sample mean was larger than the null value, μ_0 (e.g. μ_0 would represent 7 if H_0: $\mu = 7$). Then if we can flip to a one-sided test, we would use H_A: $\mu > \mu_0$. Now if we obtain any observation with a Z-score greater than 1.65, we would reject H_0. If the null hypothesis is true, we incorrectly reject the null hypothesis about 5% of the time when the sample mean is above the null value, as shown in Figure 4.19.

Suppose the sample mean was smaller than the null value. Then if we change to a one-sided test, we would use H_A: $\mu < \mu_0$. If $\bar{x}$ had a Z-score smaller than -1.65, we would reject H_0. If the null hypothesis is true, then we would observe such a case about 5% of the time.

By examining these two scenarios, we can determine that we will make a Type 1 Error $5\% + 5\% = 10\%$ of the time if we are allowed to swap to the "best" one-sided test for the data. This is twice the error rate we prescribed with our significance level: $\alpha = 0.05$ (!).

Caution: One-sided hypotheses are allowed only *before* seeing data

After observing data, it is tempting to turn a two-sided test into a one-sided test. Avoid this temptation. Hypotheses must be set up *before* observing the data. If they are not, the test should be two-sided.

4.3.6   Choosing a significance level

Choosing a significance level for a test is important in many contexts, and the traditional level is 0.05. However, it is often helpful to adjust the significance level based on the

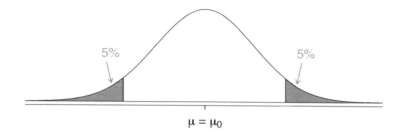

Figure 4.19: The shaded regions represent areas where we would reject H_0 under the bad practices considered in Example 4.35 when $\alpha = 0.05$.

application. We may select a level that is smaller or larger than 0.05 depending on the consequences of any conclusions reached from the test.

If making a Type 1 Error is dangerous or especially costly, we should choose a small significance level (e.g. 0.01). Under this scenario we want to be very cautious about rejecting the null hypothesis, so we demand very strong evidence favoring H_A before we would reject H_0.

If a Type 2 Error is relatively more dangerous or much more costly than a Type 1 Error, then we should choose a higher significance level (e.g. 0.10). Here we want to be cautious about failing to reject H_0 when the null is actually false.

Significance levels should reflect consequences of errors
The significance level selected for a test should reflect the consequences associated with Type 1 and Type 2 Errors.

● **Example 4.36** A car manufacturer is considering a higher quality but more expensive supplier for window parts in its vehicles. They sample a number of parts from their current supplier and also parts from the new supplier. They decide that if the high quality parts will last more than 12% longer, it makes financial sense to switch to this more expensive supplier. Is there good reason to modify the significance level in such a hypothesis test?

The null hypothesis is that the more expensive parts last no more than 12% longer while the alternative is that they do last more than 12% longer. This decision is just one of the many regular factors that have a marginal impact on the car and company. A significance level of 0.05 seems reasonable since neither a Type 1 or Type 2 Error should be dangerous or (relatively) much more expensive.

● **Example 4.37** The same car manufacturer is considering a slightly more expensive supplier for parts related to safety, not windows. If the durability of these safety components is shown to be better than the current supplier, they will switch manufacturers. Is there good reason to modify the significance level in such an evaluation?

The null hypothesis would be that the suppliers' parts are equally reliable. Because safety is involved, the car company should be eager to switch to the slightly more expensive manufacturer (reject H_0) even if the evidence of increased safety is only moderately strong. A slightly larger significance level, such as $\alpha = 0.10$, might be appropriate.

⊙ **Guided Practice 4.38** A part inside of a machine is very expensive to replace. However, the machine usually functions properly even if this part is broken, so the part is replaced only if we are extremely certain it is broken based on a series of measurements. Identify appropriate hypotheses for this test (in plain language) and suggest an appropriate significance level.[31]

4.4 Examining the Central Limit Theorem

The normal model for the sample mean tends to be very good when the sample consists of at least 30 independent observations and the population data are not strongly skewed. The Central Limit Theorem provides the theory that allows us to make this assumption.

Central Limit Theorem, informal definition

The distribution of $\bar{x}$ is approximately normal. The approximation can be poor if the sample size is small, but it improves with larger sample sizes.

The Central Limit Theorem states that when the sample size is small, the normal approximation may not be very good. However, as the sample size becomes large, the normal approximation improves. We will investigate three cases to see roughly when the approximation is reasonable.

We consider three data sets: one from a *uniform* distribution, one from an *exponential* distribution, and the other from a *log-normal* distribution. These distributions are shown in the top panels of Figure 4.20. The uniform distribution is symmetric, the exponential distribution may be considered as having moderate skew since its right tail is relatively short (few outliers), and the log-normal distribution is strongly skewed and will tend to produce more apparent outliers.

The left panel in the $n = 2$ row represents the sampling distribution of $\bar{x}$ if it is the sample mean of two observations from the uniform distribution shown. The dashed line represents the closest approximation of the normal distribution. Similarly, the center and right panels of the $n = 2$ row represent the respective distributions of $\bar{x}$ for data from exponential and log-normal distributions.

⊙ **Guided Practice 4.39** Examine the distributions in each row of Figure 4.20. What do you notice about the normal approximation for each sampling distribution as the sample size becomes larger?[32]

● **Example 4.40** Would the normal approximation be good in all applications where the sample size is at least 30?

Not necessarily. For example, the normal approximation for the log-normal example is questionable for a sample size of 30. Generally, the more skewed a population distribution or the more common the frequency of outliers, the larger the sample required to guarantee the distribution of the sample mean is nearly normal.

---

[31]Here the null hypothesis is that the part is not broken, and the alternative is that it is broken. If we don't have sufficient evidence to reject H_0, we would not replace the part. It sounds like failing to fix the part if it is broken (H_0 false, H_A true) is not very problematic, and replacing the part is expensive. Thus, we should require very strong evidence against H_0 before we replace the part. Choose a small significance level, such as $\alpha = 0.01$.

[32]The normal approximation becomes better as larger samples are used.

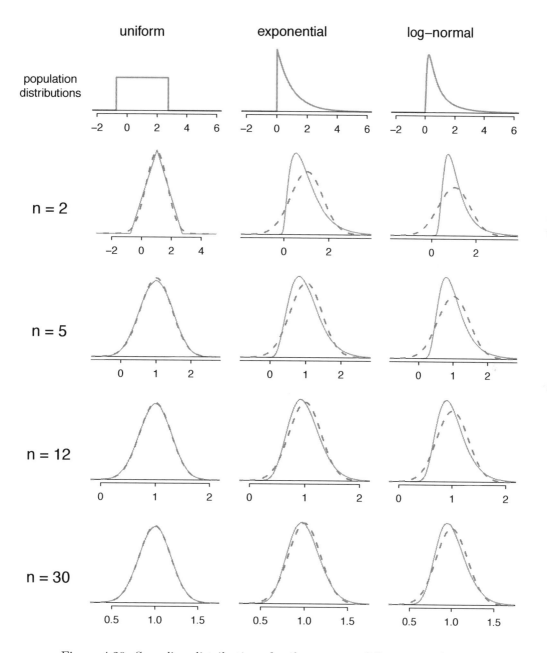

Figure 4.20: Sampling distributions for the mean at different sample sizes and for three different distributions. The dashed red lines show normal distributions.

> **TIP: With larger n, the sampling distribution of $\bar{x}$ becomes more normal**
> As the sample size increases, the normal model for $\bar{x}$ becomes more reasonable. We can also relax our condition on skew when the sample size is very large.

We discussed in Section 4.1.3 that the sample standard deviation, s, could be used as a substitute of the population standard deviation, σ, when computing the standard error. This estimate tends to be reasonable when $n \geq 30$. We will encounter alternative distributions for smaller sample sizes in Chapters 5 and 6.

● **Example 4.41**   Figure 4.21 shows a histogram of 50 observations. These represent winnings and losses from 50 consecutive days of a professional poker player. Can the normal approximation be applied to the sample mean, 90.69?

We should consider each of the required conditions.

(1) These are referred to as **time series data**, because the data arrived in a particular sequence. If the player wins on one day, it may influence how she plays the next. To make the assumption of independence we should perform careful checks on such data. While the supporting analysis is not shown, no evidence was found to indicate the observations are not independent.

(2) The sample size is 50, satisfying the sample size condition.

(3) There are two outliers, one very extreme, which suggests the data are very strongly skewed or very distant outliers may be common for this type of data. Outliers can play an important role and affect the distribution of the sample mean and the estimate of the standard error.

Since we should be skeptical of the independence of observations and the very extreme upper outlier poses a challenge, we should not use the normal model for the sample mean of these 50 observations. If we can obtain a much larger sample, perhaps several hundred observations, then the concerns about skew and outliers would no longer apply.

> **Caution: Examine data structure when considering independence**
> Some data sets are collected in such a way that they have a natural underlying structure between observations, e.g. when observations occur consecutively. Be especially cautious about independence assumptions regarding such data sets.

> **Caution: Watch out for strong skew and outliers**
> Strong skew is often identified by the presence of clear outliers. If a data set has prominent outliers, or such observations are somewhat common for the type of data under study, then it is useful to collect a sample with many more than 30 observations if the normal model will be used for $\bar{x}$.

You won't be a pro at assessing skew by the end of this book, so just use your best judgement and continue learning. As you develop your statistics skills and encounter tough situations, also consider learning about better ways to analyze skewed data, such as the studentized bootstrap (bootstrap-t), or consult a more experienced statistician.

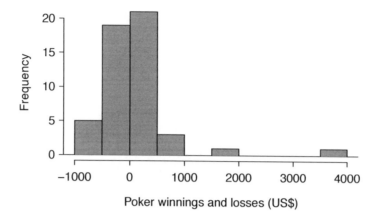

Figure 4.21: Sample distribution of poker winnings. These data include some very clear outliers. These are problematic when considering the normality of the sample mean. For example, outliers are often an indicator of very strong skew.

4.5 Inference for other estimators

The sample mean is not the only point estimate for which the sampling distribution is nearly normal. For example, the sampling distribution of sample proportions closely resembles the normal distribution when the sample size is sufficiently large. In this section, we introduce a number of examples where the normal approximation is reasonable for the point estimate. Chapters 5 and 6 will revisit each of the point estimates you see in this section along with some other new statistics.

We make another important assumption about each point estimate encountered in this section: the estimate is unbiased. A point estimate is **unbiased** if the sampling distribution of the estimate is centered at the parameter it estimates. That is, an unbiased estimate does not naturally over or underestimate the parameter. Rather, it tends to provide a "good" estimate. The sample mean is an example of an unbiased point estimate, as are each of the examples we introduce in this section.

Finally, we will discuss the general case where a point estimate may follow some distribution other than the normal distribution. We also provide guidance about how to handle scenarios where the statistical techniques you are familiar with are insufficient for the problem at hand.

4.5.1 Confidence intervals for nearly normal point estimates

In Section 4.2, we used the point estimate $\bar{x}$ with a standard error $SE_{\bar{x}}$ to create a 95% confidence interval for the population mean:

$$\bar{x} \ \pm \ 1.96 \times SE_{\bar{x}} \tag{4.42}$$

We constructed this interval by noting that the sample mean is within 1.96 standard errors of the actual mean about 95% of the time. This same logic generalizes to any unbiased point estimate that is nearly normal. We may also generalize the confidence level by using a place-holder $z^\star$.

General confidence interval for the normal sampling distribution case

A confidence interval based on an unbiased and nearly normal point estimate is

$$\text{point estimate} \ \pm \ z^{\star}SE \qquad\qquad (4.43)$$

where $z^{\star}$ is selected to correspond to the confidence level, and SE represents the standard error. The value $z^{\star}SE$ is called the *margin of error*.

Generally the standard error for a point estimate is estimated from the data and computed using a formula. For example, the standard error for the sample mean is

$$SE_{\bar{x}} = \frac{s}{\sqrt{n}}$$

In this section, we provide the computed standard error for each example and exercise without detailing where the values came from. In future chapters, you will learn to fill in these and other details for each situation.

● **Example 4.44** In Guided Practice 4.1 on page 170, we computed a point estimate for the difference in the average days active per week between male and female students: $\bar{x}_{female} - \bar{x}_{male} = 1.1$ days. This point estimate is associated with a nearly normal distribution with standard error $SE = 0.5$ days. What is a reasonable 95% confidence interval for the difference in average days active per week?

The normal approximation is said to be valid, so we apply Equation (4.43):

$$\text{point estimate} \ \pm \ z^{\star}SE \quad \rightarrow \quad 1.1 \pm 1.96 \times 0.5 \quad \rightarrow \quad (0.12, 2.08)$$

We are 95% confident that the male students, on average, were physically active 0.12 to 2.08 days more than female students in YRBSS each week. That is, the actual average difference is plausibly between 0.12 and 2.08 days per week with 95% confidence.

● **Example 4.45** Does Example 4.44 guarantee that if a male and female student are selected at random from YRBSS, the male student would be active 0.12 to 2.08 days more than the female student?

Our confidence interval says absolutely nothing about individual observations. It only makes a statement about a plausible range of values for the *average* difference between all male and female students who participated in YRBSS.

⊙ **Guided Practice 4.46** What $z^{\star}$ would be appropriate for a 99% confidence level? For help, see Figure 4.10 on page 178.[33]

---

[33]We seek $z^{\star}$ such that 99% of the area under the normal curve will be between the Z-scores $-z^{\star}$ and $z^{\star}$. Because the remaining 1% is found in the tails, each tail has area 0.5%, and we can identify $-z^{\star}$ by looking up 0.0050 in the normal probability table: $z^{\star} = 2.58$. See also Figure 4.10 on page 178.

⊙ **Guided Practice 4.47** The proportion of students who are male in the yrbss_samp sample is $\hat{p} = 0.48$. This sample meets certain conditions that ensure $\hat{p}$ will be nearly normal, and the standard error of the estimate is $SE_{\hat{p}} = 0.05$. Create a 90% confidence interval for the proportion of students in the 2013 YRBSS survey who are male.[34]

4.5.2 Hypothesis testing for nearly normal point estimates

Just as the confidence interval method works with many other point estimates, we can generalize our hypothesis testing methods to new point estimates. Here we only consider the p-value approach, introduced in Section 4.3.4, since it is the most commonly used technique and also extends to non-normal cases.

Hypothesis testing using the normal model

1. First write the hypotheses in plain language, then set them up in mathematical notation.

2. Identify an appropriate point estimate of the parameter of interest.

3. Verify conditions to ensure the standard error estimate is reasonable and the point estimate is nearly normal and unbiased.

4. Compute the standard error. Draw a picture depicting the distribution of the estimate under the idea that H_0 is true. Shade areas representing the p-value.

5. Using the picture and normal model, compute the *test statistic* (Z-score) and identify the p-value to evaluate the hypotheses. Write a conclusion in plain language.

⊙ **Guided Practice 4.48** A drug called sulphinpyrazone was under consideration for use in reducing the death rate in heart attack patients. To determine whether the drug was effective, a set of 1,475 patients were recruited into an experiment and randomly split into two groups: a control group that received a placebo and a treatment group that received the new drug. What would be an appropriate null hypothesis? And the alternative?[35]

---

[34]We use $z^\star = 1.65$ (see Guided Practice 4.17 on page 180), and apply the general confidence interval formula:

$$\hat{p} \pm z^\star SE_{\hat{p}} \quad \rightarrow \quad 0.48 \pm 1.65 \times 0.05 \quad \rightarrow \quad (0.3975, 0.5625)$$

Thus, we are 90% confident that between 40% and 56% of the YRBSS students were male.

[35]The skeptic's perspective is that the drug does not work at reducing deaths in heart attack patients (H_0), while the alternative is that the drug does work (H_A).

We can formalize the hypotheses from Guided Practice 4.48 by letting $p_{control}$ and $p_{treatment}$ represent the proportion of patients who died in the control and treatment groups, respectively. Then the hypotheses can be written as

$$H_0 : p_{control} = p_{treatment} \quad \text{(the drug doesn't work)}$$
$$H_A : p_{control} > p_{treatment} \quad \text{(the drug works)}$$

or equivalently,

$$H_0 : p_{control} - p_{treatment} = 0 \quad \text{(the drug doesn't work)}$$
$$H_A : p_{control} - p_{treatment} > 0 \quad \text{(the drug works)}$$

Strong evidence against the null hypothesis and in favor of the alternative would correspond to an observed difference in death rates,

$$\text{point estimate} = \hat{p}_{control} - \hat{p}_{treatment}$$

being larger than we would expect from chance alone. This difference in sample proportions represents a point estimate that is useful in evaluating the hypotheses.

● **Example 4.49** We want to evaluate the hypothesis setup from Guided Practice 4.48 using data from the actual study.[36] In the control group, 60 of 742 patients died. In the treatment group, 41 of 733 patients died. The sample difference in death rates can be summarized as

$$\text{point estimate} = \hat{p}_{control} - \hat{p}_{treatment} = \frac{60}{742} - \frac{41}{733} = 0.025$$

This point estimate is nearly normal and is an unbiased estimate of the actual difference in death rates. The standard error of this sample difference is $SE = 0.013$. Evaluate the hypothesis test at a 5% significance level: $\alpha = 0.05$.

We would like to identify the p-value to evaluate the hypotheses. If the null hypothesis is true, then the point estimate would have come from a nearly normal distribution, like the one shown in Figure 4.22. The distribution is centered at zero since $p_{control} - p_{treatment} = 0$ under the null hypothesis. Because a large positive difference provides evidence against the null hypothesis and in favor of the alternative, the upper tail has been shaded to represent the p-value. We need not shade the lower tail since this is a one-sided test: an observation in the lower tail does not support the alternative hypothesis.

The p-value can be computed by using the Z-score of the point estimate and the normal probability table.

$$Z = \frac{\text{point estimate} - \text{null value}}{SE_{\text{point estimate}}} = \frac{0.025 - 0}{0.013} = 1.92 \qquad (4.50)$$

Examining Z in the normal probability table, we find that the lower unshaded tail is about 0.973. Thus, the upper shaded tail representing the p-value is

$$\text{p-value} = 1 - 0.973 = 0.027$$

---

[36] Anturane Reinfarction Trial Research Group. 1980. Sulfinpyrazone in the prevention of sudden death after myocardial infarction. New England Journal of Medicine 302(5):250-256.

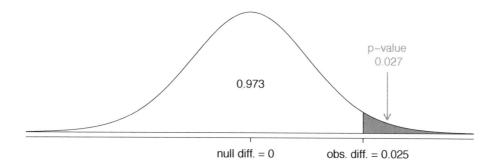

Figure 4.22: The distribution of the sample difference if the null hypothesis is true.

Because the p-value is less than the significance level ($\alpha = 0.05$), we say the null hypothesis is implausible. That is, we reject the null hypothesis in favor of the alternative and conclude that the drug is effective at reducing deaths in heart attack patients.

The Z-score in Equation (4.50) is called a **test statistic**. In most hypothesis tests, a test statistic is a particular data summary that is especially useful for computing the p-value and evaluating the hypothesis test. In the case of point estimates that are nearly normal, the test statistic is the Z-score.

Test statistic

A *test statistic* is a summary statistic that is particularly useful for evaluating a hypothesis test or identifying the p-value. When a point estimate is nearly normal, we use the Z-score of the point estimate as the test statistic. In later chapters we encounter situations where other test statistics are helpful.

4.5.3 Non-normal point estimates

We may apply the ideas of confidence intervals and hypothesis testing to cases where the point estimate or test statistic is not necessarily normal. There are many reasons why such a situation may arise:

- the sample size is too small for the normal approximation to be valid;
- the standard error estimate may be poor; or
- the point estimate tends towards some distribution that is not the normal distribution.

For each case where the normal approximation is not valid, our first task is always to understand and characterize the sampling distribution of the point estimate or test statistic. Next, we can apply the general frameworks for confidence intervals and hypothesis testing to these alternative distributions.

4.5.4 When to retreat

Statistical tools rely on conditions. When the conditions are not met, these tools are unreliable and drawing conclusions from them is treacherous. The conditions for these tools typically come in two forms.

- **The individual observations must be independent.** A random sample from less than 10% of the population ensures the observations are independent. In experiments, we generally require that subjects are randomized into groups. If independence fails, then advanced techniques must be used, and in some such cases, inference may not be possible.

- **Other conditions focus on sample size and skew.** For example, if the sample size is too small, the skew too strong, or extreme outliers are present, then the normal model for the sample mean will fail.

Verification of conditions for statistical tools is always necessary. Whenever conditions are not satisfied for a statistical technique, there are three options. The first is to learn new methods that are appropriate for the data. The second route is to consult a statistician.[37] The third route is to ignore the failure of conditions. This last option effectively invalidates any analysis and may discredit novel and interesting findings.

Finally, we caution that there may be no inference tools helpful when considering data that include unknown biases, such as convenience samples. For this reason, there are books, courses, and researchers devoted to the techniques of sampling and experimental design. See Sections 1.3-1.5 for basic principles of data collection.

4.5.5 Statistical significance versus practical significance

When the sample size becomes larger, point estimates become more precise and any real differences in the mean and null value become easier to detect and recognize. Even a very small difference would likely be detected if we took a large enough sample. Sometimes researchers will take such large samples that even the slightest difference is detected. While we still say that difference is **statistically significant**, it might not be **practically significant**.

Statistically significant differences are sometimes so minor that they are not practically relevant. This is especially important to research: if we conduct a study, we want to focus on finding a meaningful result. We don't want to spend lots of money finding results that hold no practical value.

The role of a statistician in conducting a study often includes planning the size of the study. The statistician might first consult experts or scientific literature to learn what would be the smallest meaningful difference from the null value. She also would obtain some reasonable estimate for the standard deviation. With these important pieces of information, she would choose a sufficiently large sample size so that the power for the meaningful difference is perhaps 80% or 90%. While larger sample sizes may still be used, she might advise against using them in some cases, especially in sensitive areas of research.

---

[37]If you work at a university, then there may be campus consulting services to assist you. Alternatively, there are many private consulting firms that are also available for hire.

4.6 Exercises

4.6.1 Variability in estimates

4.1 Identify the parameter, Part I. For each of the following situations, state whether the parameter of interest is a mean or a proportion. It may be helpful to examine whether individual responses are numerical or categorical.

(a) In a survey, one hundred college students are asked how many hours per week they spend on the Internet.

(b) In a survey, one hundred college students are asked: "What percentage of the time you spend on the Internet is part of your course work?"

(c) In a survey, one hundred college students are asked whether or not they cited information from Wikipedia in their papers.

(d) In a survey, one hundred college students are asked what percentage of their total weekly spending is on alcoholic beverages.

(e) In a sample of one hundred recent college graduates, it is found that 85 percent expect to get a job within one year of their graduation date.

4.2 Identify the parameter, Part II. For each of the following situations, state whether the parameter of interest is a mean or a proportion.

(a) A poll shows that 64% of Americans personally worry a great deal about federal spending and the budget deficit.

(b) A survey reports that local TV news has shown a 17% increase in revenue between 2009 and 2011 while newspaper revenues decreased by 6.4% during this time period.

(c) In a survey, high school and college students are asked whether or not they use geolocation services on their smart phones.

(d) In a survey, smart phone users are asked whether or not they use a web-based taxi service.

(e) In a survey, smart phone users are asked how many times they used a web-based taxi service over the last year.

4.3 College credits. A college counselor is interested in estimating how many credits a student typically enrolls in each semester. The counselor decides to randomly sample 100 students by using the registrar's database of students. The histogram below shows the distribution of the number of credits taken by these students. Sample statistics for this distribution are also provided.

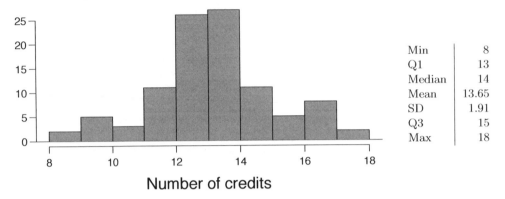

Min	8
Q1	13
Median	14
Mean	13.65
SD	1.91
Q3	15
Max	18

Number of credits

(a) What is the point estimate for the average number of credits taken per semester by students at this college? What about the median?

(b) What is the point estimate for the standard deviation of the number of credits taken per semester by students at this college? What about the IQR?

(c) Is a load of 16 credits unusually high for this college? What about 18 credits? Explain your reasoning. *Hint:* Observations farther than two standard deviations from the mean are usually considered to be unusual.

(d) The college counselor takes another random sample of 100 students and this time finds a sample mean of 14.02 units. Should she be surprised that this sample statistic is slightly different than the one from the original sample? Explain your reasoning.

(e) The sample means given above are point estimates for the mean number of credits taken by all students at that college. What measures do we use to quantify the variability of this estimate (Hint: recall that $SD_{\bar{x}} = \frac{\sigma}{\sqrt{n}}$)? Compute this quantity using the data from the original sample.

4.4 Heights of adults. Researchers studying anthropometry collected body girth measurements and skeletal diameter measurements, as well as age, weight, height and gender, for 507 physically active individuals. The histogram below shows the sample distribution of heights in centimeters.[38]

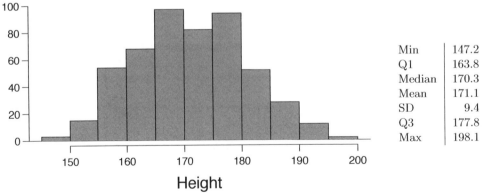

Min	147.2
Q1	163.8
Median	170.3
Mean	171.1
SD	9.4
Q3	177.8
Max	198.1

Height

(a) What is the point estimate for the average height of active individuals? What about the median? **(See the next page for parts (b)-(e).)**

---

[38] G. Heinz et al. "Exploring relationships in body dimensions". In: *Journal of Statistics Education* 11.2 (2003).

(b) What is the point estimate for the standard deviation of the heights of active individuals? What about the IQR?

(c) Is a person who is 1m 80cm (180 cm) tall considered unusually tall? And is a person who is 1m 55cm (155cm) considered unusually short? Explain your reasoning.

(d) The researchers take another random sample of physically active individuals. Would you expect the mean and the standard deviation of this new sample to be the ones given above? Explain your reasoning.

(e) The sample means obtained are point estimates for the mean height of all active individuals, if the sample of individuals is equivalent to a simple random sample. What measure do we use to quantify the variability of such an estimate (Hint: recall that $SD_{\bar{x}} = \frac{\sigma}{\sqrt{n}}$)? Compute this quantity using the data from the original sample under the condition that the data are a simple random sample.

4.5 Hen eggs. The distribution of the number of eggs laid by a certain species of hen during their breeding period is 35 eggs with a standard deviation of 18.2. Suppose a group of researchers randomly samples 45 hens of this species, counts the number of eggs laid during their breeding period, and records the sample mean. They repeat this 1,000 times, and build a distribution of sample means.

(a) What is this distribution called?

(b) Would you expect the shape of this distribution to be symmetric, right skewed, or left skewed? Explain your reasoning.

(c) Calculate the variability of this distribution and state the appropriate term used to refer to this value.

(d) Suppose the researchers' budget is reduced and they are only able to collect random samples of 10 hens. The sample mean of the number of eggs is recorded, and we repeat this 1,000 times, and build a new distribution of sample means. How will the variability of this new distribution compare to the variability of the original distribution?

4.6 Art after school. Elijah and Tyler, two high school juniors, conducted a survey on 15 students at their school, asking the students whether they would like the school to offer an after-school art program, counted the number of "yes" answers, and recorded the sample proportion. 14 out of the 15 students responded "yes". They repeated this 100 times and built a distribution of sample means.

(a) What is this distribution called?

(b) Would you expect the shape of this distribution to be symmetric, right skewed, or left skewed? Explain your reasoning.

(c) Calculate the variability of this distribution and state the appropriate term used to refer to this value.

(d) Suppose that the students were able to recruit a few more friends to help them with sampling, and are now able to collect data from random samples of 25 students. Once again, they record the number of "yes" answers, and record the sample proportion, and repeat this 100 times to build a new distribution of sample proportions. How will the variability of this new distribution compare to the variability of the original distribution?

4.6.2 Confidence intervals

4.7 Chronic illness, Part I. In 2013, the Pew Research Foundation reported that "45% of U.S. adults report that they live with one or more chronic conditions".[39] However, this value was based on a sample, so it may not be a perfect estimate for the population parameter of interest on its own. The study reported a standard error of about 1.2%, and a normal model may reasonably be used in this setting. Create a 95% confidence interval for the proportion of U.S. adults who live with one or more chronic conditions. Also interpret the confidence interval in the context of the study.

4.8 Twitter users and news, Part I. A poll conducted in 2013 found that 52% of U.S. adult Twitter users get at least some news on Twitter.[40]. The standard error for this estimate was 2.4%, and a normal distribution may be used to model the sample proportion. Construct a 99% confidence interval for the fraction of U.S. adult Twitter users who get some news on Twitter, and interpret the confidence interval in context.

4.9 Chronic illness, Part II. In 2013, the Pew Research Foundation reported that "45% of U.S. adults report that they live with one or more chronic conditions", and the standard error for this estimate is 1.2%. Identify each of the following statements as true or false. Provide an explanation to justify each of your answers.

(a) We can say with certainty that the confidence interval from Exercise 4.7 contains the true percentage of U.S. adults who suffer from a chronic illness.

(b) If we repeated this study 1,000 times and constructed a 95% confidence interval for each study, then approximately 950 of those confidence intervals would contain the true fraction of U.S. adults who suffer from chronic illnesses.

(c) The poll provides statistically significant evidence (at the $\alpha = 0.05$ level) that the percentage of U.S. adults who suffer from chronic illnesses is below 50%.

(d) Since the standard error is 1.2%, only 1.2% of people in the study communicated uncertainty about their answer.

4.10 Twitter users and news, Part II. A poll conducted in 2013 found that 52% of U.S. adult Twitter users get at least some news on Twitter, and the standard error for this estimate was 2.4%. Identify each of the following statements as true or false. Provide an explanation to justify each of your answers.

(a) The data provide statistically significant evidence that more than half of U.S. adult Twitter users get some news through Twitter. Use a significance level of $\alpha = 0.01$.

(b) Since the standard error is 2.4%, we can conclude that 97.6% of all U.S. adult Twitter users were included in the study.

(c) If we want to reduce the standard error of the estimate, we should collect less data.

(d) If we construct a 90% confidence interval for the percentage of U.S. adults Twitter users who get some news through Twitter, this confidence interval will be wider than a corresponding 99% confidence interval.

---

[39]Pew Research Center, Washington, D.C. The Diagnosis Difference, November 26, 2013.

[40]Pew Research Center, Washington, D.C. Twitter News Consumers: Young, Mobile and Educated, November 4, 2013.

4.11 Relaxing after work. The 2010 General Social Survey asked the question: "After an average work day, about how many hours do you have to relax or pursue activities that you enjoy?" to a random sample of 1,155 Americans.[41] A 95% confidence interval for the mean number of hours spent relaxing or pursuing activities they enjoy was (1.38, 1.92).

(a) Interpret this interval in context of the data.

(b) Suppose another set of researchers reported a confidence interval with a larger margin of error based on the same sample of 1,155 Americans. How does their confidence level compare to the confidence level of the interval stated above?

(c) Suppose next year a new survey asking the same question is conducted, and this time the sample size is 2,500. Assuming that the population characteristics, with respect to how much time people spend relaxing after work, have not changed much within a year. How will the margin of error of the 95% confidence interval constructed based on data from the new survey compare to the margin of error of the interval stated above?

4.12 Mental health. The 2010 General Social Survey asked the question: "For how many days during the past 30 days was your mental health, which includes stress, depression, and problems with emotions, not good?" Based on responses from 1,151 US residents, the survey reported a 95% confidence interval of 3.40 to 4.24 days in 2010.

(a) Interpret this interval in context of the data.

(b) What does "95% confident" mean? Explain in the context of the application.

(c) Suppose the researchers think a 99% confidence level would be more appropriate for this interval. Will this new interval be smaller or larger than the 95% confidence interval?

(d) If a new survey were to be done with 500 Americans, would the standard error of the estimate be larger, smaller, or about the same. Assume the standard deviation has remained constant since 2010.

4.13 Waiting at an ER, Part I. A hospital administrator hoping to improve wait times decides to estimate the average emergency room waiting time at her hospital. She collects a simple random sample of 64 patients and determines the time (in minutes) between when they checked in to the ER until they were first seen by a doctor. A 95% confidence interval based on this sample is (128 minutes, 147 minutes), which is based on the normal model for the mean. Determine whether the following statements are true or false, and explain your reasoning.

(a) This confidence interval is not valid since we do not know if the population distribution of the ER wait times is nearly Normal.

(b) We are 95% confident that the average waiting time of these 64 emergency room patients is between 128 and 147 minutes.

(c) We are 95% confident that the average waiting time of all patients at this hospital's emergency room is between 128 and 147 minutes.

(d) 95% of random samples have a sample mean between 128 and 147 minutes.

(e) A 99% confidence interval would be narrower than the 95% confidence interval since we need to be more sure of our estimate.

(f) The margin of error is 9.5 and the sample mean is 137.5.

(g) In order to decrease the margin of error of a 95% confidence interval to half of what it is now, we would need to double the sample size.

[41]National Opinion Research Center, General Social Survey, 2010.

4.14   Thanksgiving spending, Part I. The 2009 holiday retail season, which kicked off on November 27, 2009 (the day after Thanksgiving), had been marked by somewhat lower self-reported consumer spending than was seen during the comparable period in 2008. To get an estimate of consumer spending, 436 randomly sampled American adults were surveyed. Daily consumer spending for the six-day period after Thanksgiving, spanning the Black Friday weekend and Cyber Monday, averaged $84.71. A 95% confidence interval based on this sample is ($80.31, $89.11). Determine whether the following statements are true or false, and explain your reasoning.

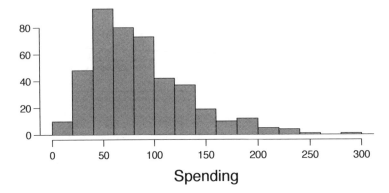

(a) We are 95% confident that the average spending of these 436 American adults is between $80.31 and $89.11.

(b) This confidence interval is not valid since the distribution of spending in the sample is right skewed.

(c) 95% of random samples have a sample mean between $80.31 and $89.11.

(d) We are 95% confident that the average spending of all American adults is between $80.31 and $89.11.

(e) A 90% confidence interval would be narrower than the 95% confidence interval since we don't need to be as sure about our estimate.

(f) In order to decrease the margin of error of a 95% confidence interval to a third of what it is now, we would need to use a sample 3 times larger.

(g) The margin of error is 4.4.

4.15   Exclusive relationships. A survey conducted on a reasonably random sample of 203 undergraduates asked, among many other questions, about the number of exclusive relationships these students have been in. The histogram below shows the distribution of the data from this sample. The sample average is 3.2 with a standard deviation of 1.97.

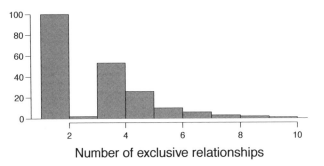

Estimate the average number of exclusive relationships Duke students have been in using a 90% confidence interval and interpret this interval in context. Check any conditions required for inference, and note any assumptions you must make as you proceed with your calculations and conclusions.

4.16 Age at first marriage, Part I. The National Survey of Family Growth conducted by the Centers for Disease Control gathers information on family life, marriage and divorce, pregnancy, infertility, use of contraception, and men's and women's health. One of the variables collected on this survey is the age at first marriage. The histogram below shows the distribution of ages at first marriage of 5,534 randomly sampled women between 2006 and 2010. The average age at first marriage among these women is 23.44 with a standard deviation of 4.72.[42]

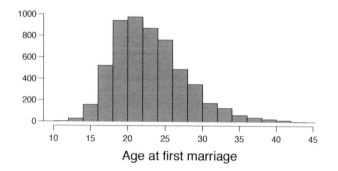

Age at first marriage

Estimate the average age at first marriage of women using a 95% confidence interval, and interpret this interval in context. Discuss any relevant assumptions.

4.6.3 Hypothesis testing

4.17 Identify hypotheses, Part I. Write the null and alternative hypotheses in words and then symbols for each of the following situations.

(a) New York is known as "the city that never sleeps". A random sample of 25 New Yorkers were asked how much sleep they get per night. Do these data provide convincing evidence that New Yorkers on average sleep less than 8 hours a night?

(b) Employers at a firm are worried about the effect of March Madness, a basketball championship held each spring in the US, on employee productivity. They estimate that on a regular business day employees spend on average 15 minutes of company time checking personal email, making personal phone calls, etc. They also collect data on how much company time employees spend on such non- business activities during March Madness. They want to determine if these data provide convincing evidence that employee productivity decreases during March Madness.

4.18 Identify hypotheses, Part II. Write the null and alternative hypotheses in words and using symbols for each of the following situations.

(a) Since 2008, chain restaurants in California have been required to display calorie counts of each menu item. Prior to menus displaying calorie counts, the average calorie intake of diners at a restaurant was 1100 calories. After calorie counts started to be displayed on menus, a nutritionist collected data on the number of calories consumed at this restaurant from a random sample of diners. Do these data provide convincing evidence of a difference in the average calorie intake of a diners at this restaurant?

(b) Based on the performance of those who took the GRE exam between July 1, 2004 and June 30, 2007, the average Verbal Reasoning score was calculated to be 462. In 2011 the average verbal score was slightly higher. Do these data provide convincing evidence that the average GRE Verbal Reasoning score has changed since 2004?

---

[42]Centers for Disease Control and Prevention, National Survey of Family Growth, 2010.

4.19 Online communication. A study suggests that the average college student spends 10 hours per week communicating with others online. You believe that this is an underestimate and decide to collect your own sample for a hypothesis test. You randomly sample 60 students from your dorm and find that on average they spent 13.5 hours a week communicating with others online. A friend of yours, who offers to help you with the hypothesis test, comes up with the following set of hypotheses. Indicate any errors you see.

$$H_0 : \bar{x} < 10 \; hours$$
$$H_A : \bar{x} > 13.5 \; hours$$

4.20 Age at first marriage, Part II. Exercise 4.16 presents the results of a 2006 - 2010 survey showing that the average age of women at first marriage is 23.44. Suppose a social scientist believes that this value has increased in 2012, but she would also be interested if she found a decrease. Below is how she set up her hypotheses. Indicate any errors you see.

$$H_0 : \bar{x} = 23.44 \; years \; old$$
$$H_A : \bar{x} > 23.44 \; years \; old$$

4.21 Waiting at an ER, Part II. Exercise 4.13 provides a 95% confidence interval for the mean waiting time at an emergency room (ER) of (128 minutes, 147 minutes). Answer the following questions based on this interval.

(a) A local newspaper claims that the average waiting time at this ER exceeds 3 hours. Is this claim supported by the confidence interval? Explain your reasoning.

(b) The Dean of Medicine at this hospital claims the average wait time is 2.2 hours. Is this claim supported by the confidence interval? Explain your reasoning.

(c) Without actually calculating the interval, determine if the claim of the Dean from part (b) would be supported based on a 99% confidence interval?

4.22 Thanksgiving spending, Part II. Exercise 4.14 provides a 95% confidence interval for the average spending by American adults during the six-day period after Thanksgiving 2009: ($80.31, $89.11).

(a) A local news anchor claims that the average spending during this period in 2009 was $100. What do you think of her claim?

(b) Would the news anchor's claim be considered reasonable based on a 90% confidence interval? Why or why not? (Do not actually calculate the interval.)

4.23 Nutrition labels. The nutrition label on a bag of potato chips says that a one ounce (28 gram) serving of potato chips has 130 calories and contains ten grams of fat, with three grams of saturated fat. A random sample of 35 bags yielded a sample mean of 134 calories with a standard deviation of 17 calories. Is there evidence that the nutrition label does not provide an accurate measure of calories in the bags of potato chips? We have verified the independence, sample size, and skew conditions are satisfied.

4.24 Gifted children, Part I. Researchers investigating characteristics of gifted children collected data from schools in a large city on a random sample of thirty-six children who were identified as gifted children soon after they reached the age of four. The following histogram shows the distribution of the ages (in months) at which these children first counted to 10 successfully. Also provided are some sample statistics.[43]

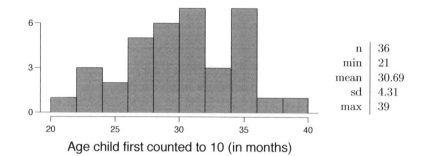

n	36
min	21
mean	30.69
sd	4.31
max	39

Age child first counted to 10 (in months)

(a) Are conditions for inference satisfied?

(b) Suppose you read online that children first count to 10 successfully when they are 32 months old, on average. Perform a hypothesis test to evaluate if these data provide convincing evidence that the average age at which gifted children fist count to 10 successfully is less than the general average of 32 months. Use a significance level of 0.10.

(c) Interpret the p-value in context of the hypothesis test and the data.

(d) Calculate a 90% confidence interval for the average age at which gifted children first count to 10 successfully.

(e) Do your results from the hypothesis test and the confidence interval agree? Explain.

4.25 Waiting at an ER, Part III. The hospital administrator mentioned in Exercise 4.13 randomly selected 64 patients and measured the time (in minutes) between when they checked in to the ER and the time they were first seen by a doctor. The average time is 137.5 minutes and the standard deviation is 39 minutes. She is getting grief from her supervisor on the basis that the wait times in the ER has increased greatly from last year's average of 127 minutes. However, she claims that the increase is probably just due to chance.

(a) Are conditions for inference met? Note any assumptions you must make to proceed.

(b) Using a significance level of $\alpha = 0.05$, is the change in wait times statistically significant? Use a two-sided test since it seems the supervisor had to inspect the data before she suggested an increase occurred.

(c) Would the conclusion of the hypothesis test change if the significance level was changed to $\alpha = 0.01$?

---

[43]F.A. Graybill and H.K. Iyer. *Regression Analysis: Concepts and Applications.* Duxbury Press, 1994, pp. 511–516.

4.26 Gifted children, Part II. Exercise 4.24 describes a study on gifted children. In this study, along with variables on the children, the researchers also collected data on the mother's and father's IQ of the 36 randomly sampled gifted children. The histogram below shows the distribution of mother's IQ. Also provided are some sample statistics.

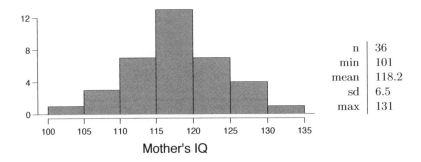

n	36
min	101
mean	118.2
sd	6.5
max	131

(a) Perform a hypothesis test to evaluate if these data provide convincing evidence that the average IQ of mothers of gifted children is different than the average IQ for the population at large, which is 100. Use a significance level of 0.10.

(b) Calculate a 90% confidence interval for the average IQ of mothers of gifted children.

(c) Do your results from the hypothesis test and the confidence interval agree? Explain.

4.27 Working backwards, one-sided. You are given the following hypotheses:

$$H_0 : \mu = 30$$
$$H_A : \mu > 30$$

We know that the sample standard deviation is 10 and the sample size is 70. For what sample mean would the p-value be equal to 0.05? Assume that all conditions necessary for inference are satisfied.

4.28 Working backwards, two-sided. You are given the following hypotheses:

$$H_0 : \mu = 30$$
$$H_A : \mu \neq 30$$

We know that the sample standard deviation is 10 and the sample size is 70. For what sample mean would the p-value be equal to 0.05? Assume that all conditions necessary for inference are satisfied.

4.29 Testing for Fibromyalgia. A patient named Diana was diagnosed with Fibromyalgia, a long-term syndrome of body pain, and was prescribed anti-depressants. Being the skeptic that she is, Diana didn't initially believe that anti-depressants would help her symptoms. However after a couple months of being on the medication she decides that the anti-depressants are working, because she feels like her symptoms are in fact getting better.

(a) Write the hypotheses in words for Diana's skeptical position when she started taking the anti-depressants.

(b) What is a Type 1 Error in this context?

(c) What is a Type 2 Error in this context?

4.30 Testing for food safety. A food safety inspector is called upon to investigate a restaurant with a few customer reports of poor sanitation practices. The food safety inspector uses a hypothesis testing framework to evaluate whether regulations are not being met. If he decides the restaurant is in gross violation, its license to serve food will be revoked.

(a) Write the hypotheses in words.

(b) What is a Type 1 Error in this context?

(c) What is a Type 2 Error in this context?

(d) Which error is more problematic for the restaurant owner? Why?

(e) Which error is more problematic for the diners? Why?

(f) As a diner, would you prefer that the food safety inspector requires strong evidence or very strong evidence of health concerns before revoking a restaurant's license? Explain your reasoning.

4.31 Which is higher? In each part below, there is a value of interest and two scenarios (I and II). For each part, report if the value of interest is larger under scenario I, scenario II, or whether the value is equal under the scenarios.

(a) The standard error of $\bar{x}$ when $s = 120$ and (I) n = 25 or (II) n = 125.

(b) The margin of error of a confidence interval when the confidence level is (I) 90% or (II) 80%.

(c) The p-value for a Z-statistic of 2.5 when (I) n = 500 or (II) n = 1000.

(d) The probability of making a Type 2 Error when the alternative hypothesis is true and the significance level is (I) 0.05 or (II) 0.10.

4.32 True or false. Determine if the following statements are true or false, and explain your reasoning. If false, state how it could be corrected.

(a) If a given value (for example, the null hypothesized value of a parameter) is within a 95% confidence interval, it will also be within a 99% confidence interval.

(b) Decreasing the significance level (α) will increase the probability of making a Type 1 Error.

(c) Suppose the null hypothesis is $\mu = 5$ and we fail to reject H_0. Under this scenario, the true population mean is 5.

(d) If the alternative hypothesis is true, then the probability of making a Type 2 Error and the power of a test add up to 1.

(e) With large sample sizes, even small differences between the null value and the true value of the parameter, a difference often called the effect size , will be identified as statistically significant.

4.6.4   Examining the Central Limit Theorem

4.33   Ages of pennies. The histogram below shows the distribution of ages of pennies at a bank.

(a) Describe the distribution.

(b) Sampling distributions for means from simple random samples of 5, 30, and 100 pennies is shown in the histograms below. Describe the shapes of these distributions and comment on whether they look like what you would expect to see based on the Central Limit Theorem.

(c) The mean age of the pennies is 10.44 years, with a standard deviation of 9.2 years. Using the Central Limit Theorem, calculate the means and standard deviations of the distribution of means from random samples of size 5, 30, and 100. Comment on whether the sampling distributions shown in part (b) agree with the values you compute.

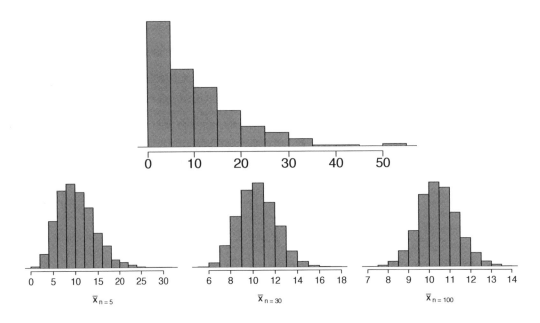

4.34   CLT. Define the term "sampling distribution" of the mean, and describe how the shape, center, and spread of the sampling distribution of the mean change as sample size increases.

4.35   Housing prices. A housing survey was conducted to determine the price of a typical home in Topanga, CA. The mean price of a house was roughly $1.3 million with a standard deviation of $300,000. There were no houses listed below $600,000 but a few houses above $3 million.

(a) Is the distribution of housing prices in Topanga symmetric, right skewed, or left skewed? *Hint:* Sketch the distribution.

(b) Would you expect most houses in Topanga to cost more or less than $1.3 million?

(c) Can we estimate the probability that a randomly chosen house in Topanga costs more than $1.4 million using the normal distribution?

(d) What is the probability that the mean of 60 randomly chosen houses in Topanga is more than $1.4 million?

(e) How would doubling the sample size affect the standard deviation of the mean?

4.36 Stats final scores. Each year about 1500 students take the introductory statistics course at a large university. This year scores on the final exam are distributed with a median of 74 points, a mean of 70 points, and a standard deviation of 10 points. There are no students who scored above 100 (the maximum score attainable on the final) but a few students scored below 20 points.

(a) Is the distribution of scores on this final exam symmetric, right skewed, or left skewed?

(b) Would you expect most students to have scored above or below 70 points?

(c) Can we calculate the probability that a randomly chosen student scored above 75 using the normal distribution?

(d) What is the probability that the average score for a random sample of 40 students is above 75?

(e) How would cutting the sample size in half affect the standard deviation of the mean?

4.37 Identify distributions, Part I. Four plots are presented below. The plot at the top is a distribution for a population. The mean is 10 and the standard deviation is 3. Also shown below is a distribution of (1) a single random sample of 100 values from this population, (2) a distribution of 100 sample means from random samples with size 5, and (3) a distribution of 100 sample means from random samples with size 25. Determine which plot (A, B, or C) is which and explain your reasoning.

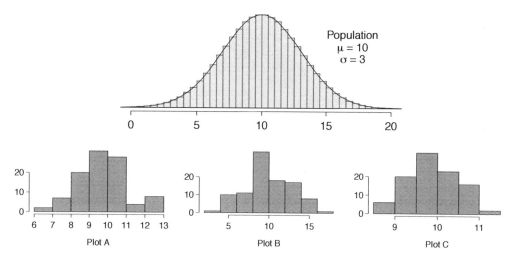

4.38 Identify distributions, Part II. Four plots are presented below. The plot at the top is a distribution for a population. The mean is 60 and the standard deviation is 18. Also shown below is a distribution of (1) a single random sample of 500 values from this population, (2) a distribution of 500 sample means from random samples of each size 18, and (3) a distribution of 500 sample means from random samples of each size 81. Determine which plot (A, B, or C) is which and explain your reasoning.

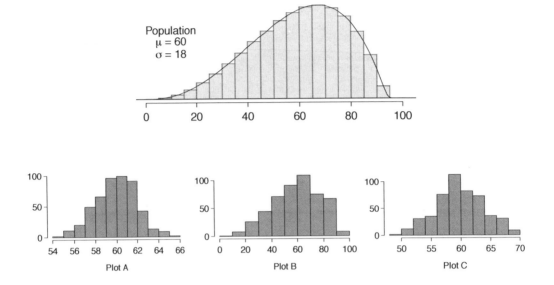

4.39 Weights of pennies. The distribution of weights of United States pennies is approximately normal with a mean of 2.5 grams and a standard deviation of 0.03 grams.

(a) What is the probability that a randomly chosen penny weighs less than 2.4 grams?

(b) Describe the sampling distribution of the mean weight of 10 randomly chosen pennies.

(c) What is the probability that the mean weight of 10 pennies is less than 2.4 grams?

(d) Sketch the two distributions (population and sampling) on the same scale.

(e) Could you estimate the probabilities from (a) and (c) if the weights of pennies had a skewed distribution?

4.40 CFLBs. A manufacturer of compact fluorescent light bulbs advertises that the distribution of the lifespans of these light bulbs is nearly normal with a mean of 9,000 hours and a standard deviation of 1,000 hours.

(a) What is the probability that a randomly chosen light bulb lasts more than 10,500 hours?

(b) Describe the distribution of the mean lifespan of 15 light bulbs.

(c) What is the probability that the mean lifespan of 15 randomly chosen light bulbs is more than 10,500 hours?

(d) Sketch the two distributions (population and sampling) on the same scale.

(e) Could you estimate the probabilities from parts (a) and (c) if the lifespans of light bulbs had a skewed distribution?

4.41 Songs on an iPod. Suppose an iPod has 3,000 songs. The histogram below shows the distribution of the lengths of these songs. We also know that, for this iPod, the mean length is 3.45 minutes and the standard deviation is 1.63 minutes.

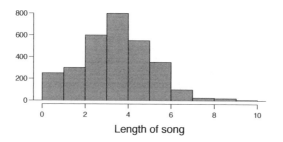

(a) Calculate the probability that a randomly selected song lasts more than 5 minutes.

(b) You are about to go for an hour run and you make a random playlist of 15 songs. What is the probability that your playlist lasts for the entire duration of your run? *Hint:* If you want the playlist to last 60 minutes, what should be the minimum average length of a song?

(c) You are about to take a trip to visit your parents and the drive is 6 hours. You make a random playlist of 100 songs. What is the probability that your playlist lasts the entire drive?

4.42 Spray paint. Suppose the area that can be painted using a single can of spray paint is slightly variable and follows a nearly normal distribution with a mean of 25 square feet and a standard deviation of 3 square feet.

(a) What is the probability that the area covered by a can of spray paint is more than 27 square feet?

(b) Suppose you want to spray paint an area of 540 square feet using 20 cans of spray paint. On average, how many square feet must each can be able to cover to spray paint all 540 square feet?

(c) What is the probability that you can cover a 540 square feet area using 20 cans of spray paint?

(d) If the area covered by a can of spray paint had a slightly skewed distribution, could you still calculate the probabilities in parts (a) and (c) using the normal distribution?

4.6.5 Inference for other estimators

4.43 Spam mail counts. The 2004 National Technology Readiness Survey sponsored by the Smith School of Business at the University of Maryland surveyed 418 randomly sampled Americans, asking them how many spam emails they receive per day. The survey was repeated on a new random sample of 499 Americans in 2009.[44]

(a) What are the hypotheses for evaluating if the average spam emails per day has changed from 2004 to 2009.

(b) In 2004 the mean was 18.5 spam emails per day, and in 2009 this value was 14.9 emails per day. What is the point estimate for the difference between the two population means?

(c) A report on the survey states that the observed difference between the sample means is not statistically significant. Explain what this means in context of the hypothesis test and data.

(d) Would you expect a confidence interval for the difference between the two population means to contain 0? Explain your reasoning.

---

[44]Rockbridge, 2009 National Technology Readiness Survey SPAM Report.

4.44 Nearsighted. It is believed that nearsightedness affects about 8% of all children. In a random sample of 194 children, 21 are nearsighted.

(a) Construct hypotheses appropriate for the following question: do these data provide evidence that the 8% value is inaccurate?

(b) What proportion of children in this sample are nearsighted?

(c) Given that the standard error of the sample proportion is 0.0195 and the point estimate follows a nearly normal distribution, calculate the test statistic (the Z-statistic).

(d) What is the p-value for this hypothesis test?

(e) What is the conclusion of the hypothesis test?

4.45 Spam mail percentages. The National Technology Readiness Survey sponsored by the Smith School of Business at the University of Maryland surveyed 418 randomly sampled Americans, asking them how often they delete spam emails. In 2004, 23% of the respondents said they delete their spam mail once a month or less, and in 2009 this value was 16%.

(a) What are the hypotheses for evaluating if the proportion of those who delete their email once a month or less has changed from 2004 to 2009?

(b) What is the point estimate for the difference between the two population proportions?

(c) A report on the survey states that the observed decrease from 2004 to 2009 is statistically significant. Explain what this means in context of the hypothesis test and the data.

(d) Would you expect a confidence interval for the difference between the two population proportions to contain 0? Explain your reasoning.

4.46 Unemployment and relationship problems. A USA Today/Gallup poll conducted between 2010 and 2011 asked a group of unemployed and underemployed Americans if they have had major problems in their relationships with their spouse or another close family member as a result of not having a job (if unemployed) or not having a full-time job (if underemployed). 27% of the 1,145 unemployed respondents and 25% of the 675 underemployed respondents said they had major problems in relationships as a result of their employment status.

(a) What are the hypotheses for evaluating if the proportions of unemployed and underemployed people who had relationship problems were different?

(b) The p-value for this hypothesis test is approximately 0.35. Explain what this means in context of the hypothesis test and the data.

4.47 Practical vs. statistical. Determine whether the following statement is true or false, and explain your reasoning: "With large sample sizes, even small differences between the null value and the point estimate can be statistically significant."

4.48 Same observation, different sample size. Suppose you conduct a hypothesis test based on a sample where the sample size is $n = 50$, and arrive at a p-value of 0.08. You then refer back to your notes and discover that you made a careless mistake, the sample size should have been $n = 500$. Will your p-value increase, decrease, or stay the same? Explain.

Chapter 5

Inference for numerical data

Chapter 4 introduced a framework for statistical inference based on confidence intervals and hypotheses. In this chapter, we encounter several new point estimates and scenarios. In each case, the inference ideas remain the same:

1. Determine which point estimate or test statistic is useful.

2. Identify an appropriate distribution for the point estimate or test statistic.

3. Apply the ideas from Chapter 4 using the distribution from step 2.

5.1 One-sample means with the t-distribution

We required a large sample in Chapter 4 for two reasons:

1. The sampling distribution of $\bar{x}$ tends to be more normal when the sample is large.

2. The calculated standard error is typically very accurate when using a large sample.

So what should we do when the sample size is small? As we'll discuss in Section 5.1.1, if the population data are nearly normal, then $\bar{x}$ will also follow a normal distribution, which addresses the first problem. The accuracy of the standard error is trickier, and for this challenge we'll introduce a new distribution called the t-distribution.

While we emphasize the use of the t-distribution for small samples, this distribution is also generally used for large samples, where it produces similar results to those from the normal distribution.

5.1.1 The normality condition

A special case of the Central Limit Theorem ensures the distribution of sample means will be nearly normal, regardless of sample size, when the data come from a nearly normal distribution.

Central Limit Theorem for normal data

The sampling distribution of the mean is nearly normal when the sample observations are independent and come from a nearly normal distribution. This is true for any sample size.

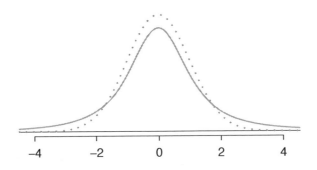

Figure 5.1: Comparison of a t-distribution (solid line) and a normal distribution (dotted line).

While this seems like a very helpful special case, there is one small problem. It is inherently difficult to verify normality in small data sets.

Caution: Checking the normality condition

We should exercise caution when verifying the normality condition for small samples. It is important to not only examine the data but also think about where the data come from. For example, ask: would I expect this distribution to be symmetric, and am I confident that outliers are rare?

You may relax the normality condition as the sample size goes up. If the sample size is 10 or more, slight skew is not problematic. Once the sample size hits about 30, then moderate skew is reasonable. Data with strong skew or outliers require a more cautious analysis.

5.1.2   Introducing the t-distribution

In the cases where we will use a small sample to calculate the standard error, it will be useful to rely on a new distribution for inference calculations: the t-distribution. A t-distribution, shown as a solid line in Figure 5.1, has a bell shape. However, its tails are thicker than the normal model's. This means observations are more likely to fall beyond two standard deviations from the mean than under the normal distribution.[1] While our estimate of the standard error will be a little less accurate when we are analyzing a small data set, these extra thick tails of the t-distribution are exactly the correction we need to resolve the problem of a poorly estimated standard error.

The t-distribution, always centered at zero, has a single parameter: degrees of freedom. The **degrees of freedom (df)** describe the precise form of the bell-shaped t-distribution. Several t-distributions are shown in Figure 5.2. When there are more degrees of freedom, the t-distribution looks very much like the standard normal distribution.

---

[1]The standard deviation of the t-distribution is actually a little more than 1. However, it is useful to always think of the t-distribution as having a standard deviation of 1 in all of our applications.

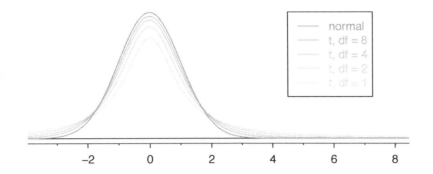

Figure 5.2: The larger the degrees of freedom, the more closely the *t*-distribution resembles the standard normal model.

Degrees of freedom (df)

The degrees of freedom describe the shape of the *t*-distribution. The larger the degrees of freedom, the more closely the distribution approximates the normal model.

When the degrees of freedom is about 30 or more, the *t*-distribution is nearly indistinguishable from the normal distribution. In Section 5.1.3, we relate degrees of freedom to sample size.

It's very useful to become familiar with the *t*-distribution, because it allows us greater flexibility than the normal distribution when analyzing numerical data. We use a **t-table**, partially shown in Table 5.3, in place of the normal probability table. A larger *t*-table is in Appendix B.2 on page 430. In practice, it's more common to use statistical software instead of a table, and you can see some of these options at

www.openintro.org/stat/prob-tables

Each row in the *t*-table represents a *t*-distribution with different degrees of freedom. The columns correspond to tail probabilities. For instance, if we know we are working with the *t*-distribution with $df = 18$, we can examine row 18, which is highlighted in Table 5.3. If we want the value in this row that identifies the cutoff for an upper tail of 10%, we can look in the column where *one tail* is 0.100. This cutoff is 1.33. If we had wanted the cutoff for the lower 10%, we would use -1.33. Just like the normal distribution, all *t*-distributions are symmetric.

● **Example 5.1** What proportion of the *t*-distribution with 18 degrees of freedom falls below -2.10?

Just like a normal probability problem, we first draw the picture in Figure 5.4 and shade the area below -2.10. To find this area, we identify the appropriate row: $df = 18$. Then we identify the column containing the absolute value of -2.10; it is the third column. Because we are looking for just one tail, we examine the top line of the table, which shows that a one tail area for a value in the third row corresponds to 0.025. About 2.5% of the distribution falls below -2.10. In the next example we encounter a case where the exact *t* value is not listed in the table.

one tail	0.100	0.050	0.025	0.010	0.005
two tails	0.200	0.100	0.050	0.020	0.010
df 1	3.08	6.31	12.71	31.82	63.66
2	1.89	2.92	4.30	6.96	9.92
3	1.64	2.35	3.18	4.54	5.84
⋮	⋮	⋮	⋮	⋮	⋮
17	1.33	1.74	2.11	2.57	2.90
18	**1.33**	**1.73**	**2.10**	**2.55**	**2.88**
19	1.33	1.73	2.09	2.54	2.86
20	1.33	1.72	2.09	2.53	2.85
⋮	⋮	⋮	⋮	⋮	⋮
400	1.28	1.65	1.97	2.34	2.59
500	1.28	1.65	1.96	2.33	2.59
∞	1.28	1.64	1.96	2.33	2.58

Table 5.3: An abbreviated look at the *t*-table. Each row represents a different *t*-distribution. The columns describe the cutoffs for specific tail areas. The row with *df* = 18 has been **highlighted**.

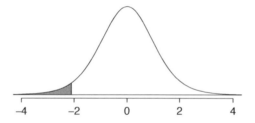

Figure 5.4: The *t*-distribution with 18 degrees of freedom. The area below -2.10 has been shaded.

● **Example 5.2** A *t*-distribution with 20 degrees of freedom is shown in the left panel of Figure 5.5. Estimate the proportion of the distribution falling above 1.65.

We identify the row in the *t*-table using the degrees of freedom: *df* = 20. Then we look for 1.65; it is not listed. It falls between the first and second columns. Since these values bound 1.65, their tail areas will bound the tail area corresponding to 1.65. We identify the one tail area of the first and second columns, 0.050 and 0.10, and we conclude that between 5% and 10% of the distribution is more than 1.65 standard deviations above the mean. If we like, we can identify the precise area using statistical software: 0.0573.

● **Example 5.3** A *t*-distribution with 2 degrees of freedom is shown in the right panel of Figure 5.5. Estimate the proportion of the distribution falling more than 3 units from the mean (above or below).

As before, first identify the appropriate row: *df* = 2. Next, find the columns that capture 3; because 2.92 < 3 < 4.30, we use the second and third columns. Finally, we find bounds for the tail areas by looking at the two tail values: 0.05 and 0.10. We use the two tail values because we are looking for two (symmetric) tails.

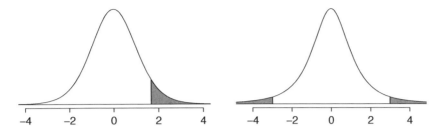

Figure 5.5: Left: The t-distribution with 20 degrees of freedom, with the area above 1.65 shaded. Right: The t-distribution with 2 degrees of freedom, with the area further than 3 units from 0 shaded.

⊙ **Guided Practice 5.4**   What proportion of the t-distribution with 19 degrees of freedom falls above -1.79 units?[2]

5.1.3   Conditions for using the t-distribution for inference on a sample mean

To proceed with the t-distribution for inference about a single mean, we first check two conditions.

Independence of observations. We verify this condition just as we did before. We collect a simple random sample from less than 10% of the population, or if the data are from an experiment or random process, we check to the best of our abilities that the observations were independent.

Observations come from a nearly normal distribution. This second condition is difficult to verify with small data sets. We often (i) take a look at a plot of the data for obvious departures from the normal model, and (ii) consider whether any previous experiences alert us that the data may not be nearly normal.

When examining a sample mean and estimated standard error from a sample of n independent and nearly normal observations, we use a t-distribution with $n - 1$ degrees of freedom (df). For example, if the sample size was 19, then we would use the t-distribution with $df = 19 - 1 = 18$ degrees of freedom and proceed exactly as we did in Chapter 4, except that *now we use the t-distribution*.

TIP: When to use the t-distribution

Use the t-distribution for inference of the sample mean when observations are independent and nearly normal. You may relax the nearly normal condition as the sample size increases. For example, the data distribution may be moderately skewed when the sample size is at least 30.

---

[2]We find the shaded area *above* -1.79 (we leave the picture to you). The small left tail is between 0.025 and 0.05, so the larger upper region must have an area between 0.95 and 0.975.

5.1.4   One sample *t*-confidence intervals

Dolphins are at the top of the oceanic food chain, which causes dangerous substances such as mercury to concentrate in their organs and muscles. This is an important problem for both dolphins and other animals, like humans, who occasionally eat them. For instance, this is particularly relevant in Japan where school meals have included dolphin at times.

Figure 5.6: A Risso's dolphin.

Photo by Mike Baird (www.bairdphotos.com). CC BY 2.0 license.

Here we identify a confidence interval for the average mercury content in dolphin muscle using a sample of 19 Risso's dolphins from the Taiji area in Japan.[3] The data are summarized in Table 5.7. The minimum and maximum observed values can be used to evaluate whether or not there are obvious outliers or skew.

n	$\bar{x}$	s	minimum	maximum
19	4.4	2.3	1.7	9.2

Table 5.7: Summary of mercury content in the muscle of 19 Risso's dolphins from the Taiji area. Measurements are in μg/wet g (micrograms of mercury per wet gram of muscle).

● **Example 5.5** Are the independence and normality conditions satisfied for this data set?

The observations are a simple random sample and consist of less than 10% of the population, therefore independence is reasonable. The summary statistics in Table 5.7 do not suggest any skew or outliers; all observations are within 2.5 standard deviations of the mean. Based on this evidence, the normality assumption seems reasonable.

---

[3]Taiji was featured in the movie *The Cove*, and it is a significant source of dolphin and whale meat in Japan. Thousands of dolphins pass through the Taiji area annually, and we will assume these 19 dolphins represent a simple random sample from those dolphins. Data reference: Endo T and Haraguchi K. 2009. High mercury levels in hair samples from residents of Taiji, a Japanese whaling town. Marine Pollution Bulletin 60(5):743-747.

In the normal model, we used $z^\star$ and the standard error to determine the width of a confidence interval. We revise the confidence interval formula slightly when using the t-distribution:

$$\bar{x} \pm t_{df}^\star SE$$

$t_{df}^\star$

Multiplication factor for t conf. interval

The sample mean and estimated standard error are computed just as before ($\bar{x} = 4.4$ and $SE = s/\sqrt{n} = 0.528$). The value $t_{df}^\star$ is a cutoff we obtain based on the confidence level and the t-distribution with df degrees of freedom. Before determining this cutoff, we will first need the degrees of freedom.

Degrees of freedom for a single sample

If the sample has n observations and we are examining a single mean, then we use the t-distribution with $df = n - 1$ degrees of freedom.

In our current example, we should use the t-distribution with $df = 19 - 1 = 18$ degrees of freedom. Then identifying $t_{18}^\star$ is similar to how we found $z^\star$.

- For a 95% confidence interval, we want to find the cutoff $t_{18}^\star$ such that 95% of the t-distribution is between -$t_{18}^\star$ and $t_{18}^\star$.

- We look in the t-table on page 222, find the column with area totaling 0.05 in the two tails (third column), and then the row with 18 degrees of freedom: $t_{18}^\star = 2.10$.

Generally the value of $t_{df}^\star$ is slightly larger than what we would get under the normal model with $z^\star$.

Finally, we can substitute all our values into the confidence interval equation to create the 95% confidence interval for the average mercury content in muscles from Risso's dolphins that pass through the Taiji area:

$$\bar{x} \pm t_{18}^\star SE \quad \rightarrow \quad 4.4 \pm 2.10 \times 0.528 \quad \rightarrow \quad (3.29, 5.51)$$

We are 95% confident the average mercury content of muscles in Risso's dolphins is between 3.29 and 5.51 μg/wet gram, which is considered extremely high.

Finding a t-confidence interval for the mean

Based on a sample of n independent and nearly normal observations, a confidence interval for the population mean is

$$\bar{x} \pm t_{df}^\star SE$$

where $\bar{x}$ is the sample mean, $t_{df}^\star$ corresponds to the confidence level and degrees of freedom, and SE is the standard error as estimated by the sample.

⊙ **Guided Practice 5.6**   The FDA's webpage provides some data on mercury content of fish.[4] Based on a sample of 15 croaker white fish (Pacific), a sample mean and standard deviation were computed as 0.287 and 0.069 ppm (parts per million), respectively. The 15 observations ranged from 0.18 to 0.41 ppm. We will assume these observations are independent. Based on the summary statistics of the data, do you have any objections to the normality condition of the individual observations?[5]

● **Example 5.7**   Estimate the standard error of $\bar{x} = 0.287$ ppm using the data summaries in Guided Practice 5.6. If we are to use the t-distribution to create a 90% confidence interval for the actual mean of the mercury content, identify the degrees of freedom we should use and also find $t_{df}^{\star}$.

The standard error: $SE = \frac{0.069}{\sqrt{15}} = 0.0178$. Degrees of freedom: $df = n - 1 = 14$.

Looking in the column where two tails is 0.100 (for a 90% confidence interval) and row $df = 14$, we identify $t_{14}^{\star} = 1.76$.

⊙ **Guided Practice 5.8**   Using the results of Guided Practice 5.6 and Example 5.7, compute a 90% confidence interval for the average mercury content of croaker white fish (Pacific).[6]

5.1.5   One sample t-tests

Is the typical US runner getting faster or slower over time? We consider this question in the context of the Cherry Blossom Race, which is a 10-mile race in Washington, DC each spring.[7]

The average time for all runners who finished the Cherry Blossom Race in 2006 was 93.29 minutes (93 minutes and about 17 seconds). We want to determine using data from 100 participants in the 2012 Cherry Blossom Race whether runners in this race are getting faster or slower, versus the other possibility that there has been no change.

⊙ **Guided Practice 5.9**   What are appropriate hypotheses for this context?[8]

⊙ **Guided Practice 5.10**   The data come from a simple random sample from less than 10% of all participants, so the observations are independent. However, should we be worried about skew in the data? See Figure 5.8 for a histogram of the differences.[9]

With independence satisfied and slight skew not a concern for this large of a sample, we can proceed with performing a hypothesis test using the t-distribution.

---

[4]www.fda.gov/food/foodborneillnesscontaminants/metals/ucm115644.htm

[5]There are no obvious outliers; all observations are within 2 standard deviations of the mean. If there is skew, it is not evident. There are no red flags for the normal model based on this (limited) information, and we do not have reason to believe the mercury content is not nearly normal in this type of fish.

[6]$\bar{x} \pm t_{14}^{\star}SE \rightarrow 0.287 \pm 1.76 \times 0.0178 \rightarrow (0.256, 0.318)$. We are 90% confident that the average mercury content of croaker white fish (Pacific) is between 0.256 and 0.318 ppm.

[7]www.cherryblossom.org

[8]H_0: The average 10 mile run time was the same for 2006 and 2012. $\mu = 93.29$ minutes. H_A: The average 10 mile run time for 2012 was *different* than that of 2006. $\mu \neq 93.29$ minutes.

[9]With a sample of 100, we should only be concerned if there is very strong skew. The histogram of the data suggests, at worst, slight skew.

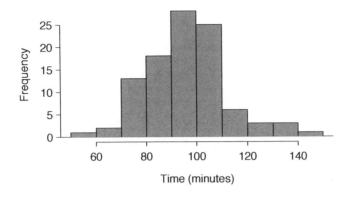

Figure 5.8: A histogram of `time` for the sample Cherry Blossom Race data.

⊙ **Guided Practice 5.11** The sample mean and sample standard deviation of the sample of 100 runners from the 2012 Cherry Blossom Race are 95.61 and 15.78 minutes, respectively. Recall that the sample size is 100. What is the p-value for the test, and what is your conclusion?[10]

When using a t-distribution, we use a T-score (same as Z-score)

To help us remember to use the t-distribution, we use a T to represent the test statistic, and we often call this a **T-score**. The Z-score and T-score are computed in the exact same way and are conceptually identical: each represents how many standard errors the observed value is from the null value.

▶ Calculator videos

Videos covering confidence intervals and hypothesis tests for a single mean using TI and Casio graphing calculators are available at openintro.org/videos.

---

[10]With the conditions satisfied for the t-distribution, we can compute the standard error ($SE = 15.78/\sqrt{100} = 1.58$ and the *T-score*: $T = \frac{95.61-93.29}{1.58} = 1.47$. (There is more on this after the guided practice, but a T-score and Z-score are calculated in the same way.) For $df = 100 - 1 = 99$, we would find $T = 1.47$ to fall between the first and second column, which means the p-value is between 0.10 and 0.20 (use $df = 90$ and consider two tails since the test is two-sided). The p-value could also have been calculated more precisely with statistical software: 0.1447. Because the p-value is greater than 0.05, we do not reject the null hypothesis. That is, the data do not provide strong evidence that the average run time for the Cherry Blossom Run in 2012 is any different than the 2006 average.

5.2   Paired data

Are textbooks actually cheaper online? Here we compare the price of textbooks at the University of California, Los Angeles' (UCLA's) bookstore and prices at Amazon.com. Seventy-three UCLA courses were randomly sampled in Spring 2010, representing less than 10% of all UCLA courses.[11] A portion of the data set is shown in Table 5.9.

	dept	course	ucla	amazon	diff
1	Am Ind	C170	27.67	27.95	-0.28
2	Anthro	9	40.59	31.14	9.45
3	Anthro	135T	31.68	32.00	-0.32
4	Anthro	191HB	16.00	11.52	4.48
⋮	⋮	⋮	⋮	⋮	⋮
72	Wom Std	M144	23.76	18.72	5.04
73	Wom Std	285	27.70	18.22	9.48

Table 5.9: Six cases of the `textbooks` data set.

5.2.1   Paired observations

Each textbook has two corresponding prices in the data set: one for the UCLA bookstore and one for Amazon. Therefore, each textbook price from the UCLA bookstore has a natural correspondence with a textbook price from Amazon. When two sets of observations have this special correspondence, they are said to be **paired**.

Paired data

Two sets of observations are *paired* if each observation in one set has a special correspondence or connection with exactly one observation in the other data set.

To analyze paired data, it is often useful to look at the difference in outcomes of each pair of observations. In the `textbook` data set, we look at the differences in prices, which is represented as the `diff` variable in the `textbooks` data. Here the differences are taken as

$$\text{UCLA price} - \text{Amazon price}$$

for each book. It is important that we always subtract using a consistent order; here Amazon prices are always subtracted from UCLA prices. A histogram of these differences is shown in Figure 5.10. Using differences between paired observations is a common and useful way to analyze paired data.

⊙ **Guided Practice 5.12**   The first difference shown in Table 5.9 is computed as $27.67 - 27.95 = -0.28$. Verify the differences are calculated correctly for observations 2 and 3.[12]

---

[11]When a class had multiple books, only the most expensive text was considered.
[12]Observation 2: $40.59 - 31.14 = 9.45$. Observation 3: $31.68 - 32.00 = -0.32$.

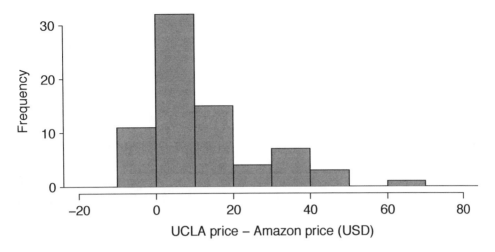

Figure 5.10: Histogram of the difference in price for each book sampled. These data are strongly skewed.

5.2.2 Inference for paired data

To analyze a paired data set, we simply analyze the differences. We can use the same t-distribution techniques we applied in the last section.

n_{diff}	$\bar{x}_{diff}$	s_{diff}
73	12.76	14.26

Table 5.11: Summary statistics for the price differences. There were 73 books, so there are 73 differences.

● **Example 5.13** Set up and implement a hypothesis test to determine whether, on average, there is a difference between Amazon's price for a book and the UCLA bookstore's price.

We are considering two scenarios: there is no difference or there is some difference in average prices.

H_0: $\mu_{diff} = 0$. There is no difference in the average textbook price.
H_A: $\mu_{diff} \neq 0$. There is a difference in average prices.

Can the t-distribution be used for this application? The observations are based on a simple random sample from less than 10% of all books sold at the bookstore, so independence is reasonable. While the distribution is strongly skewed, the sample is reasonably large ($n = 73$), so we can proceed. Because the conditions are reasonably satisfied, we can apply the t-distribution to this setting.

We compute the standard error associated with $\bar{x}_{diff}$ using the standard deviation of the differences ($s_{diff} = 14.26$) and the number of differences ($n_{diff} = 73$):

$$SE_{\bar{x}_{diff}} = \frac{s_{diff}}{\sqrt{n_{diff}}} = \frac{14.26}{\sqrt{73}} = 1.67$$

To visualize the p-value, the sampling distribution of $\bar{x}_{diff}$ is drawn as though H_0 is true, which is shown in Figure 5.12. The p-value is represented by the two (very) small tails.

To find the tail areas, we compute the test statistic, which is the T-score of $\bar{x}_{diff}$ under the null condition that the actual mean difference is 0:

$$T = \frac{\bar{x}_{diff} - 0}{SE_{x_{diff}}} = \frac{12.76 - 0}{1.67} = 7.65$$

The degrees of freedom are $df = 73 - 1 = 72$. If we examined Appendix B.2 on page 430, we would see that this value is larger than any in the 70 df row (we round down for df when using the table), meaning the two-sided p-value is less than 0.01. If we used statistical software, we would find the p-value is less than 1-in-10 billion! Because the p-value is less than 0.05, we reject the null hypothesis. We have found convincing evidence that Amazon was, on average, cheaper than the UCLA bookstore for UCLA course textbooks.

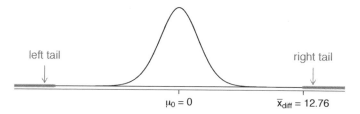

Figure 5.12: Sampling distribution for the mean difference in book prices, if the true average difference is zero.

⊙ **Guided Practice 5.14**   Create a 95% confidence interval for the average price difference between books at the UCLA bookstore and books on Amazon.[13]

5.3   Difference of two means

In this section we consider a difference in two population means, $\mu_1 - \mu_2$, under the condition that the data are not paired. Just as with a single sample, we identify conditions to ensure we can use the t-distribution with a point estimate of the difference, $\bar{x}_1 - \bar{x}_2$.

We apply these methods in three contexts: determining whether stem cells can improve heart function, exploring the impact of pregnant womens' smoking habits on birth weights of newborns, and exploring whether there is statistically significant evidence that one variations of an exam is harder than another variation. This section is motivated by questions like "Is there convincing evidence that newborns from mothers who smoke have a different average birth weight than newborns from mothers who don't smoke?"

---

[13]Conditions have already verified and the standard error computed in Example 5.13. To find the interval, identify $t_{72}^\star$ (use $df = 70$ in the table, $t_{70}^\star = 1.99$) and plug it, the point estimate, and the standard error into the confidence interval formula:

$$\text{point estimate} \pm z^\star SE \quad \rightarrow \quad 12.76 \pm 1.99 \times 1.67 \quad \rightarrow \quad (9.44, 16.08)$$

We are 95% confident that Amazon is, on average, between $9.44 and $16.08 cheaper than the UCLA bookstore for UCLA course books.

5.3.1   Confidence interval for a difference of means

Does treatment using embryonic stem cells (ESCs) help improve heart function following a heart attack? Table 5.13 contains summary statistics for an experiment to test ESCs in sheep that had a heart attack. Each of these sheep was randomly assigned to the ESC or control group, and the change in their hearts' pumping capacity was measured in the study. A positive value corresponds to increased pumping capacity, which generally suggests a stronger recovery. Our goal will be to identify a 95% confidence interval for the effect of ESCs on the change in heart pumping capacity relative to the control group.

A point estimate of the difference in the heart pumping variable can be found using the difference in the sample means:

$$\bar{x}_{esc} - \bar{x}_{control} \; = \; 3.50 - (-4.33) \; = \; 7.83$$

	n	$\bar{x}$	s
ESCs	9	3.50	5.17
control	9	-4.33	2.76

Table 5.13: Summary statistics of the embryonic stem cell study.

Using the t-distribution for a difference in means

The t-distribution can be used for inference when working with the standardized difference of two means if (1) each sample meets the conditions for using the t-distribution and (2) the samples are independent.

● **Example 5.15** Can the t-distribution be used to make inference using the point estimate, $\bar{x}_{esc} - \bar{x}_{control} = 7.83$?

We check the two required conditions:

1. In this study, the sheep were independent of each other. Additionally, the distributions in Figure 5.14 don't show any clear deviations from normality, where we watch for prominent outliers in particular for such small samples. These findings imply each sample mean could itself be modeled using a t-distribution.

2. The sheep in each group were also independent of each other.

Because both conditions are met, we can use the t-distribution to model the difference of the two sample means.

We can quantify the variability in the point estimate, $\bar{x}_{esc} - \bar{x}_{control}$, using the following formula for its standard error:

$$SE_{\bar{x}_{esc} - \bar{x}_{control}} = \sqrt{\frac{\sigma_{esc}^2}{n_{esc}} + \frac{\sigma_{control}^2}{n_{control}}}$$

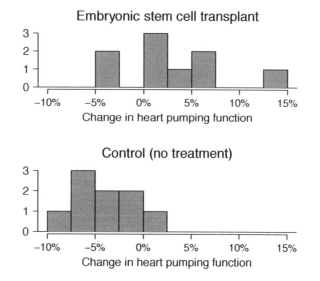

Figure 5.14: Histograms for both the embryonic stem cell group and the control group. Higher values are associated with greater improvement. We don't see any evidence of skew in these data; however, it is worth noting that skew would be difficult to detect with such a small sample.

We usually estimate this standard error using standard deviation estimates based on the samples:

$$SE_{\bar{x}_{esc} - \bar{x}_{control}} = \sqrt{\frac{\sigma^2_{esc}}{n_{esc}} + \frac{\sigma^2_{control}}{n_{control}}}$$

$$\approx \sqrt{\frac{s^2_{esc}}{n_{esc}} + \frac{s^2_{control}}{n_{control}}} = \sqrt{\frac{5.17^2}{9} + \frac{2.76^2}{9}} = 1.95$$

Because we will use the t-distribution, we also must identify the appropriate degrees of freedom. This can be done using computer software. An alternative technique is to use the smaller of $n_1 - 1$ and $n_2 - 1$, which is the method we will typically apply in the examples and guided practice.[14]

Distribution of a difference of sample means

The sample difference of two means, $\bar{x}_1 - \bar{x}_2$, can be modeled using the t-distribution and the standard error

$$SE_{\bar{x}_1 - \bar{x}_2} = \sqrt{\frac{s^2_1}{n_1} + \frac{s^2_2}{n_2}} \qquad (5.16)$$

when each sample mean can itself be modeled using a t-distribution and the samples are independent. To calculate the degrees of freedom, use statistical software or the smaller of $n_1 - 1$ and $n_2 - 1$.

---

[14]This technique for degrees of freedom is conservative with respect to a Type 1 Error; it is more difficult to reject the null hypothesis using this df method. In this example, computer software would have provided us a more precise degrees of freedom of $df = 12.225$.

● **Example 5.17** Calculate a 95% confidence interval for the effect of ESCs on the change in heart pumping capacity of sheep after they've suffered a heart attack.

We will use the sample difference and the standard error for that point estimate from our earlier calculations:

$$\bar{x}_{esc} - \bar{x}_{control} = 7.83$$

$$SE = \sqrt{\frac{5.17^2}{9} + \frac{2.76^2}{9}} = 1.95$$

Using $df = 8$, we can identify the appropriate $t^{\star}_{df} = t^{\star}_8$ for a 95% confidence interval as 2.31. Finally, we can enter the values into the confidence interval formula:

$$\text{point estimate } \pm \ t^{\star} SE \quad \rightarrow \quad 7.83 \ \pm \ 2.31 \times 1.95 \quad \rightarrow \quad (3.32, 12.34)$$

We are 95% confident that embryonic stem cells improve the heart's pumping function in sheep that have suffered a heart attack by 3.32% to 12.34%.

5.3.2 Hypothesis tests based on a difference in means

A data set called `baby_smoke` represents a random sample of 150 cases of mothers and their newborns in North Carolina over a year. Four cases from this data set are represented in Table 5.15. We are particularly interested in two variables: `weight` and `smoke`. The `weight` variable represents the weights of the newborns and the `smoke` variable describes which mothers smoked during pregnancy. We would like to know, is there convincing evidence that newborns from mothers who smoke have a different average birth weight than newborns from mothers who don't smoke? We will use the North Carolina sample to try to answer this question. The smoking group includes 50 cases and the nonsmoking group contains 100 cases, represented in Figure 5.16.

	fAge	mAge	weeks	weight	sexBaby	smoke
1	NA	13	37	5.00	female	nonsmoker
2	NA	14	36	5.88	female	nonsmoker
3	19	15	41	8.13	male	smoker
⋮	⋮	⋮	⋮	⋮	⋮	
150	45	50	36	9.25	female	nonsmoker

Table 5.15: Four cases from the `baby_smoke` data set. The value "NA", shown for the first two entries of the first variable, indicates that piece of data is missing.-2mm

● **Example 5.18** Set up appropriate hypotheses to evaluate whether there is a relationship between a mother smoking and average birth weight.

The null hypothesis represents the case of no difference between the groups.

H_0: There is no difference in average birth weight for newborns from mothers who did and did not smoke. In statistical notation: $\mu_n - \mu_s = 0$, where μ_n represents non-smoking mothers and μ_s represents mothers who smoked.

H_A: There is some difference in average newborn weights from mothers who did and did not smoke ($\mu_n - \mu_s \neq 0$).

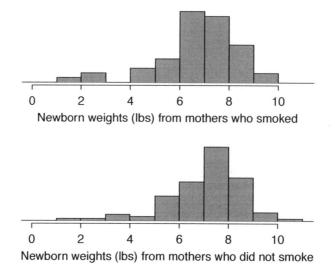

Figure 5.16: The top panel represents birth weights for infants whose mothers smoked. The bottom panel represents the birth weights for infants whose mothers who did not smoke. The distributions exhibit moderate-to-strong and strong skew, respectively.

We check the two conditions necessary to apply the t-distribution to the difference in sample means. (1) Because the data come from a simple random sample and consist of less than 10% of all such cases, the observations are independent. Additionally, while each distribution is strongly skewed, the sample sizes of 50 and 100 would make it reasonable to model each mean separately using a t-distribution. The skew is reasonable for these sample sizes of 50 and 100. (2) The independence reasoning applied in (1) also ensures the observations in each sample are independent. Since both conditions are satisfied, the difference in sample means may be modeled using a t-distribution.

	smoker	nonsmoker
mean	6.78	7.18
st. dev.	1.43	1.60
samp. size	50	100

Table 5.17: Summary statistics for the baby_smoke data set.

⊙ **Guided Practice 5.19**   The summary statistics in Table 5.17 may be useful for this exercise. (a) What is the point estimate of the population difference, $\mu_n - \mu_s$? (b) Compute the standard error of the point estimate from part (a).[15]

---

[15](a) The difference in sample means is an appropriate point estimate: $\bar{x}_n - \bar{x}_s = 0.40$. (b) The standard error of the estimate can be estimated using Equation (5.16):

$$SE = \sqrt{\frac{\sigma_n^2}{n_n} + \frac{\sigma_s^2}{n_s}} \approx \sqrt{\frac{s_n^2}{n_n} + \frac{s_s^2}{n_s}} = \sqrt{\frac{1.60^2}{100} + \frac{1.43^2}{50}} = 0.26$$

● **Example 5.20** Draw a picture to represent the p-value for the hypothesis test from Example 5.18.

To depict the p-value, we draw the distribution of the point estimate as though H_0 were true and shade areas representing at least as much evidence against H_0 as what was observed. Both tails are shaded because it is a two-sided test.

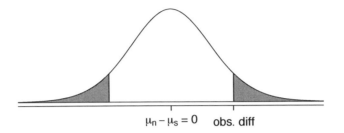

● **Example 5.21** Compute the p-value of the hypothesis test using the figure in Example 5.20, and evaluate the hypotheses using a significance level of $\alpha = 0.05$.

We start by computing the T-score:

$$T = \frac{0.40 - 0}{0.26} = 1.54$$

Next, we compare this value to values in the t-table in Appendix B.2 on page 430, where we use the smaller of $n_n - 1 = 99$ and $n_s - 1 = 49$ as the degrees of freedom: $df = 49$. The T-score falls between the first and second columns in the $df = 49$ row of the t-table, meaning the two-sided p-value falls between 0.10 and 0.20 (reminder, find tail areas along the top of the table). This p-value is larger than the significance value, 0.05, so we fail to reject the null hypothesis. There is insufficient evidence to say there is a difference in average birth weight of newborns from North Carolina mothers who did smoke during pregnancy and newborns from North Carolina mothers who did not smoke during pregnancy.

⊙ **Guided Practice 5.22** Does the conclusion to Example 5.21 mean that smoking and average birth weight are unrelated?[16]

⊙ **Guided Practice 5.23** If we made a Type 2 Error and there is a difference, what could we have done differently in data collection to be more likely to detect the difference?[17]

---

[16]Absolutely not. It is possible that there is some difference but we did not detect it. If there is a difference, we made a Type 2 Error. Notice: we also don't have enough information to, if there is an actual difference, confidently say which direction that difference would be in.

[17]We could have collected more data. If the sample sizes are larger, we tend to have a better shot at finding a difference if one exists.

Public service announcement: while we have used this relatively small data set as an example, larger data sets show that women who smoke tend to have smaller newborns. In fact, some in the tobacco industry actually had the audacity to tout that as a *benefit* of smoking:

> *It's true. The babies born from women who smoke are smaller, but they're just as healthy as the babies born from women who do not smoke. And some women would prefer having smaller babies.*
>
> - Joseph Cullman, Philip Morris' Chairman of the Board
> on CBS' *Face the Nation*, Jan 3, 1971

Fact check: the babies from women who smoke are not actually as healthy as the babies from women who do not smoke.[18]

5.3.3   Case study: two versions of a course exam

An instructor decided to run two slight variations of the same exam. Prior to passing out the exams, she shuffled the exams together to ensure each student received a random version. Summary statistics for how students performed on these two exams are shown in Table 5.18. Anticipating complaints from students who took Version B, she would like to evaluate whether the difference observed in the groups is so large that it provides convincing evidence that Version B was more difficult (on average) than Version A.

Version	n	$\bar{x}$	s	min	max
A	30	79.4	14	45	100
B	27	74.1	20	32	100

Table 5.18: Summary statistics of scores for each exam version.

⊙ **Guided Practice 5.24**   Construct a hypotheses to evaluate whether the observed difference in sample means, $\bar{x}_A - \bar{x}_B = 5.3$, is due to chance.[19]

⊙ **Guided Practice 5.25**   To evaluate the hypotheses in Guided Practice 5.24 using the t-distribution, we must first verify assumptions. (a) Does it seem reasonable that the scores are independent within each group? (b) What about the normality / skew condition for observations in each group? (c) Do you think scores from the two groups would be independent of each other, i.e. the two samples are independent?[20]

After verifying the conditions for each sample and confirming the samples are independent of each other, we are ready to conduct the test using the t-distribution. In this case,

---

[18]You can watch an episode of John Oliver on *This Week Tonight* to explore the present day offenses of the tobacco industry. Please be aware that there is some adult language: youtu.be/6UsHHOCH4q8.

[19]Because the teacher did not expect one exam to be more difficult prior to examining the test results, she should use a two-sided hypothesis test. H_0: the exams are equally difficult, on average. $\mu_A - \mu_B = 0$. H_A: one exam was more difficult than the other, on average. $\mu_A - \mu_B \neq 0$.

[20](a) It is probably reasonable to conclude the scores are independent, provided there was no cheating. (b) The summary statistics suggest the data are roughly symmetric about the mean, and it doesn't seem unreasonable to suggest the data might be normal. Note that since these samples are each nearing 30, moderate skew in the data would be acceptable. (c) It seems reasonable to suppose that the samples are independent since the exams were handed out randomly.

we are estimating the true difference in average test scores using the sample data, so the point estimate is $\bar{x}_A - \bar{x}_B = 5.3$. The standard error of the estimate can be calculated as

$$SE = \sqrt{\frac{s_A^2}{n_A} + \frac{s_B^2}{n_B}} = \sqrt{\frac{14^2}{30} + \frac{20^2}{27}} = 4.62$$

Finally, we construct the test statistic:

$$T = \frac{\text{point estimate} - \text{null value}}{SE} = \frac{(79.4 - 74.1) - 0}{4.62} = 1.15$$

If we have a computer handy, we can identify the degrees of freedom as 45.97. Otherwise we use the smaller of $n_1 - 1$ and $n_2 - 1$: $df = 26$.

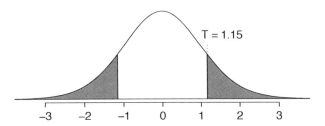

Figure 5.19: The t-distribution with 26 degrees of freedom. The shaded right tail represents values with $T \geq 1.15$. Because it is a two-sided test, we also shade the corresponding lower tail.

● **Example 5.26** Identify the p-value using $df = 26$ and provide a conclusion in the context of the case study.

We examine row $df = 26$ in the t-table. Because this value is smaller than the value in the left column, the p-value is larger than 0.200 (two tails!). Because the p-value is so large, we do not reject the null hypothesis. That is, the data do not convincingly show that one exam version is more difficult than the other, and the teacher should not be convinced that she should add points to the Version B exam scores.

5.3.4 Summary for inference using the t-distribution

Hypothesis tests. When applying the t-distribution for a hypothesis test, we proceed as follows:

- Write appropriate hypotheses.
- Verify conditions for using the t-distribution.
  - One-sample or differences from paired data: the observations (or differences) must be independent and nearly normal. For larger sample sizes, we can relax the nearly normal requirement, e.g. slight skew is okay for sample sizes of 15, moderate skew for sample sizes of 30, and strong skew for sample sizes of 60.
  - For a difference of means when the data are not paired: each sample mean must separately satisfy the one-sample conditions for the t-distribution, and the data in the groups must also be independent.

- Compute the point estimate of interest, the standard error, and the degrees of freedom. For df, use $n-1$ for one sample, and for two samples use either statistical software or the smaller of $n_1 - 1$ and $n_2 - 1$.

- Compute the T-score and p-value.

- Make a conclusion based on the p-value, and write a conclusion in context and in plain language so anyone can understand the result.

Confidence intervals. Similarly, the following is how we generally computed a confidence interval using a t-distribution:

- Verify conditions for using the t-distribution. (See above.)

- Compute the point estimate of interest, the standard error, the degrees of freedom, and $t_{df}^\star$.

- Calculate the confidence interval using the general formula, point estimate $\pm t_{df}^\star SE$.

- Put the conclusions in context and in plain language so even non-statisticians can understand the results.

▶ Calculator videos

Videos covering confidence intervals and hypothesis tests for a difference of means using TI and Casio graphing calculators are available at openintro.org/videos.

5.3.5 Examining the standard error formula (special topic)

The formula for the standard error of the difference in two means is similar to the formula for other standard errors. Recall that the standard error of a single mean, $\bar{x}_1$, can be approximated by

$$SE_{\bar{x}_1} = \frac{s_1}{\sqrt{n_1}}$$

where s_1 and n_1 represent the sample standard deviation and sample size.

The standard error of the difference of two sample means can be constructed from the standard errors of the separate sample means:

$$SE_{\bar{x}_1 - \bar{x}_2} = \sqrt{SE_{\bar{x}_1}^2 + SE_{\bar{x}_2}^2} = \sqrt{\frac{s_1^2}{n_1} + \frac{s_2^2}{n_2}} \qquad (5.27)$$

This special relationship follows from probability theory.

⊙ **Guided Practice 5.28** Prerequisite: Section 2.4. We can rewrite Equation (5.27) in a different way:

$$SE_{\bar{x}_1 - \bar{x}_2}^2 = SE_{\bar{x}_1}^2 + SE_{\bar{x}_2}^2$$

Explain where this formula comes from using the ideas of probability theory.[21]

---

[21] The standard error squared represents the variance of the estimate. If X and Y are two random variables with variances σ_x^2 and σ_y^2, then the variance of $X - Y$ is $\sigma_x^2 + \sigma_y^2$. Likewise, the variance corresponding to $\bar{x}_1 - \bar{x}_2$ is $\sigma_{\bar{x}_1}^2 + \sigma_{\bar{x}_2}^2$. Because $\sigma_{\bar{x}_1}^2$ and $\sigma_{\bar{x}_2}^2$ are just another way of writing $SE_{\bar{x}_1}^2$ and $SE_{\bar{x}_2}^2$, the variance associated with $\bar{x}_1 - \bar{x}_2$ may be written as $SE_{\bar{x}_1}^2 + SE_{\bar{x}_2}^2$.

5.3.6   Pooled standard deviation estimate (special topic)

Occasionally, two populations will have standard deviations that are so similar that they can be treated as identical. For example, historical data or a well-understood biological mechanism may justify this strong assumption. In such cases, we can make the *t*-distribution approach slightly more precise by using a pooled standard deviation.

The **pooled standard deviation** of two groups is a way to use data from both samples to better estimate the standard deviation and standard error. If s_1 and s_2 are the standard deviations of groups 1 and 2 and there are good reasons to believe that the population standard deviations are equal, then we can obtain an improved estimate of the group variances by pooling their data:

$$s^2_{pooled} = \frac{s_1^2 \times (n_1 - 1) + s_2^2 \times (n_2 - 1)}{n_1 + n_2 - 2}$$

where n_1 and n_2 are the sample sizes, as before. To use this new statistic, we substitute s^2_{pooled} in place of s_1^2 and s_2^2 in the standard error formula, and we use an updated formula for the degrees of freedom:

$$df = n_1 + n_2 - 2$$

The benefits of pooling the standard deviation are realized through obtaining a better estimate of the standard deviation for each group and using a larger degrees of freedom parameter for the *t*-distribution. Both of these changes may permit a more accurate model of the sampling distribution of $\bar{x}_1 - \bar{x}_2$, if the standard deviations of the two groups are equal.

Caution: Pool standard deviations only after careful consideration

A pooled standard deviation is only appropriate when background research indicates the population standard deviations are nearly equal. When the sample size is large and the condition may be adequately checked with data, the benefits of pooling the standard deviations greatly diminishes.

5.4   Power calculations for a difference of means (special topic)

Often times in experiment planning, there are two competing considerations:

- We want to collect enough data that we can detect important effects.
- Collecting data can be expensive, and in experiments involving people, there may be some risk to patients.

In this section, we focus on the context of a clinical trial, which is a health-related experiment where the subject are people, and we will determine an appropriate sample size where we can be 80% sure that we would detect any practically important effects.[22]

---

[22]Even though we don't cover it explicitly, similar sample size planning is also helpful for observational studies.

5.4.1   Going through the motions of a test

We're going to go through the motions of a hypothesis test. This will help us frame our calculations for determining an appropriate sample size for the study.

● **Example 5.29** Suppose a pharmaceutical company has developed a new drug for lowering blood pressure, and they are preparing a clinical trial (experiment) to test the drug's effectiveness. They recruit people who are taking a particular standard blood pressure medication. People in the control group will continue to take their current medication through generic-looking pills to ensure blinding. Write down the hypotheses for a two-sided hypothesis test in this context.

Generally, clinical trials use a two-sided alternative hypothesis, so below are suitable hypotheses for this context:

H_0: The new drug performs exactly as well as the standard medication.
   $\mu_{trmt} - \mu_{ctrl} = 0$.
H_A: The new drug's performance differs from the standard medication.
   $\mu_{trmt} - \mu_{ctrl} \neq 0$.

Some researchers might argue for a one-sided test here, where the alternative would consider only whether the new drug performs better than the standard medication. However, it would be very informative to know whether the new drug performs worse than the standard medication, so we use a two-sided test to consider this possibility during the analysis.

● **Example 5.30** The researchers would like to run the clinical trial on patients with systolic blood pressures between 140 and 180 mmHg. Suppose previously published studies suggest that the standard deviation of the patients' blood pressures will be about 12 mmHg and the distribution of patient blood pressures will be approximately symmetric.[23] If we had 100 patients per group, what would be the approximate standard error for $\bar{x}_{trmt} - \bar{x}_{ctrl}$?

The standard error is calculated as follows:

$$SE_{\bar{x}_{trmt} - \bar{x}_{ctrl}} = \sqrt{\frac{s_{trmt}^2}{n_{trmt}} + \frac{s_{ctrl}^2}{n_{ctrl}}} = \sqrt{\frac{12^2}{100} + \frac{12^2}{100}} = 1.70$$

This may be an imperfect estimate of $SE_{\bar{x}_{trmt} - \bar{x}_{ctrl}}$, since the standard deviation estimate we used may not be correct for this group of patients. However, it is sufficient for our purposes.

---

[23]In this particular study, we'd generally measure each patient's blood pressure at the beginning and end of the study, and then the outcome measurement for the study would be the average change in blood pressure. That is, both μ_{trmt} and μ_{ctrl} would represent average differences. This is what you might think of as a 2-sample paired testing structure, and we'd analyze it exactly just like a hypothesis test for a difference in the average change for patients. In the calculations we perform here, we'll suppose that 12 mmHg is the predicted standard deviation of a patient's blood pressure difference over the course of the study.

● **Example 5.31**   What does the null distribution of $\bar{x}_{trmt} - \bar{x}_{ctrl}$ look like?

The degrees of freedom are greater than 30, so the distribution of $\bar{x}_{trmt} - \bar{x}_{ctrl}$ will be approximately normal. The standard deviation of this distribution (the standard error) would be about 1.70, and under the null hypothesis, its mean would be 0.

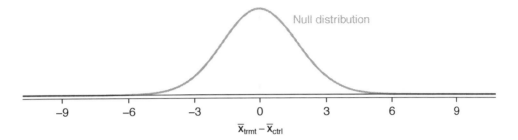

● **Example 5.32**   For what values of $\bar{x}_{trmt} - \bar{x}_{ctrl}$ would we reject the null hypothesis?

For $\alpha = 0.05$, we would reject H_0 if the difference is in the lower 2.5% or upper 2.5% tail:

Lower 2.5%: For the normal model, this is 1.96 standard errors below 0, so any difference smaller than $-1.96 \times 1.70 = -3.332$ mmHg.

Upper 2.5%: For the normal model, this is 1.96 standard errors above 0, so any difference larger than $1.96 \times 1.70 = 3.332$ mmHg.

The boundaries of these **rejection regions** are shown below:

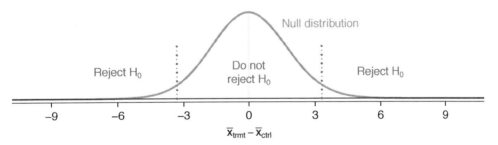

Next, we'll perform some hypothetical calculations to determine the probability we reject the null hypothesis, if the alternative hypothesis were actually true.

5.4.2   Computing the power for a 2-sample test

When planning a study, we want to know how likely we are to detect an effect we care about. In other words, if there is a real effect, and that effect is large enough that it has practical value, then what's the probability that we detect that effect? This probability is called the **power**, and we can compute it for different sample sizes or for different *effect sizes*.

We first determine what is a practically significant result. Suppose that the company researchers care about finding any effect on blood pressure that is 3 mmHg or larger vs the standard medication. Here, 3 mmHg is the minimum **effect size** of interest, and we want to know how likely we are to detect this size of an effect in the study.

● **Example 5.33** Suppose we decided to move forward with 100 patients per treatment group and the new drug reduces blood pressure by an additional 3 mmHg relative to the standard medication. What is the probability that we detect a drop?

Before we even do any calculations, notice that if $\bar{x}_{trmt} - \bar{x}_{ctrl} = -3$ mmHg, there wouldn't even be sufficient evidence to reject H_0. That's not a good sign.

To calculate the probability that we will reject H_0, we need to determine a few things:

- The sampling distribution for $\bar{x}_{trmt} - \bar{x}_{ctrl}$ when the true difference is -3 mmHg. This is the same as the null distribution, except it is shifted to the left by 3:

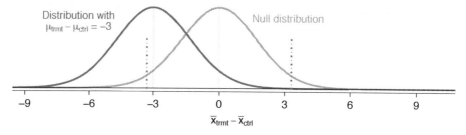

- The rejection regions, which are outside of the dotted lines above.
- The fraction of the distribution that falls in the rejection region.

In short, we need to calculate the probability that $x < -3.332$ for a normal distribution with mean -3 and standard deviation 1.7. To do so, we first shade the area we want to calculate:

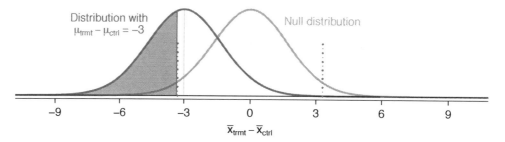

Then we calculate the Z-score and find the tail area using either the normal probability table or statistical software:

$$Z = \frac{-3.332 - (-3)}{1.7} = -0.20 \quad \rightarrow \quad 0.4207$$

The power for the test is about 42% when $\mu_{trmt} - \mu_{ctrl} = -3$ and each group has a sample size of 100.

In Example 5.33, we ignored the upper rejection region in the calculation, which was in the opposite direction of the hypothetical truth, i.e. -3. The reasoning? There wouldn't be any value in rejecting the null hypothesis and concluding there was an increase when in fact there was a decrease.

5.4.3 Determining a proper sample size

In the last example, we found that if we have a sample size of 100 in each group, we can only detect an effect size of 3 mmHg with a probability of about 0.42. Suppose the researchers moved forward and only used 100 patients per group, and the data did not support the alternative hypothesis, i.e. the researchers did not reject H_0. This is a very bad situation to be in for a few reasons:

- In the back of the researchers' minds, they'd all be wondering, *maybe there is a real and meaningful difference, but we weren't able to detect it with such a small sample.*

- The company probably invested hundreds of millions of dollars in developing the new drug, so now they are left with great uncertainty about its potential since the experiment didn't have a great shot at detecting effects that could still be important.

- Patients were subjected to the drug, and we can't even say with much certainty that the drug doesn't help (or harm) patients.

- Another clinical trial may need to be run to get a more conclusive answer as to whether the drug does hold any practical value, and conducting a second clinical trial may take years and many millions of dollars.

We want to avoid this situation, so we need to determine an appropriate sample size to ensure we can be pretty confident that we'll detect any effects that are practically important. As mentioned earlier, a change of 3 mmHg was deemed to be the minimum difference that was practically important. As a first step, we could calculate power for several different sample sizes. For instance, let's try 500 patients per group.

⊙ **Guided Practice 5.34** Calculate the power to detect a change of -3 mmHg when using a sample size of 500 per group.[24]

(a) Determine the standard error (recall that the standard deviation for patients was expected to be about 12 mmHg).

(b) Identify the null distribution and rejection regions.

(c) Identify the alternative distribution when $\mu_{trmt} - \mu_{ctrl} = -3$.

(d) Compute the probability we reject the null hypothesis.

The researchers decided 3 mmHg was the minimum difference that was practically important, and with a sample size of 500, we can be very certain (97.7% or better) that we will detect any such difference. We now have moved to another extreme where we are

---

[24](a) The standard error is given as $SE = \sqrt{\frac{12^2}{500} + \frac{12^2}{500}} = 0.76$.

(b) & (c) The null distribution, rejection boundaries, and alternative distribution are shown below:

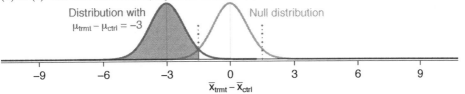

The rejection regions are the areas on the outside of the two dotted lines and are at $\pm 0.76 \times 1.96 = \pm 1.49$. (d) The area of the alternative distribution where $\mu_{trmt} - \mu_{ctrl} = -3$ has been shaded. We compute the Z-score and find the tail area: $Z = \frac{-1.49 - (-3)}{0.76} = 1.99 \rightarrow 0.9767$, which is the power of the test for a difference of 3 mmHg. With 500 patients per group, we would be about 97.7% sure (or more) that we'd detect any effects that are at least 3 mmHg in size.

exposing an unnecessary number of patients to the new drug in the clinical trial. Not only is this ethically questionable, but it would also cost a lot more money than is necessary to be quite sure we'd detect any important effects.

The most common practice is to identify the sample size where the power is around 80%, and sometimes 90%. Other values may be reasonable for a specific context, but 80% and 90% are most commonly targeted as a good balance between high power and not exposing too many patients to a new treatment (or wasting too much money). We could compute the power of the test at several other possible sample sizes until we find one that's close to 80%, but that's inefficient. Instead, we should solve the problem backwards.

● **Example 5.35** What sample size will lead to a power of 80%?

We start by identifying the Z-score that would give us a lower tail of 80%: it would be about 0.84:

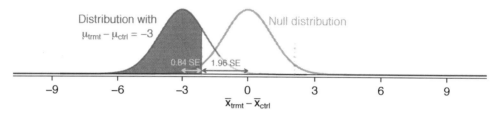

Additionally, the rejection region always extends $1.96 \times SE$ from the center of the null distribution for $\alpha = 0.05$. This allows us to calculate the target distance between the center of the null and alternative distributions in terms of the standard error:

$$0.84 \times SE + 1.96 \times SE = 2.8 \times SE$$

In our example, we also want the distance between the null and alternative distributions' centers to equal the minimum effect size of interest, 3 mmHg, which allows us to set up an equation between this difference and the standard error:

$$3 = 2.8 \times SE$$
$$3 = 2.8 \times \sqrt{\frac{12^2}{n} + \frac{12^2}{n}}$$
$$n = \frac{2.8^2}{3^2} \times \left(12^2 + 12^2\right) = 250.88$$

We should target about 251 patients per group.

The standard error difference of $2.8 \times SE$ is specific to a context where the targeted power is 80% and the significance level is $\alpha = 0.05$. If the targeted power is 90% or if we use a difference significance level, then we'll use something a little different than $2.8 \times SE$.

⊙ **Guided Practice 5.36** Suppose the targeted power was 90% and we were using $\alpha = 0.01$. How many standard errors should separate the centers of the null and alternative distribution, where the alternative distribution is centered at the minimum effect size of interest? Assume the test is two-sided.[25]

---

[25]First, find the Z-score such that 90% of the distribution is below it: $Z = 1.28$. Next, find the cutoffs for the rejection regions: ± 2.58. Then the difference in centers should be about $1.28 \times SE + 2.58 \times SE = 3.86 \times SE$.

⊙ **Guided Practice 5.37** List some considerations that are important in determining what the power should be for an experiment.[26]

Figure 5.20 shows the power for sample sizes from 20 patients to 5,000 patients when $\alpha = 0.05$ and the true difference is -3. This curve was constructed by writing a program to compute the power for many different sample sizes.

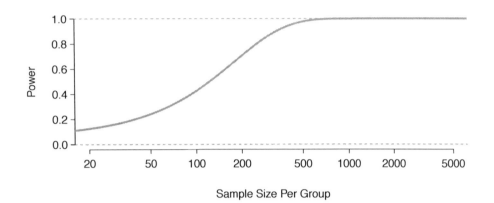

Figure 5.20: The curve shows the power for different sample sizes in the context of the blood pressure example when the true difference is -3. Having more than about 250 to 350 observations doesn't provide much additional value in detecting an effect when $\alpha = 0.05$.

Power calculations for expensive or risky experiments are critical. However, what about experiments that are inexpensive and where the ethical considerations are minimal? For example, if we are doing final testing on a new feature on a popular website, how would our sample size considerations change? As before, we'd want to make sure the sample is big enough. However, if the feature has undergone some testing and is known to perform well (i.e. not frustrate many site users), then we may run a much larger experiment than is necessary to detect the minimum effects of interest. The reason is that there may be additional benefits to having an even more precise estimate of the effect of the new feature. We may even conduct a large experiment as part of the rollout of the new feature.

---

[26]Answers will vary, but here are a few important considerations:

- Whether there is any risk to patients in the study.
- The cost of enrolling more patients.
- The potential downside of not detecting an effect of interest.

5.5 Comparing many means with ANOVA (special topic)

Sometimes we want to compare means across many groups. We might initially think to do pairwise comparisons; for example, if there were three groups, we might be tempted to compare the first mean with the second, then with the third, and then finally compare the second and third means for a total of three comparisons. However, this strategy can be treacherous. If we have many groups and do many comparisons, it is likely that we will eventually find a difference just by chance, even if there is no difference in the populations.

In this section, we will learn a new method called **analysis of variance (ANOVA)** and a new test statistic called F. ANOVA uses a single hypothesis test to check whether the means across many groups are equal:

H_0: The mean outcome is the same across all groups. In statistical notation, $\mu_1 = \mu_2 = \cdots = \mu_k$ where μ_i represents the mean of the outcome for observations in category i.

H_A: At least one mean is different.

Generally we must check three conditions on the data before performing ANOVA:

- the observations are independent within and across groups,
- the data within each group are nearly normal, and
- the variability across the groups is about equal.

When these three conditions are met, we may perform an ANOVA to determine whether the data provide strong evidence against the null hypothesis that all the μ_i are equal.

⬤ **Example 5.38** College departments commonly run multiple lectures of the same introductory course each semester because of high demand. Consider a statistics department that runs three lectures of an introductory statistics course. We might like to determine whether there are statistically significant differences in first exam scores in these three classes (A, B, and C). Describe appropriate hypotheses to determine whether there are any differences between the three classes.

The hypotheses may be written in the following form:

H_0: The average score is identical in all lectures. Any observed difference is due to chance. Notationally, we write $\mu_A = \mu_B = \mu_C$.

H_A: The average score varies by class. We would reject the null hypothesis in favor of the alternative hypothesis if there were larger differences among the class averages than what we might expect from chance alone.

Strong evidence favoring the alternative hypothesis in ANOVA is described by unusually large differences among the group means. We will soon learn that assessing the variability of the group means relative to the variability among individual observations within each group is key to ANOVA's success.

● **Example 5.39** Examine Figure 5.21. Compare groups I, II, and III. Can you visually determine if the differences in the group centers is due to chance or not? Now compare groups IV, V, and VI. Do these differences appear to be due to chance?

Any real difference in the means of groups I, II, and III is difficult to discern, because the data within each group are very volatile relative to any differences in the average outcome. On the other hand, it appears there are differences in the centers of groups IV, V, and VI. For instance, group V appears to have a higher mean than that of the other two groups. Investigating groups IV, V, and VI, we see the differences in the groups' centers are noticeable because those differences are large *relative to the variability in the individual observations within each group.*

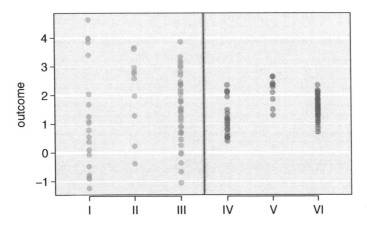

Figure 5.21: Side-by-side dot plot for the outcomes for six groups.

5.5.1 Is batting performance related to player position in MLB?

We would like to discern whether there are real differences between the batting performance of baseball players according to their position: outfielder (OF), infielder (IF), designated hitter (DH), and catcher (C). We will use a data set called bat10, which includes batting records of 327 Major League Baseball (MLB) players from the 2010 season. Six of the 327 cases represented in bat10 are shown in Table 5.22, and descriptions for each variable are provided in Table 5.23. The measure we will use for the player batting performance (the outcome variable) is on-base percentage (OBP). The on-base percentage roughly represents the fraction of the time a player successfully gets on base or hits a home run.

⊙ **Guided Practice 5.40** The null hypothesis under consideration is the following: $\mu_{OF} = \mu_{IF} = \mu_{DH} = \mu_C$. Write the null and corresponding alternative hypotheses in plain language.[27]

---

[27]H_0: The average on-base percentage is equal across the four positions. H_A: The average on-base percentage varies across some (or all) groups.

	name	team	position	AB	H	HR	RBI	AVG	OBP
1	I Suzuki	SEA	OF	680	214	6	43	0.315	0.359
2	D Jeter	NYY	IF	663	179	10	67	0.270	0.340
3	M Young	TEX	IF	656	186	21	91	0.284	0.330
$\vdots$	$\vdots$	$\vdots$	$\vdots$	$\vdots$	$\vdots$	$\vdots$	$\vdots$		
325	B Molina	SF	C	202	52	3	17	0.257	0.312
326	J Thole	NYM	C	202	56	3	17	0.277	0.357
327	C Heisey	CIN	OF	201	51	8	21	0.254	0.324

Table 5.22: Six cases from the `bat10` data matrix.

variable	description
`name`	Player name
`team`	The abbreviated name of the player's team
`position`	The player's primary field position (`OF`, `IF`, `DH`, `C`)
`AB`	Number of opportunities at bat
`H`	Number of hits
`HR`	Number of home runs
`RBI`	Number of runs batted in
`AVG`	Batting average, which is equal to `H/AB`
`OBP`	On-base percentage, which is roughly equal to the fraction of times a player gets on base or hits a home run

Table 5.23: Variables and their descriptions for the `bat10` data set.

● **Example 5.41** The player positions have been divided into four groups: outfield (`OF`), infield (`IF`), designated hitter (`DH`), and catcher (`C`). What would be an appropriate point estimate of the on-base percentage by outfielders, μ_{OF}?

A good estimate of the on-base percentage by outfielders would be the sample average of `OBP` for just those players whose position is outfield: $\bar{x}_{OF} = 0.334$.

Table 5.24 provides summary statistics for each group. A side-by-side box plot for the on-base percentage is shown in Figure 5.25. Notice that the variability appears to be approximately constant across groups; nearly constant variance across groups is an important assumption that must be satisfied before we consider the ANOVA approach.

	OF	IF	DH	C
Sample size (n_i)	120	154	14	39
Sample mean ($\bar{x}_i$)	0.334	0.332	0.348	0.323
Sample SD (s_i)	0.029	0.037	0.036	0.045

Table 5.24: Summary statistics of on-base percentage, split by player position.

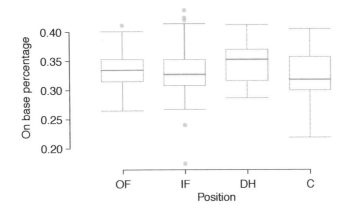

Figure 5.25: Side-by-side box plot of the on-base percentage for 327 players across four groups. There is one prominent outlier visible in the infield group, but with 154 observations in the infield group, this outlier is not a concern.

Example 5.42 The largest difference between the sample means is between the designated hitter and the catcher positions. Consider again the original hypotheses:

H_0: $\mu_{OF} = \mu_{IF} = \mu_{DH} = \mu_C$
H_A: The average on-base percentage (μ_i) varies across some (or all) groups.

Why might it be inappropriate to run the test by simply estimating whether the difference of μ_{DH} and μ_C is statistically significant at a 0.05 significance level?

The primary issue here is that we are inspecting the data before picking the groups that will be compared. It is inappropriate to examine all data by eye (informal testing) and only afterwards decide which parts to formally test. This is called **data snooping** or **data fishing**. Naturally we would pick the groups with the large differences for the formal test, leading to an inflation in the Type 1 Error rate. To understand this better, let's consider a slightly different problem.

Suppose we are to measure the aptitude for students in 20 classes in a large elementary school at the beginning of the year. In this school, all students are randomly assigned to classrooms, so any differences we observe between the classes at the start of the year are completely due to chance. However, with so many groups, we will probably observe a few groups that look rather different from each other. If we select only these classes that look so different, we will probably make the wrong conclusion that the assignment wasn't random. While we might only formally test differences for a few pairs of classes, we informally evaluated the other classes by eye before choosing the most extreme cases for a comparison.

For additional information on the ideas expressed in Example 5.42, we recommend reading about the **prosecutor's fallacy**.[28]

In the next section we will learn how to use the F statistic and ANOVA to test whether observed differences in sample means could have happened just by chance even if there was no difference in the respective population means.

---

[28]See, for example, andrewgelman.com/2007/05/18/the_prosecutors.

5.5.2   Analysis of variance (ANOVA) and the F test

The method of analysis of variance in this context focuses on answering one question: is the variability in the sample means so large that it seems unlikely to be from chance alone? This question is different from earlier testing procedures since we will *simultaneously* consider many groups, and evaluate whether their sample means differ more than we would expect from natural variation. We call this variability the **mean square between groups** (MSG), and it has an associated degrees of freedom, $df_G = k - 1$ when there are k groups. The MSG can be thought of as a scaled variance formula for means. If the null hypothesis is true, any variation in the sample means is due to chance and shouldn't be too large. Details of MSG calculations are provided in the footnote,[29] however, we typically use software for these computations.

The mean square between the groups is, on its own, quite useless in a hypothesis test. We need a benchmark value for how much variability should be expected among the sample means if the null hypothesis is true. To this end, we compute a pooled variance estimate, often abbreviated as the **mean square error** (MSE), which has an associated degrees of freedom value $df_E = n - k$. It is helpful to think of MSE as a measure of the variability within the groups. Details of the computations of the MSE are provided in the footnote[30] for interested readers.

When the null hypothesis is true, any differences among the sample means are only due to chance, and the MSG and MSE should be about equal. As a test statistic for ANOVA, we examine the fraction of MSG and MSE:

$$F = \frac{MSG}{MSE} \tag{5.43}$$

The MSG represents a measure of the between-group variability, and MSE measures the variability within each of the groups.

⊙ **Guided Practice 5.44**   For the baseball data, $MSG = 0.00252$ and $MSE = 0.00127$. Identify the degrees of freedom associated with MSG and MSE and verify the F statistic is approximately 1.994.[31]

---

[29]Let $\bar{x}$ represent the mean of outcomes across all groups. Then the mean square between groups is computed as

$$MSG = \frac{1}{df_G} SSG = \frac{1}{k-1} \sum_{i=1}^{k} n_i \left(\bar{x}_i - \bar{x} \right)^2$$

where SSG is called the **sum of squares between groups** and n_i is the sample size of group i.

[30]Let $\bar{x}$ represent the mean of outcomes across all groups. Then the **sum of squares total** (SST) is computed as

$$SST = \sum_{i=1}^{n} (x_i - \bar{x})^2$$

where the sum is over all observations in the data set. Then we compute the **sum of squared errors** (SSE) in one of two equivalent ways:

$$SSE = SST - SSG$$
$$= (n_1 - 1)s_1^2 + (n_2 - 1)s_2^2 + \cdots + (n_k - 1)s_k^2$$

where s_i^2 is the sample variance (square of the standard deviation) of the residuals in group i. Then the MSE is the standardized form of SSE: $MSE = \frac{1}{df_E} SSE$.

[31]There are $k = 4$ groups, so $df_G = k - 1 = 3$. There are $n = n_1 + n_2 + n_3 + n_4 = 327$ total observations, so $df_E = n - k = 323$. Then the F statistic is computed as the ratio of MSG and MSE: $F = \frac{MSG}{MSE} = \frac{0.00252}{0.00127} = 1.984 \approx 1.994$. ($F = 1.994$ was computed by using values for MSG and MSE that were not rounded.)

We can use the F statistic to evaluate the hypotheses in what is called an **F test**. A p-value can be computed from the F statistic using an F distribution, which has two associated parameters: df_1 and df_2. For the F statistic in ANOVA, $df_1 = df_G$ and $df_2 = df_E$. An F distribution with 3 and 323 degrees of freedom, corresponding to the F statistic for the baseball hypothesis test, is shown in Figure 5.26.

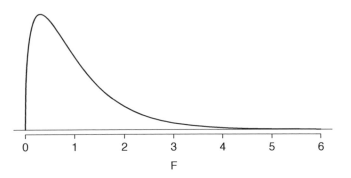

Figure 5.26: An F distribution with $df_1 = 3$ and $df_2 = 323$.

The larger the observed variability in the sample means (MSG) relative to the within-group observations (MSE), the larger F will be and the stronger the evidence against the null hypothesis. Because larger values of F represent stronger evidence against the null hypothesis, we use the upper tail of the distribution to compute a p-value.

The F statistic and the F test

Analysis of variance (ANOVA) is used to test whether the mean outcome differs across 2 or more groups. ANOVA uses a test statistic F, which represents a standardized ratio of variability in the sample means relative to the variability within the groups. If H_0 is true and the model assumptions are satisfied, the statistic F follows an F distribution with parameters $df_1 = k - 1$ and $df_2 = n - k$. The upper tail of the F distribution is used to represent the p-value.

⊙ **Guided Practice 5.45** The test statistic for the baseball example is $F = 1.994$. Shade the area corresponding to the p-value in Figure 5.26. [32]

● **Example 5.46** The p-value corresponding to the shaded area in the solution of Guided Practice 5.45 is equal to about 0.115. Does this provide strong evidence against the null hypothesis?

The p-value is larger than 0.05, indicating the evidence is not strong enough to reject the null hypothesis at a significance level of 0.05. That is, the data do not provide strong evidence that the average on-base percentage varies by player's primary field position.

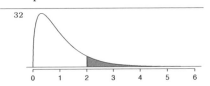

5.5.3   Reading an ANOVA table from software

The calculations required to perform an ANOVA by hand are tedious and prone to human error. For these reasons, it is common to use statistical software to calculate the F statistic and p-value.

An ANOVA can be summarized in a table very similar to that of a regression summary, which we will see in Chapters 7 and 8. Table 5.27 shows an ANOVA summary to test whether the mean of on-base percentage varies by player positions in the MLB. Many of these values should look familiar; in particular, the F test statistic and p-value can be retrieved from the last columns.

	Df	Sum Sq	Mean Sq	F value	Pr(>F)
position	3	0.0076	0.0025	1.9943	0.1147
Residuals	323	0.4080	0.0013		

$$s_{pooled} = 0.036 \text{ on } df = 323$$

Table 5.27: ANOVA summary for testing whether the average on-base percentage differs across player positions.

5.5.4   Graphical diagnostics for an ANOVA analysis

There are three conditions we must check for an ANOVA analysis: all observations must be independent, the data in each group must be nearly normal, and the variance within each group must be approximately equal.

Independence. If the data are a simple random sample from less than 10% of the population, this condition is satisfied. For processes and experiments, carefully consider whether the data may be independent (e.g. no pairing). For example, in the MLB data, the data were not sampled. However, there are not obvious reasons why independence would not hold for most or all observations.

Approximately normal. As with one- and two-sample testing for means, the normality assumption is especially important when the sample size is quite small. The normal probability plots for each group of the MLB data are shown in Figure 5.28; there is some deviation from normality for infielders, but this isn't a substantial concern since there are about 150 observations in that group and the outliers are not extreme. Sometimes in ANOVA there are so many groups or so few observations per group that checking normality for each group isn't reasonable. See the footnote[33] for guidance on how to handle such instances.

Constant variance. The last assumption is that the variance in the groups is about equal from one group to the next. This assumption can be checked by examining a side-by-side box plot of the outcomes across the groups, as in Figure 5.25 on page 249. In this case, the variability is similar in the four groups but not identical. We see in Table 5.24 on page 248 that the standard deviation varies a bit from one group to the next. Whether these differences are from natural variation is unclear, so we should report this uncertainty with the final results.

---

[33] First calculate the **residuals** of the baseball data, which are calculated by taking the observed values and subtracting the corresponding group means. For example, an outfielder with OBP of 0.405 would have a residual of $0.405 - \bar{x}_{OF} = 0.071$. Then to check the normality condition, create a normal probability plot using all the residuals simultaneously.

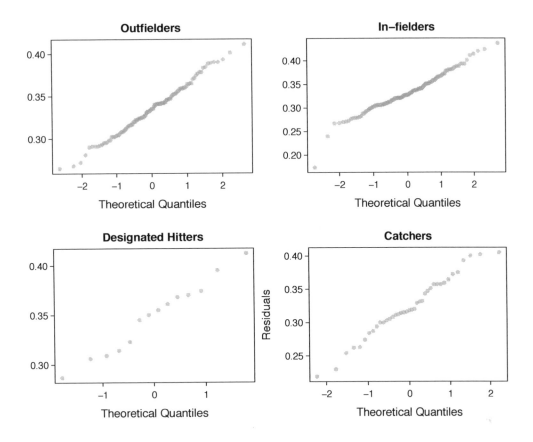

Figure 5.28: Normal probability plot of OBP for each field position.

Caution: Diagnostics for an ANOVA analysis

Independence is always important to an ANOVA analysis. The normality condition is very important when the sample sizes for each group are relatively small. The constant variance condition is especially important when the sample sizes differ between groups.

5.5.5   Multiple comparisons and controlling Type 1 Error rate

When we reject the null hypothesis in an ANOVA analysis, we might wonder, which of these groups have different means? To answer this question, we compare the means of each possible pair of groups. For instance, if there are three groups and there is strong evidence that there are some differences in the group means, there are three comparisons to make: group 1 to group 2, group 1 to group 3, and group 2 to group 3. These comparisons can be accomplished using a two-sample t-test, but we use a modified significance level and a pooled estimate of the standard deviation across groups. Usually this pooled standard deviation can be found in the ANOVA table, e.g. along the bottom of Table 5.27.

Class i	A	B	C
n_i	58	55	51
$\bar{x}_i$	75.1	72.0	78.9
s_i	13.9	13.8	13.1

Table 5.29: Summary statistics for the first midterm scores in three different lectures of the same course.

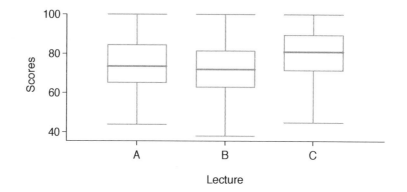

Figure 5.30: Side-by-side box plot for the first midterm scores in three different lectures of the same course.

● **Example 5.47** Example 5.38 on page 246 discussed three statistics lectures, all taught during the same semester. Table 5.29 shows summary statistics for these three courses, and a side-by-side box plot of the data is shown in Figure 5.30. We would like to conduct an ANOVA for these data. Do you see any deviations from the three conditions for ANOVA?

In this case (like many others) it is difficult to check independence in a rigorous way. Instead, the best we can do is use common sense to consider reasons the assumption of independence may not hold. For instance, the independence assumption may not be reasonable if there is a star teaching assistant that only half of the students may access; such a scenario would divide a class into two subgroups. No such situations were evident for these particular data, and we believe that independence is acceptable.

The distributions in the side-by-side box plot appear to be roughly symmetric and show no noticeable outliers.

The box plots show approximately equal variability, which can be verified in Table 5.29, supporting the constant variance assumption.

⊙ **Guided Practice 5.48** An ANOVA was conducted for the midterm data, and summary results are shown in Table 5.31. What should we conclude?[34]

There is strong evidence that the different means in each of the three classes is not simply due to chance. We might wonder, which of the classes are actually different? As

---

[34]The p-value of the test is 0.0330, less than the default significance level of 0.05. Therefore, we reject the null hypothesis and conclude that the difference in the average midterm scores are not due to chance.

	Df	Sum Sq	Mean Sq	F value	Pr(>F)
lecture	2	1290.11	645.06	3.48	0.0330
Residuals	161	29810.13	185.16		

$$s_{pooled} = 13.61 \text{ on } df = 161$$

Table 5.31: ANOVA summary table for the midterm data.

discussed in earlier chapters, a two-sample t-test could be used to test for differences in each possible pair of groups. However, one pitfall was discussed in Example 5.42 on page 249: when we run so many tests, the Type 1 Error rate increases. This issue is resolved by using a modified significance level.

Multiple comparisons and the Bonferroni correction for α

The scenario of testing many pairs of groups is called **multiple comparisons**. The **Bonferroni correction** suggests that a more stringent significance level is more appropriate for these tests:

$$\alpha^* = \alpha/K$$

where K is the number of comparisons being considered (formally or informally). If there are k groups, then usually all possible pairs are compared and $K = \frac{k(k-1)}{2}$.

● **Example 5.49** In Guided Practice 5.48, you found strong evidence of differences in the average midterm grades between the three lectures. Complete the three possible pairwise comparisons using the Bonferroni correction and report any differences.

We use a modified significance level of $\alpha^* = 0.05/3 = 0.0167$. Additionally, we use the pooled estimate of the standard deviation: $s_{pooled} = 13.61$ on $df = 161$, which is provided in the ANOVA summary table.

Lecture A versus Lecture B: The estimated difference and standard error are, respectively,

$$\bar{x}_A - \bar{x}_B = 75.1 - 72 = 3.1 \qquad SE = \sqrt{\frac{13.61^2}{58} + \frac{13.61^2}{55}} = 2.56$$

(See Section 5.3.6 on page 239 for additional details.) This results in a T score of 1.21 on $df = 161$ (we use the df associated with s_{pooled}). Statistical software was used to precisely identify the two-sided p-value since the modified significance of 0.0167 is not found in the t-table. The p-value (0.228) is larger than $\alpha^* = 0.0167$, so there is not strong evidence of a difference in the means of lectures A and B.

Lecture A versus Lecture C: The estimated difference and standard error are 3.8 and 2.61, respectively. This results in a T score of 1.46 on $df = 161$ and a two-sided p-value of 0.1462. This p-value is larger than α^*, so there is not strong evidence of a difference in the means of lectures A and C.

Lecture B versus Lecture C: The estimated difference and standard error are 6.9 and 2.65, respectively. This results in a T score of 2.60 on $df = 161$ and a two-sided p-value of 0.0102. This p-value is smaller than α^*. Here we find strong evidence of a difference in the means of lectures B and C.

We might summarize the findings of the analysis from Example 5.49 using the following notation:

$$\mu_A \overset{?}{=} \mu_B \qquad\qquad \mu_A \overset{?}{=} \mu_C \qquad\qquad \mu_B \neq \mu_C$$

The midterm mean in lecture A is not statistically distinguishable from those of lectures B or C. However, there is strong evidence that lectures B and C are different. In the first two pairwise comparisons, we did not have sufficient evidence to reject the null hypothesis. Recall that failing to reject H_0 does not imply H_0 is true.

Caution: Sometimes an ANOVA will reject the null but no groups will have statistically significant differences

It is possible to reject the null hypothesis using ANOVA and then to not subsequently identify differences in the pairwise comparisons. However, *this does not invalidate the ANOVA conclusion*. It only means we have not been able to successfully identify which groups differ in their means.

The ANOVA procedure examines the big picture: it considers all groups simultaneously to decipher whether there is evidence that some difference exists. Even if the test indicates that there is strong evidence of differences in group means, identifying with high confidence a specific difference as statistically significant is more difficult.

Consider the following analogy: we observe a Wall Street firm that makes large quantities of money based on predicting mergers. Mergers are generally difficult to predict, and if the prediction success rate is extremely high, that may be considered sufficiently strong evidence to warrant investigation by the Securities and Exchange Commission (SEC). While the SEC may be quite certain that there is insider trading taking place at the firm, the evidence against any single trader may not be very strong. It is only when the SEC considers all the data that they identify the pattern. This is effectively the strategy of ANOVA: stand back and consider all the groups simultaneously.

5.6 Exercises

5.6.1 One-sample means with the t-distribution

5.1 Identify the critical t. An independent random sample is selected from an approximately normal population with unknown standard deviation. Find the degrees of freedom and the critical t-value ($t^\star$) for the given sample size and confidence level.

(a) $n = 6$, CL = 90%

(b) $n = 21$, CL = 98%

(c) $n = 29$, CL = 95%

(d) $n = 12$, CL = 99%

5.2 t-distribution. The figure on the right shows three unimodal and symmetric curves: the standard normal (z) distribution, the t-distribution with 5 degrees of freedom, and the t-distribution with 1 degree of freedom. Determine which is which, and explain your reasoning.

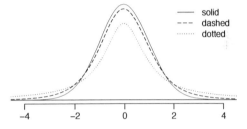

5.3 Find the p-value, Part I. An independent random sample is selected from an approximately normal population with an unknown standard deviation. Find the p-value for the given set of hypotheses and T test statistic. Also determine if the null hypothesis would be rejected at $\alpha = 0.05$.

(a) $H_A : \mu > \mu_0$, $n = 11$, $T = 1.91$

(b) $H_A : \mu < \mu_0$, $n = 17$, $T = -3.45$

(c) $H_A : \mu \neq \mu_0$, $n = 7$, $T = 0.83$

(d) $H_A : \mu > \mu_0$, $n = 28$, $T = 2.13$

5.4 Find the p-value, Part II. An independent random sample is selected from an approximately normal population with an unknown standard deviation. Find the p-value for the given set of hypotheses and T test statistic. Also determine if the null hypothesis would be rejected at $\alpha = 0.01$.

(a) $H_A : \mu > 0.5$, $n = 26$, $T = 2.485$

(b) $H_A : \mu < 3$, $n = 18$, $T = 0.5$

5.5 Working backwards, Part I. A 95% confidence interval for a population mean, μ, is given as (18.985, 21.015). This confidence interval is based on a simple random sample of 36 observations. Calculate the sample mean and standard deviation. Assume that all conditions necessary for inference are satisfied. Use the t-distribution in any calculations.

5.6 Working backwards, Part II. A 90% confidence interval for a population mean is (65, 77). The population distribution is approximately normal and the population standard deviation is unknown. This confidence interval is based on a simple random sample of 25 observations. Calculate the sample mean, the margin of error, and the sample standard deviation.

5.7   Sleep habits of New Yorkers. New York is known as "the city that never sleeps". A random sample of 25 New Yorkers were asked how much sleep they get per night. Statistical summaries of these data are shown below. Do these data provide strong evidence that New Yorkers sleep less than 8 hours a night on average?

n	$\bar{x}$	s	min	max
25	7.73	0.77	6.17	9.78

(a) Write the hypotheses in symbols and in words.

(b) Check conditions, then calculate the test statistic, T, and the associated degrees of freedom.

(c) Find and interpret the p-value in this context. Drawing a picture may be helpful.

(d) What is the conclusion of the hypothesis test?

(e) If you were to construct a 90% confidence interval that corresponded to this hypothesis test, would you expect 8 hours to be in the interval?

5.8   Fuel efficiency of Prius. Fueleconomy.gov, the official US government source for fuel economy information, allows users to share gas mileage information on their vehicles. The histogram below shows the distribution of gas mileage in miles per gallon (MPG) from 14 users who drive a 2012 Toyota Prius. The sample mean is 53.3 MPG and the standard deviation is 5.2 MPG. Note that these data are user estimates and since the source data cannot be verified, the accuracy of these estimates are not guaranteed.[35]

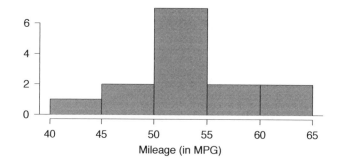

(a) We would like to use these data to evaluate the average gas mileage of all 2012 Prius drivers. Do you think this is reasonable? Why or why not?

(b) The EPA claims that a 2012 Prius gets 50 MPG (city and highway mileage combined). Do these data provide strong evidence against this estimate for drivers who participate on fueleconomy.gov? Note any assumptions you must make as you proceed with the test.

(c) Calculate a 95% confidence interval for the average gas mileage of a 2012 Prius by drivers who participate on fueleconomy.gov.

5.9   Find the mean. You are given the following hypotheses:

$$H_0 : \mu = 60$$
$$H_A : \mu < 60$$

We know that the sample standard deviation is 8 and the sample size is 20. For what sample mean would the p-value be equal to 0.05? Assume that all conditions necessary for inference are satisfied.

5.10 $t^\star$ vs. $z^\star$. For a given confidence level, $t^\star_{df}$ is larger than $z^\star$. Explain how $t^\star_{df}$ being slightly larger than $z^\star$ affects the width of the confidence interval.

---

[35]Fuelecomy.gov, Shared MPG Estimates: Toyota Prius 2012.

5.11 Play the piano. Georgianna claims that in a small city renowned for its music school, the average child takes at least 5 years of piano lessons. We have a random sample of 20 children from the city, with a mean of 4.6 years of piano lessons and a standard deviation of 2.2 years.

(a) Evaluate Georgianna's claim using a hypothesis test.

(b) Construct a 95% confidence interval for the number of years students in this city take piano lessons, and interpret it in context of the data.

(c) Do your results from the hypothesis test and the confidence interval agree? Explain your reasoning.

5.12 Auto exhaust and lead exposure. Researchers interested in lead exposure due to car exhaust sampled the blood of 52 police officers subjected to constant inhalation of automobile exhaust fumes while working traffic enforcement in a primarily urban environment. The blood samples of these officers had an average lead concentration of 124.32 μg/l and a SD of 37.74 μg/l; a previous study of individuals from a nearby suburb, with no history of exposure, found an average blood level concentration of 35 μg/l.[36]

(a) Write down the hypotheses that would be appropriate for testing if the police officers appear to have been exposed to a higher concentration of lead.

(b) Explicitly state and check all conditions necessary for inference on these data.

(c) Test the hypothesis that the downtown police officers have a higher lead exposure than the group in the previous study. Interpret your results in context.

(d) Based on your preceding result, without performing a calculation, would a 99% confidence interval for the average blood concentration level of police officers contain 35 μg/l?

5.13 Car insurance savings. A market researcher wants to evaluate car insurance savings at a competing company. Based on past studies he is assuming that the standard deviation of savings is $100. He wants to collect data such that he can get a margin of error of no more than $10 at a 95% confidence level. How large of a sample should he collect?

5.14 SAT scores. SAT scores of students at an Ivy League college are distributed with a standard deviation of 250 points. Two statistics students, Raina and Luke, want to estimate the average SAT score of students at this college as part of a class project. They want their margin of error to be no more than 25 points.

(a) Raina wants to use a 90% confidence interval. How large a sample should she collect?

(b) Luke wants to use a 99% confidence interval. Without calculating the actual sample size, determine whether his sample should be larger or smaller than Raina's, and explain your reasoning.

(c) Calculate the minimum required sample size for Luke.

5.6.2 Paired data

5.15 Air quality. Air quality measurements were collected in a random sample of 25 country capitals in 2013, and then again in the same cities in 2014. We would like to use these data to compare average air quality between the two years.

(a) Should we use a one-sided or a two-sided test? Explain your reasoning.

(b) Should we use a paired or non-paired test? Explain your reasoning.

(c) Should we use a *t*-test or a *z*-test? Explain your reasoning.

---

[36]WI Mortada et al. "Study of lead exposure from automobile exhaust as a risk for nephrotoxicity among traffic policemen." In: *American journal of nephrology* 21.4 (2000), pp. 274–279.

5.16   True / False: paired. Determine if the following statements are true or false. If false, explain.

(a) In a paired analysis we first take the difference of each pair of observations, and then we do inference on these differences.

(b) Two data sets of different sizes cannot be analyzed as paired data.

(c) Each observation in one data set has a natural correspondence with exactly one observation from the other data set.

(d) Each observation in one data set is subtracted from the average of the other data set's observations.

5.17   Paired or not, Part I? In each of the following scenarios, determine if the data are paired.

(a) Compare pre- (beginning of semester) and post-test (end of semester) scores of students.

(b) Assess gender-related salary gap by comparing salaries of randomly sampled men and women.

(c) Compare artery thicknesses at the beginning of a study and after 2 years of taking Vitamin E for the same group of patients.

(d) Assess effectiveness of a diet regimen by comparing the before and after weights of subjects.

5.18   Paired or not, Part II? In each of the following scenarios, determine if the data are paired.

(a) We would like to know if Intel's stock and Southwest Airlines' stock have similar rates of return. To find out, we take a random sample of 50 days, and record Intel's and Southwest's stock on those same days.

(b) We randomly sample 50 items from Target stores and note the price for each. Then we visit Walmart and collect the price for each of those same 50 items.

(c) A school board would like to determine whether there is a difference in average SAT scores for students at one high school versus another high school in the district. To check, they take a simple random sample of 100 students from each high school.

5.19   Global warming, Part I. Is there strong evidence of global warming? Let's consider a small scale example, comparing how temperatures have changed in the US from 1968 to 2008. The daily high temperature reading on January 1 was collected in 1968 and 2008 for 51 randomly selected locations in the continental US. Then the difference between the two readings (temperature in 2008 - temperature in 1968) was calculated for each of the 51 different locations. The average of these 51 values was 1.1 degrees with a standard deviation of 4.9 degrees. We are interested in determining whether these data provide strong evidence of temperature warming in the continental US.

(a) Is there a relationship between the observations collected in 1968 and 2008? Or are the observations in the two groups independent? Explain.

(b) Write hypotheses for this research in symbols and in words.

(c) Check the conditions required to complete this test.

(d) Calculate the test statistic and find the p-value.

(e) What do you conclude? Interpret your conclusion in context.

(f) What type of error might we have made? Explain in context what the error means.

(g) Based on the results of this hypothesis test, would you expect a confidence interval for the average difference between the temperature measurements from 1968 and 2008 to include 0? Explain your reasoning.

5.20 High School and Beyond, Part I. The National Center of Education Statistics conducted a survey of high school seniors, collecting test data on reading, writing, and several other subjects. Here we examine a simple random sample of 200 students from this survey. Side-by-side box plots of reading and writing scores as well as a histogram of the differences in scores are shown below.

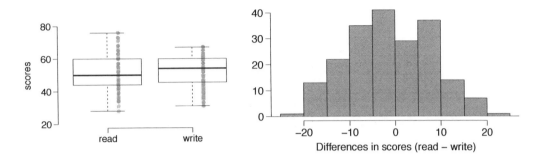

(a) Is there a clear difference in the average reading and writing scores?

(b) Are the reading and writing scores of each student independent of each other?

(c) Create hypotheses appropriate for the following research question: is there an evident difference in the average scores of students in the reading and writing exam?

(d) Check the conditions required to complete this test.

(e) The average observed difference in scores is $\bar{x}_{read-write} = -0.545$, and the standard deviation of the differences is 8.887 points. Do these data provide convincing evidence of a difference between the average scores on the two exams?

(f) What type of error might we have made? Explain what the error means in the context of the application.

(g) Based on the results of this hypothesis test, would you expect a confidence interval for the average difference between the reading and writing scores to include 0? Explain your reasoning.

5.21 Global warming, Part II. We considered the differences between the temperature readings in January 1 of 1968 and 2008 at 51 locations in the continental US in Exercise 5.19. The mean and standard deviation of the reported differences are 1.1 degrees and 4.9 degrees.

(a) Calculate a 90% confidence interval for the average difference between the temperature measurements between 1968 and 2008.

(b) Interpret this interval in context.

(c) Does the confidence interval provide convincing evidence that the temperature was higher in 2008 than in 1968 in the continental US? Explain.

5.22 High school and beyond, Part II. We considered the differences between the reading and writing scores of a random sample of 200 students who took the High School and Beyond Survey in Exercise 5.20. The mean and standard deviation of the differences are $\bar{x}_{read-write} = -0.545$ and 8.887 points.

(a) Calculate a 95% confidence interval for the average difference between the reading and writing scores of all students.

(b) Interpret this interval in context.

(c) Does the confidence interval provide convincing evidence that there is a real difference in the average scores? Explain.

5.23 Gifted children. Researchers collected a simple random sample of 36 children who had been identified as gifted in a large city. The following histograms show the distributions of the IQ scores of mothers and fathers of these children. Also provided are some sample statistics.[37]

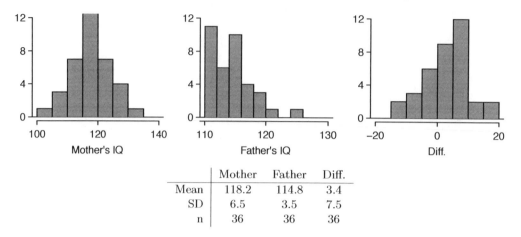

	Mother	Father	Diff.
Mean	118.2	114.8	3.4
SD	6.5	3.5	7.5
n	36	36	36

(a) Are the IQs of mothers and the IQs of fathers in this data set related? Explain.

(b) Conduct a hypothesis test to evaluate if the scores are equal on average. Make sure to clearly state your hypotheses, check the relevant conditions, and state your conclusion in the context of the data.

5.24 Sample size and pairing. Determine if the following statement is true or false, and if false, explain your reasoning: If comparing means of two groups with equal sample sizes, always use a paired test.

---

[37]F.A. Graybill and H.K. Iyer. *Regression Analysis: Concepts and Applications.* Duxbury Press, 1994, pp. 511–516.

5.6.3 Difference of two means

5.25 Cleveland vs. Sacramento. Average income varies from one region of the country to another, and it often reflects both lifestyles and regional living expenses. Suppose a new graduate is considering a job in two locations, Cleveland, OH and Sacramento, CA, and he wants to see whether the average income in one of these cities is higher than the other. He would like to conduct a hypothesis test based on two small samples from the 2000 Census, but he first must consider whether the conditions are met to implement the test. Below are histograms for each city. Should he move forward with the hypothesis test? Explain your reasoning.

	Cleveland, OH
Mean	$ 35,749
SD	$ 39,421
n	21

	Sacramento, CA
Mean	$ 35,500
SD	$ 41,512
n	17

5.26 Oscar winners. The first Oscar awards for best actor and best actress were given out in 1929. The histograms below show the age distribution for all of the best actor and best actress winners from 1929 to 2012. Summary statistics for these distributions are also provided. Is a hypothesis test appropriate for evaluating whether the difference in the average ages of best actors and actresses might be due to chance? Explain your reasoning.[38]

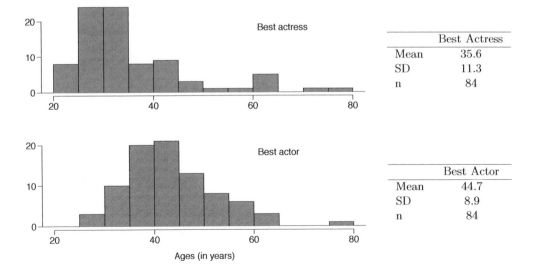

	Best Actress
Mean	35.6
SD	11.3
n	84

	Best Actor
Mean	44.7
SD	8.9
n	84

---

[38]Oscar winners from 1929 – 2012, data up to 2009 from the Journal of Statistics Education data archive and more current data from wikipedia.org.

5.27 Friday the 13th, Part I. In the early 1990's, researchers in the UK collected data on traffic flow, number of shoppers, and traffic accident related emergency room admissions on Friday the 13th and the previous Friday, Friday the 6th. The histograms below show the distribution of number of cars passing by a specific intersection on Friday the 6th and Friday the 13th for many such date pairs. Also given are some sample statistics, where the difference is the number of cars on the 6th minus the number of cars on the 13th.[39]

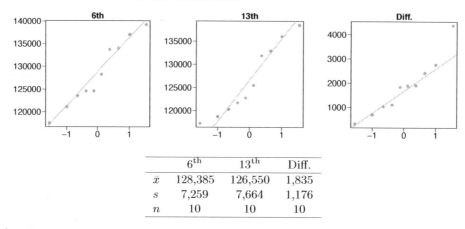

	6th	13th	Diff.
$\bar{x}$	128,385	126,550	1,835
s	7,259	7,664	1,176
n	10	10	10

(a) Are there any underlying structures in these data that should be considered in an analysis? Explain.

(b) What are the hypotheses for evaluating whether the number of people out on Friday the 6th is different than the number out on Friday the 13th?

(c) Check conditions to carry out the hypothesis test from part (b).

(d) Calculate the test statistic and the p-value.

(e) What is the conclusion of the hypothesis test?

(f) Interpret the p-value in this context.

(g) What type of error might have been made in the conclusion of your test? Explain.

5.28 Diamonds, Part I. Prices of diamonds are determined by what is known as the 4 Cs: cut, clarity, color, and carat weight. The prices of diamonds go up as the carat weight increases, but the increase is not smooth. For example, the difference between the size of a 0.99 carat diamond and a 1 carat diamond is undetectable to the naked human eye, but the price of a 1 carat diamond tends to be much higher than the price of a 0.99 diamond. In this question we use two random samples of diamonds, 0.99 carats and 1 carat, each sample of size 23, and compare the average prices of the diamonds. In order to be able to compare equivalent units, we first divide the price for each diamond by 100 times its weight in carats. That is, for a 0.99 carat diamond, we divide the price by 99. For a 1 carat diamond, we divide the price by 100. The distributions and some sample statistics are shown below.[40]

Conduct a hypothesis test to evaluate if there is a difference between the average standardized prices of 0.99 and 1 carat diamonds. Make sure to state your hypotheses clearly, check relevant conditions, and interpret your results in context of the data.

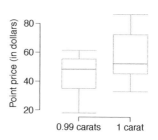

	0.99 carats	1 carat
Mean	$ 44.51	$ 56.81
SD	$ 13.32	$ 16.13
n	23	23

---

[39]T.J. Scanlon et al. "Is Friday the 13th Bad For Your Health?" In: *BMJ* 307 (1993), pp. 1584–1586.
[40]H. Wickham. *ggplot2: elegant graphics for data analysis.* Springer New York, 2009.

5.29 Friday the 13th, Part II. The Friday the 13th study reported in Exercise 5.27 also provides data on traffic accident related emergency room admissions. The distributions of these counts from Friday the 6th and Friday the 13th are shown below for six such paired dates along with summary statistics. You may assume that conditions for inference are met.

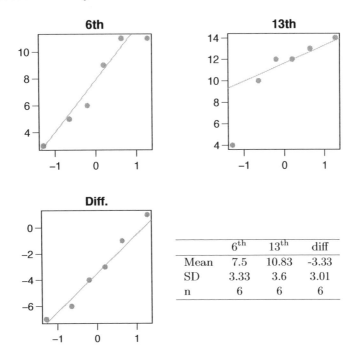

	6th	13th	diff
Mean	7.5	10.83	-3.33
SD	3.33	3.6	3.01
n	6	6	6

(a) Conduct a hypothesis test to evaluate if there is a difference between the average numbers of traffic accident related emergency room admissions between Friday the 6th and Friday the 13th.

(b) Calculate a 95% confidence interval for the difference between the average numbers of traffic accident related emergency room admissions between Friday the 6th and Friday the 13th.

(c) The conclusion of the original study states, "Friday 13th is unlucky for some. The risk of hospital admission as a result of a transport accident may be increased by as much as 52%. Staying at home is recommended." Do you agree with this statement? Explain your reasoning.

5.30 Diamonds, Part II. In Exercise 5.28, we discussed diamond prices (standardized by weight) for diamonds with weights 0.99 carats and 1 carat. See the table for summary statistics, and then construct a 95% confidence interval for the average difference between the standardized prices of 0.99 and 1 carat diamonds. You may assume the conditions for inference are met.

	0.99 carats	1 carat
Mean	$ 44.51	$ 56.81
SD	$ 13.32	$ 16.13
n	23	23

5.31 Chicken diet and weight, Part I. Chicken farming is a multi-billion dollar industry, and any methods that increase the growth rate of young chicks can reduce consumer costs while increasing company profits, possibly by millions of dollars. An experiment was conducted to measure and compare the effectiveness of various feed supplements on the growth rate of chickens. Newly hatched chicks were randomly allocated into six groups, and each group was given a different feed supplement. Below are some summary statistics from this data set along with box plots showing the distribution of weights by feed type.[41]

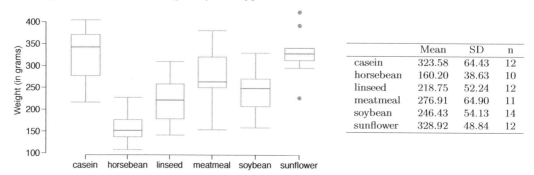

	Mean	SD	n
casein	323.58	64.43	12
horsebean	160.20	38.63	10
linseed	218.75	52.24	12
meatmeal	276.91	64.90	11
soybean	246.43	54.13	14
sunflower	328.92	48.84	12

(a) Describe the distributions of weights of chickens that were fed linseed and horsebean.

(b) Do these data provide strong evidence that the average weights of chickens that were fed linseed and horsebean are different? Use a 5% significance level.

(c) What type of error might we have committed? Explain.

(d) Would your conclusion change if we used $\alpha = 0.01$?

5.32 Fuel efficiency of manual and automatic cars, Part I. Each year the US Environmental Protection Agency (EPA) releases fuel economy data on cars manufactured in that year. Below are summary statistics on fuel efficiency (in miles/gallon) from random samples of cars with manual and automatic transmissions manufactured in 2012. Do these data provide strong evidence of a difference between the average fuel efficiency of cars with manual and automatic transmissions in terms of their average city mileage? Assume that conditions for inference are satisfied.[42]

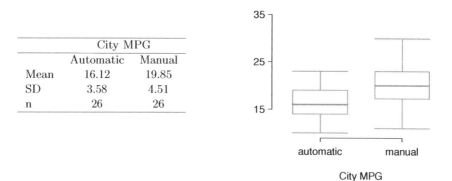

	City MPG	
	Automatic	Manual
Mean	16.12	19.85
SD	3.58	4.51
n	26	26

City MPG

5.33 Chicken diet and weight, Part II. Casein is a common weight gain supplement for humans. Does it have an effect on chickens? Using data provided in Exercise 5.31, test the hypothesis that the average weight of chickens that were fed casein is different than the average weight of chickens that were fed soybean. If your hypothesis test yields a statistically significant result, discuss whether or not the higher average weight of chickens can be attributed to the casein diet. Assume that conditions for inference are satisfied.

---

[41]Chicken Weights by Feed Type, from the `datasets` package in R..
[42]U.S. Department of Energy, Fuel Economy Data, 2012 Datafile.

5.34 Fuel efficiency of manual and automatic cars, Part II. The table provides summary statistics on highway fuel economy of cars manufactured in 2012 (from Exercise 5.32). Use these statistics to calculate a 98% confidence interval for the difference between average highway mileage of manual and automatic cars, and interpret this interval in the context of the data.[43]

	Hwy MPG	
	Automatic	Manual
Mean	22.92	27.88
SD	5.29	5.01
n	26	26

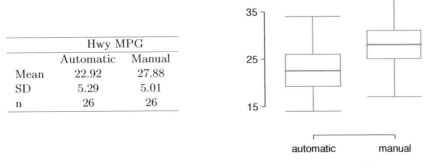

5.35 Gaming and distracted eating, Part I. A group of researchers are interested in the possible effects of distracting stimuli during eating, such as an increase or decrease in the amount of food consumption. To test this hypothesis, they monitored food intake for a group of 44 patients who were randomized into two equal groups. The treatment group ate lunch while playing solitaire, and the control group ate lunch without any added distractions. Patients in the treatment group ate 52.1 grams of biscuits, with a standard deviation of 45.1 grams, and patients in the control group ate 27.1 grams of biscuits, with a standard deviation of 26.4 grams. Do these data provide convincing evidence that the average food intake (measured in amount of biscuits consumed) is different for the patients in the treatment group? Assume that conditions for inference are satisfied.[44]

5.36 Gaming and distracted eating, Part II. The researchers from Exercise 5.35 also investigated the effects of being distracted by a game on how much people eat. The 22 patients in the treatment group who ate their lunch while playing solitaire were asked to do a serial-order recall of the food lunch items they ate. The average number of items recalled by the patients in this group was 4.9, with a standard deviation of 1.8. The average number of items recalled by the patients in the control group (no distraction) was 6.1, with a standard deviation of 1.8. Do these data provide strong evidence that the average number of food items recalled by the patients in the treatment and control groups are different?

---

[43]U.S. Department of Energy, Fuel Economy Data, 2012 Datafile.

[44]R.E. Oldham-Cooper et al. "Playing a computer game during lunch affects fullness, memory for lunch, and later snack intake". In: *The American Journal of Clinical Nutrition* 93.2 (2011), p. 308.

5.37 Prison isolation experiment, Part I. Subjects from Central Prison in Raleigh, NC, volunteered for an experiment involving an "isolation" experience. The goal of the experiment was to find a treatment that reduces subjects' psychopathic deviant T scores. This score measures a person's need for control or their rebellion against control, and it is part of a commonly used mental health test called the Minnesota Multiphasic Personality Inventory (MMPI) test. The experiment had three treatment groups:

(1) Four hours of sensory restriction plus a 15 minute "therapeutic" tape advising that professional help is available.

(2) Four hours of sensory restriction plus a 15 minute "emotionally neutral" tape on training hunting dogs.

(3) Four hours of sensory restriction but no taped message.

Forty-two subjects were randomly assigned to these treatment groups, and an MMPI test was administered before and after the treatment. Distributions of the differences between pre and post treatment scores (pre - post) are shown below, along with some sample statistics. Use this information to independently test the effectiveness of each treatment. Make sure to clearly state your hypotheses, check conditions, and interpret results in the context of the data.[45]

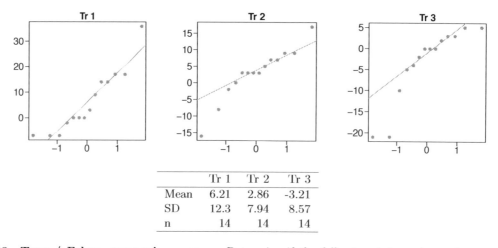

	Tr 1	Tr 2	Tr 3
Mean	6.21	2.86	-3.21
SD	12.3	7.94	8.57
n	14	14	14

5.38 True / False: comparing means. Determine if the following statements are true or false, and explain your reasoning for statements you identify as false.

(a) When comparing means of two samples where $n_1 = 20$ and $n_2 = 40$, we can use the normal model for the difference in means since $n_2 \geq 30$.

(b) As the degrees of freedom increases, the t-distribution approaches normality.

(c) We use a pooled standard error for calculating the standard error of the difference between means when sample sizes of groups are equal to each other.

---

[45]Prison isolation experiment.

5.6.4 Power calculations for a difference of means

5.39 Increasing corn yield. A large farm wants to try out a new type of fertilizer to evaluate whether it will improve the farm's corn production. The land is broken into plots that produce an average of 1,215 pounds of corn with a standard deviation of 94 pounds per plot. The owner is interested in detecting any average difference of at least 40 pounds per plot. How many plots of land would be needed for the experiment if the desired power level is 90%? Assume each plot of land gets treated with either the current fertilizer or the new fertilizer.

5.40 Email outreach efforts. A medical research group is recruiting people to complete short surveys about their medical history. For example, one survey asks for information on a person's family history in regards to cancer. Another survey asks about what topics were discussed during the person's last visit to a hospital. So far, as people sign up, they complete an average of just 4 surveys, and the standard deviation of the number of surveys is about 2.2. The research group wants to try a new interface that they think will encourage new enrollees to complete more surveys, where they will randomize each enrollee to either get the new interface or the current interface. How many new enrollees do they need for each interface to detect an effect size of 0.5 surveys per enrollee, if the desired power level is 80%?

5.6.5 Comparing many means with ANOVA

5.41 Fill in the blank. When doing an ANOVA, you observe large differences in means between groups. Within the ANOVA framework, this would most likely be interpreted as evidence strongly favoring the _____ hypothesis.

5.42 Which test? We would like to test if students who are in the social sciences, natural sciences, arts and humanities, and other fields spend the same amount of time studying for this course. What type of test should we use? Explain your reasoning.

5.43 Chicken diet and weight, Part III. In Exercises 5.31 and 5.33 we compared the effects of two types of feed at a time. A better analysis would first consider all feed types at once: casein, horsebean, linseed, meat meal, soybean, and sunflower. The ANOVA output below can be used to test for differences between the average weights of chicks on different diets.

	Df	Sum Sq	Mean Sq	F value	Pr(>F)
feed	5	231,129.16	46,225.83	15.36	0.0000
Residuals	65	195,556.02	3,008.55		

Conduct a hypothesis test to determine if these data provide convincing evidence that the average weight of chicks varies across some (or all) groups. Make sure to check relevant conditions. Figures and summary statistics are shown below.

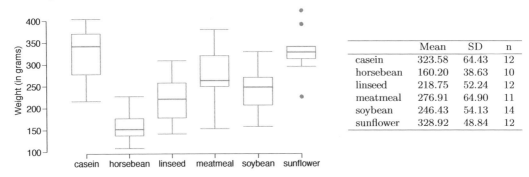

	Mean	SD	n
casein	323.58	64.43	12
horsebean	160.20	38.63	10
linseed	218.75	52.24	12
meatmeal	276.91	64.90	11
soybean	246.43	54.13	14
sunflower	328.92	48.84	12

5.44 Teaching descriptive statistics. A study compared five different methods for teaching descriptive statistics. The five methods were traditional lecture and discussion, programmed text-book instruction, programmed text with lectures, computer instruction, and computer instruction with lectures. 45 students were randomly assigned, 9 to each method. After completing the course, students took a 1-hour exam.

(a) What are the hypotheses for evaluating if the average test scores are different for the different teaching methods?

(b) What are the degrees of freedom associated with the F-test for evaluating these hypotheses?

(c) Suppose the p-value for this test is 0.0168. What is the conclusion?

5.45 Coffee, depression, and physical activity. Caffeine is the world's most widely used stimulant, with approximately 80% consumed in the form of coffee. Participants in a study investigating the relationship between coffee consumption and exercise were asked to report the number of hours they spent per week on moderate (e.g., brisk walking) and vigorous (e.g., strenuous sports and jogging) exercise. Based on these data the researchers estimated the total hours of metabolic equivalent tasks (MET) per week, a value always greater than 0. The table below gives summary statistics of MET for women in this study based on the amount of coffee consumed.[46]

Caffeinated coffee consumption

	≤ 1 cup/week	2-6 cups/week	1 cup/day	2-3 cups/day	≥ 4 cups/day	Total
Mean	18.7	19.6	19.3	18.9	17.5	
SD	21.1	25.5	22.5	22.0	22.0	
n	12,215	6,617	17,234	12,290	2,383	50,739

(a) Write the hypotheses for evaluating if the average physical activity level varies among the different levels of coffee consumption.

(b) Check conditions and describe any assumptions you must make to proceed with the test.

(c) Below is part of the output associated with this test. Fill in the empty cells.

	Df	Sum Sq	Mean Sq	F value	Pr(>F)
coffee					0.0003
Residuals		25,564,819			
Total		25,575,327			

(d) What is the conclusion of the test?

---

[46]M. Lucas et al. "Coffee, caffeine, and risk of depression among women". In: *Archives of internal medicine* 171.17 (2011), p. 1571.

5.46 Student performance across discussion sections. A professor who teaches a large introductory statistics class (197 students) with eight discussion sections would like to test if student performance differs by discussion section, where each discussion section has a different teaching assistant. The summary table below shows the average final exam score for each discussion section as well as the standard deviation of scores and the number of students in each section.

	Sec 1	Sec 2	Sec 3	Sec 4	Sec 5	Sec 6	Sec 7	Sec 8
n_i	33	19	10	29	33	10	32	31
$\bar{x}_i$	92.94	91.11	91.80	92.45	89.30	88.30	90.12	93.35
s_i	4.21	5.58	3.43	5.92	9.32	7.27	6.93	4.57

The ANOVA output below can be used to test for differences between the average scores from the different discussion sections.

	Df	Sum Sq	Mean Sq	F value	Pr(>F)
section	7	525.01	75.00	1.87	0.0767
Residuals	189	7584.11	40.13		

Conduct a hypothesis test to determine if these data provide convincing evidence that the average score varies across some (or all) groups. Check conditions and describe any assumptions you must make to proceed with the test.

5.47 GPA and major. Undergraduate students taking an introductory statistics course at Duke University conducted a survey about GPA and major. The side-by-side box plots show the distribution of GPA among three groups of majors. Also provided is the ANOVA output.

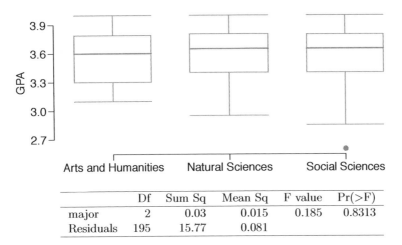

	Df	Sum Sq	Mean Sq	F value	Pr(>F)
major	2	0.03	0.015	0.185	0.8313
Residuals	195	15.77	0.081		

(a) Write the hypotheses for testing for a difference between average GPA across majors.

(b) What is the conclusion of the hypothesis test?

(c) How many students answered these questions on the survey, i.e. what is the sample size?

5.48 Work hours and education. The General Social Survey collects data on demographics, education, and work, among many other characteristics of US residents.[47] Using ANOVA, we can consider educational attainment levels for all 1,172 respondents at once. Below are the distributions of hours worked by educational attainment and relevant summary statistics that will be helpful in carrying out this analysis.

	Less than HS	HS	Jr Coll	Bachelor's	Graduate	Total
			Educational attainment			
Mean	38.67	39.6	41.39	42.55	40.85	40.45
SD	15.81	14.97	18.1	13.62	15.51	15.17
n	121	546	97	253	155	1,172

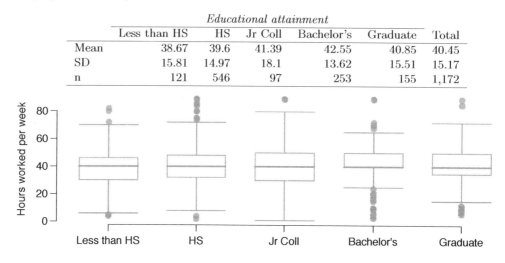

(a) Write hypotheses for evaluating whether the average number of hours worked varies across the five groups.

(b) Check conditions and describe any assumptions you must make to proceed with the test.

(c) Below is part of the output associated with this test. Fill in the empty cells.

	Df	Sum Sq	Mean Sq	F value	Pr(>F)
degree			501.54		0.0682
Residuals		267,382			
Total					

(d) What is the conclusion of the test?

5.49 True / False: ANOVA, Part I. Determine if the following statements are true or false in ANOVA, and explain your reasoning for statements you identify as false.

(a) As the number of groups increases, the modified significance level for pairwise tests increases as well.

(b) As the total sample size increases, the degrees of freedom for the residuals increases as well.

(c) The constant variance condition can be somewhat relaxed when the sample sizes are relatively consistent across groups.

(d) The independence assumption can be relaxed when the total sample size is large.

---

[47]National Opinion Research Center, General Social Survey, 2010.

5.50 Child care hours. The China Health and Nutrition Survey aims to examine the effects of the health, nutrition, and family planning policies and programs implemented by national and local governments.[48] It, for example, collects information on number of hours Chinese parents spend taking care of their children under age 6. The side-by-side box plots below show the distribution of this variable by educational attainment of the parent. Also provided below is the ANOVA output for comparing average hours across educational attainment categories.

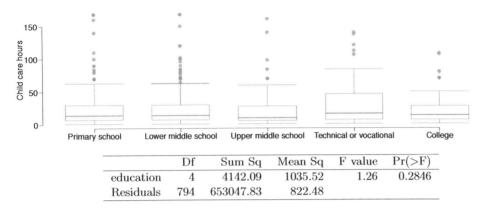

	Df	Sum Sq	Mean Sq	F value	Pr(>F)
education	4	4142.09	1035.52	1.26	0.2846
Residuals	794	653047.83	822.48		

(a) Write the hypotheses for testing for a difference between the average number of hours spent on child care across educational attainment levels.

(b) What is the conclusion of the hypothesis test?

5.51 Prison isolation experiment, Part II. Exercise 5.37 introduced an experiment that was conducted with the goal of identifying a treatment that reduces subjects' psychopathic deviant T scores, where this score measures a person's need for control or his rebellion against control. In Exercise 5.37 you evaluated the success of each treatment individually. An alternative analysis involves comparing the success of treatments. The relevant ANOVA output is given below.

	Df	Sum Sq	Mean Sq	F value	Pr(>F)
treatment	2	639.48	319.74	3.33	0.0461
Residuals	39	3740.43	95.91		

$s_{pooled} = 9.793$ on $df = 39$

(a) What are the hypotheses?

(b) What is the conclusion of the test? Use a 5% significance level.

(c) If in part (b) you determined that the test is significant, conduct pairwise tests to determine which groups are different from each other. If you did not reject the null hypothesis in part (b), recheck your answer.

5.52 True / False: ANOVA, Part II. Determine if the following statements are true or false, and explain your reasoning for statements you identify as false.

If the null hypothesis that the means of four groups are all the same is rejected using ANOVA at a 5% significance level, then ...

(a) we can then conclude that all the means are different from one another.

(b) the standardized variability between groups is higher than the standardized variability within groups.

(c) the pairwise analysis will identify at least one pair of means that are significantly different.

(d) the appropriate α to be used in pairwise comparisons is 0.05 / 4 = 0.0125 since there are four groups.

---

[48]UNC Carolina Population Center, China Health and Nutrition Survey, 2006.

Chapter 6

Inference for categorical data

Chapter 6 introduces inference in the setting of categorical data. We use these methods to answer questions like the following:

- What proportion of the American public approves of the job the Supreme Court is doing?

- The Pew Research Center conducted a poll about support for the 2010 health care law, and they used two forms of the survey question. Each respondent was randomly given one of the two questions. What is the difference in the support for respondents under the two question orderings?

The methods we learned in previous chapters will continue to be useful in these settings. For example, sample proportions are well characterized by a nearly normal distribution when certain conditions are satisfied, making it possible to employ the usual confidence interval and hypothesis testing tools. In other instances, such as those with contingency tables or when sample size conditions are not met, we will use a different distribution, though the core ideas remain the same.

6.1 Inference for a single proportion

In New York City on October 23rd, 2014, a doctor who had recently been treating Ebola patients in Guinea went to the hospital with a slight fever and was subsequently diagnosed with Ebola. Soon thereafter, an NBC 4 New York/The Wall Street Journal/Marist Poll found that 82% of New Yorkers favored a "mandatory 21-day quarantine for anyone who has come in contact with an Ebola patient".[1] This poll included responses of 1,042 New York adults between October 26th and 28th, 2014.

---

[1] Poll ID NY141026 on maristpoll.marist.edu.

6.1.1 Identifying when the sample proportion is nearly normal

A sample proportion can be described as a sample mean. If we represent each "success" as a 1 and each "failure" as a 0, then the sample proportion is the mean of these numerical outcomes:

$$\hat{p} = \frac{0 + 1 + 1 + \cdots + 0}{1042} = 0.82$$

The distribution of $\hat{p}$ is nearly normal when the distribution of 0's and 1's is not too strongly skewed for the sample size. The most common guideline for sample size and skew when working with proportions is to ensure that we expect to observe a minimum number of successes (1's) and failures (0's), typically at least 10 of each. The labels **success** and **failure** need not mean something positive or negative. These terms are just convenient words that are frequently used when discussing proportions.

Conditions for the sampling distribution of $\hat{p}$ being nearly normal

The sampling distribution for $\hat{p}$, taken from a sample of size n from a population with a true proportion p, is nearly normal when

1. the sample observations are independent and

2. we expected to see at least 10 successes and 10 failures in our sample, i.e. $np \geq 10$ and $n(1 - p) \geq 10$. This is called the **success-failure condition**.

If these conditions are met, then the sampling distribution of $\hat{p}$ is nearly normal with mean p and standard error

$$SE_{\hat{p}} = \sqrt{\frac{p(1 - p)}{n}} \tag{6.1}$$

$\hat{p}$
sample
proportion

p
population
proportion

Typically we don't know the true proportion, p, so we substitute some value to check conditions and to estimate the standard error. For confidence intervals, usually the sample proportion $\hat{p}$ is used to check the success-failure condition and compute the standard error. For hypothesis tests, typically the null value – that is, the proportion claimed in the null hypothesis – is used in place of p. Examples are presented for each of these cases in Sections 6.1.2 and 6.1.3.

TIP: Reminder on checking independence of observations
If data come from a simple random sample and consist of less than 10% of the population, then the independence assumption is reasonable. Alternatively, if the data come from a random process, we must evaluate the independence condition more carefully.

6.1.2   Confidence intervals for a proportion

We may want a confidence interval for the proportion of New York adults who favored a mandatory quarantine of anyone who had been in contact with an Ebola patient. Our point estimate, based on a sample of size $n = 1042$, is $\hat{p} = 0.82$. We would like to use the general confidence interval formula from Section 4.5. However, first we must verify that the sampling distribution of $\hat{p}$ is nearly normal and calculate the standard error of $\hat{p}$.

Observations are independent. The poll is based on a simple random sample and consists of fewer than 10% of the New York adult population, which verifies independence.

Success-failure condition. The sample size must also be sufficiently large, which is checked using the success-failure condition. There were $1042 \times \hat{p} \approx 854$ "successes" and $1042 \times (1 - \hat{p}) \approx 188$ "failures" in the sample, both easily greater than 10.

With the conditions met, we are assured that the sampling distribution of $\hat{p}$ is nearly normal. Next, a standard error for $\hat{p}$ is needed, and then we can employ the usual method to construct a confidence interval.

⊙ **Guided Practice 6.2**   Estimate the standard error of $\hat{p} = 0.82$ using Equation (6.1). Because p is unknown and the standard error is for a confidence interval, use $\hat{p}$ in place of p in the formula.[2]

● **Example 6.3**   Construct a 95% confidence interval for p, the proportion of New York adults who supported a quarantine for anyone who has come into contact with an Ebola patient.

Using the standard error $SE = 0.012$ from Guided Practice 6.2, the point estimate 0.82, and $z^\star = 1.96$ for a 95% confidence interval, the confidence interval is

$$\text{point estimate} \pm z^\star SE \quad \rightarrow \quad 0.82 \pm 1.96 \times 0.012 \quad \rightarrow \quad (0.796, 0.844)$$

We are 95% confident that the true proportion of New York adults in October 2014 who supported a quarantine for anyone who had come into contact with an Ebola patient was between 0.796 and 0.844.

Notice that since the poll was around the time where a doctor in New York had come down with Ebola, the results may not be as applicable today as they were at the time the poll was taken. This highlights an important detail about polls: they provide data about public opinion at a single point in time.

Constructing a confidence interval for a proportion

- Verify the observations are independent and also verify the success-failure condition using $\hat{p}$ and n.

- If the conditions are met, the sampling distribution of $\hat{p}$ may be well-approximated by the normal model.

- Construct the standard error using $\hat{p}$ in place of p and apply the general confidence interval formula.

---

[2] $SE = \sqrt{\frac{p(1-p)}{n}} \approx \sqrt{\frac{0.82(1-0.82)}{1042}} = 0.012$.

6.1.3 Hypothesis testing for a proportion

To apply the normal distribution framework in the context of a hypothesis test for a proportion, the independence and success-failure conditions must be satisfied. In a hypothesis test, the success-failure condition is checked using the null proportion: we verify np_0 and $n(1 - p_0)$ are at least 10, where p_0 is the null value.

⊙ **Guided Practice 6.4** Do a majority of American support nuclear arms reduction? Set up a one-sided hypothesis test to evaluate this question.[3]

● **Example 6.5** A simple random sample of 1,028 US adults in March 2013 found that 56% support nuclear arms reduction.[4] Does this provide convincing evidence that a majority of Americans supported nuclear arms reduction at the 5% significance level?

The poll was of a simple random sample that includes fewer than 10% of US adults, meaning the observations are independent. In a one-proportion hypothesis test, the success-failure condition is checked using the null proportion, which is $p_0 = 0.5$ in this context: $np_0 = n(1 - p_0) = 1028 \times 0.5 = 514 > 10$. With these conditions verified, the normal model may be applied to $\hat{p}$.

Next the standard error can be computed. The null value p_0 is used again here, because this is a hypothesis test for a single proportion.

$$SE = \sqrt{\frac{p_0(1 - p_0)}{n}} = \sqrt{\frac{0.5(1 - 0.5)}{1028}} = 0.016$$

A picture of the normal model is shown in Figure 6.1 with the p-value represented by the shaded region. Based on the normal model, the test statistic can be computed as the Z-score of the point estimate:

$$Z = \frac{\text{point estimate} - \text{null value}}{SE} = \frac{0.56 - 0.50}{0.016} = 3.75$$

The upper tail area, representing the p-value, is about 0.0001. Because the p-value is smaller than 0.05, we reject H_0. The poll provides convincing evidence that a majority of Americans supported nuclear arms reduction efforts in March 2013.

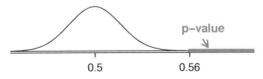

Figure 6.1: Sampling distribution for Example 6.5.

Hypothesis test for a proportion

Set up hypotheses and verify the conditions using the null value, p_0, to ensure $\hat{p}$ is nearly normal under H_0. If the conditions hold, construct the standard error, again using p_0, and show the p-value in a drawing. Lastly, compute the p-value and evaluate the hypotheses.

---

[3]$H_0 : p = 0.50$. $H_A : p > 0.50$.
[4]www.gallup.com/poll/161198/favor-russian-nuclear-arms-reductions.aspx

> **▶ Calculator videos**
> Videos covering confidence intervals and hypothesis tests for a single proportion using TI and Casio graphing calculators are available at openintro.org/videos.

6.1.4 Choosing a sample size when estimating a proportion

When collecting data, we choose a sample size suitable for the purpose of the study. Often times this means choosing a sample size large enough that the **margin of error** – which is the part we add and subtract from the point estimate in a confidence interval – is sufficiently small that the sample is useful. More explicitly, our task is to find a sample size n so that the sample proportion is within some margin of error m of the actual proportion with a certain level of confidence.

● **Example 6.6** A university newspaper is conducting a survey to determine what fraction of students support a \$200 per year increase in fees to pay for a new football stadium. How big of a sample is required to ensure the margin of error is smaller than 0.04 using a 95% confidence level?

The margin of error for a sample proportion is

$$z^{\star}\sqrt{\frac{p(1-p)}{n}}$$

Our goal is to find the smallest sample size n so that this margin of error is smaller than $m = 0.04$. For a 95% confidence level, the value $z^{\star}$ corresponds to 1.96:

$$1.96 \times \sqrt{\frac{p(1-p)}{n}} < 0.04$$

There are two unknowns in the equation: p and n. If we have an estimate of p, perhaps from a similar survey, we could enter in that value and solve for n. If we have no such estimate, we must use some other value for p. It turns out that the margin of error is largest when p is 0.5, so we typically use this *worst case value* if no estimate of the proportion is available:

$$1.96 \times \sqrt{\frac{0.5(1-0.5)}{n}} < 0.04$$
$$1.96^2 \times \frac{0.5(1-0.5)}{n} < 0.04^2$$
$$1.96^2 \times \frac{0.5(1-0.5)}{0.04^2} < n$$
$$600.25 < n$$

We would need over 600.25 participants, which means we need 601 participants or more, to ensure the sample proportion is within 0.04 of the true proportion with 95% confidence.

When an estimate of the proportion is available, we use it in place of the worst case proportion value, 0.5.

● **Example 6.7** A manager is about to oversee the mass production of a new tire model in her factory, and she would like to estimate what proportion of these tires will be rejected through quality control. The quality control team has monitored the last three tire models produced by the factory, failing 1.7% of tires in the first model, 6.2% of the second model, and 1.3% of the third model. The manager would like to examine enough tires to estimate the failure rate of the new tire model to within about 2% with a 90% confidence level.

(a) There are three different failure rates to choose from. Perform the sample size computation for each separately, and identify three sample sizes to consider.

(b) The sample sizes vary widely. Which of the three would you suggest using? What would influence your choice?

(a) For a 90% confidence interval, $z^\star = 1.65$, and since an estimate of the proportion 0.017 is available, we'll use it in the margin of error formula:

$$1.65 \times \sqrt{\frac{0.017(1 - 0.017)}{n}} < 0.02$$
$$113.7 < n$$

For sample size calculations, we always round up, so the first tire model suggests 114 tires would be sufficient.

A similar computation can be accomplished using 0.062 and 0.013 for p, and you should verify that using these proportions results in minimum sample sizes of 396 and 88 tires, respectively.

(b) We could examine which of the old models is most like the new model, then choose the corresponding sample size. Or if two of the previous estimates are based on small samples while the other is based on a larger sample, we should consider the value corresponding to the larger sample. There are also other reasonable approaches.

It should also be noted that the success-failure condition is not met with $n = 114$ or $n = 88$. That is, we would need additional methods than what we've covered so far to analyze results based on those sample sizes.

⊙ **Guided Practice 6.8** A recent estimate of Congress' approval rating was 19%.[5] What sample size does this estimate suggest we should use for a margin of error of 0.04 with 95% confidence?[6]

---

[5]www.gallup.com/poll/183128/five-months-gop-congress-approval-remains-low.aspx
[6]We complete the same computations as before, except now we use 0.19 instead of 0.5 for p:

$$1.96 \times \sqrt{\frac{p(1 - p)}{n}} \approx 1.96 \times \sqrt{\frac{0.19(1 - 0.19)}{n}} \leq 0.04 \quad \rightarrow \quad n \geq 369.5$$

A sample size of 370 or more would be reasonable. (Reminder: always round up for sample size calculations!)

6.2   Difference of two proportions

We would like to make conclusions about the difference in two population proportions: $p_1 - p_2$. We consider three examples. In the first, we compare the approval of the 2010 healthcare law under two different question phrasings. In the second application, we examine the efficacy of mammograms in reducing deaths from breast cancer. In the last example, a quadcopter company weighs whether to switch to a higher quality manufacturer of rotor blades.

In our investigations, we first identify a reasonable point estimate of $p_1 - p_2$ based on the sample. You may have already guessed its form: $\hat{p}_1 - \hat{p}_2$. Next, in each example we verify that the point estimate follows the normal model by checking certain conditions. Finally, we compute the estimate's standard error and apply our inferential framework.

6.2.1   Sample distribution of the difference of two proportions

We must check two conditions before applying the normal model to $\hat{p}_1 - \hat{p}_2$. First, the sampling distribution for each sample proportion must be nearly normal, and secondly, the samples must be independent. Under these two conditions, the sampling distribution of $\hat{p}_1 - \hat{p}_2$ may be well approximated using the normal model.

Conditions for the sampling distribution of $\hat{p}_1 - \hat{p}_2$ to be normal

The difference $\hat{p}_1 - \hat{p}_2$ tends to follow a normal model when

- each proportion separately follows a normal model, and
- the two samples are independent of each other.

The standard error of the difference in sample proportions is

$$SE_{\hat{p}_1 - \hat{p}_2} = \sqrt{SE_{\hat{p}_1}^2 + SE_{\hat{p}_2}^2} = \sqrt{\frac{p_1(1 - p_1)}{n_1} + \frac{p_2(1 - p_2)}{n_2}} \qquad (6.9)$$

where p_1 and p_2 represent the population proportions, and n_1 and n_2 represent the sample sizes.

For the difference in two means, the standard error formula took the following form:

$$SE_{\bar{x}_1 - \bar{x}_2} = \sqrt{SE_{\bar{x}_1}^2 + SE_{\bar{x}_2}^2}$$

The standard error for the difference in two proportions takes a similar form. The reasons behind this similarity are rooted in the probability theory of Section 2.4, which is described for this context in Guided Practice 5.28 on page 238.

6.2.2   Confidence intervals for $p_1 - p_2$

In the setting of confidence intervals for a difference of two proportions, the two sample proportions are used to verify the success-failure condition and also compute the standard error, just as was the case with a single proportion.

	Sample size (n_i)	Approve law (%)	Disapprove law (%)	Other
"people who cannot afford it will receive financial help from the government" is given second	771	47	49	3
"people who do not buy it will pay a penalty" is given second	732	34	63	3

Table 6.2: Results for a Pew Research Center poll where the ordering of two statements in a question regarding healthcare were randomized.

● **Example 6.10** The way a question is phrased can influence a person's response. For example, Pew Research Center conducted a survey with the following question:[7]

> As you may know, by 2014 nearly all Americans will be required to have health insurance. [People who do not buy insurance will pay a penalty] while [People who cannot afford it will receive financial help from the government]. Do you approve or disapprove of this policy?

For each randomly sampled respondent, the statements in brackets were randomized: either they were kept in the order given above, or the two statements were reversed. Table 6.2 shows the results of this experiment. Create and interpret a 90% confidence interval of the difference in approval.

First the conditions must be verified. Because each group is a simple random sample from less than 10% of the population, the observations are independent, both within the samples and between the samples. The success-failure condition also holds for each sample. Because all conditions are met, the normal model can be used for the point estimate of the difference in support, where p_1 corresponds to the original ordering and p_2 to the reversed ordering:

$$\hat{p}_1 - \hat{p}_2 = 0.47 - 0.34 = 0.13$$

The standard error may be computed from Equation (6.9) using the sample proportions:

$$SE \approx \sqrt{\frac{0.47(1 - 0.47)}{771} + \frac{0.34(1 - 0.34)}{732}} = 0.025$$

For a 90% confidence interval, we use $z^\star = 1.65$:

$$\text{point estimate} \pm z^\star SE \quad \to \quad 0.13 \pm 1.65 \times 0.025 \quad \to \quad (0.09, 0.17)$$

We are 90% confident that the approval rating for the 2010 healthcare law changes between 9% and 17% due to the ordering of the two statements in the survey question. The Pew Research Center reported that this modestly large difference suggests that the opinions of much of the public are still fluid on the health insurance mandate.

---

[7]www.people-press.org/2012/03/26/public-remains-split-on-health-care-bill-opposed-to-mandate. Sample sizes for each polling group are approximate.

6.2.3 Hypothesis tests for $p_1 - p_2$

A mammogram is an X-ray procedure used to check for breast cancer. Whether mammograms should be used is part of a controversial discussion, and it's the topic of our next example where we examine 2-proportion hypothesis test when H_0 is $p_1 - p_2 = 0$ (or equivalently, $p_1 = p_2$).

A 30-year study was conducted with nearly 90,000 female participants.[8] During a 5-year screening period, each woman was randomized to one of two groups: in the first group, women received regular mammograms to screen for breast cancer, and in the second group, women received regular non-mammogram breast cancer exams. No intervention was made during the following 25 years of the study, and we'll consider death resulting from breast cancer over the full 30-year period. Results from the study are summarized in Table 6.3.

If mammograms are much more effective than non-mammogram breast cancer exams, then we would expect to see additional deaths from breast cancer in the control group. On the other hand, if mammograms are not as effective as regular breast cancer exams, we would expect to see an increase in breast cancer deaths in the mammogram group.

	Death from breast cancer?	
	Yes	No
Mammogram	500	44,425
Control	505	44,405

Table 6.3: Summary results for breast cancer study.

⊙ **Guided Practice 6.11**   Is this study an experiment or an observational study?[9]

⊙ **Guided Practice 6.12**   Set up hypotheses to test whether there was a difference in breast cancer deaths in the mammogram and control groups.[10]

In Example 6.13, we will check the conditions for using the normal model to analyze the results of the study. The details are very similar to that of confidence intervals. However, this time we use a special proportion called the **pooled proportion** to check the success-failure condition:

$$\hat{p} = \frac{\text{\# of patients who died from breast cancer in the entire study}}{\text{\# of patients in the entire study}}$$
$$= \frac{500 + 505}{500 + 44{,}425 + 505 + 44{,}405}$$
$$= 0.0112$$

This proportion is an estimate of the breast cancer death rate across the entire study, and it's our best estimate of the proportions p_{mgm} and p_{ctrl} *if the null hypothesis is true that $p_{mgm} = p_{ctrl}$.* We will also use this pooled proportion when computing the standard error.

---

[8]Miller AB. 2014. *Twenty five year follow-up for breast cancer incidence and mortality of the Canadian National Breast Screening Study: randomised screening trial.* BMJ 2014;348:g366.

[9]This is an experiment. Patients were randomized to receive mammograms or a standard breast cancer exam. We will be able to make causal conclusions based on this study.

[10]H_0: the breast cancer death rate for patients screened using mammograms is the same as the breast cancer death rate for patients in the control, $p_{mgm} - p_{ctrl} = 0$.
H_A: the breast cancer death rate for patients screened using mammograms is different than the breast cancer death rate for patients in the control, $p_{mgm} - p_{ctrl} \neq 0$.

● **Example 6.13** Can we use the normal model to analyze this study?

Because the patients are randomized, they can be treated as independent.

We also must check the success-failure condition for each group. Under the null hypothesis, the proportions p_{mgm} and p_{ctrl} are equal, so we check the success-failure condition with our best estimate of these values under H_0, the pooled proportion from the two samples, $\hat{p} = 0.0112$:

$$\hat{p} \times n_{mgm} = 0.0112 \times 44{,}925 = 503 \quad (1 - \hat{p}) \times n_{mgm} = 0.9888 \times 44{,}925 = 44{,}422$$
$$\hat{p} \times n_{ctrl} = 0.0112 \times 44{,}910 = 503 \quad (1 - \hat{p}) \times n_{ctrl} = 0.9888 \times 44{,}910 = 44{,}407$$

The success-failure condition is satisfied since all values are at least 10, and we can safely apply the normal model.

Use the pooled proportion estimate when H_0 is $p_1 - p_2 = 0$

When the null hypothesis is that the proportions are equal, use the pooled proportion ($\hat{p}$) to verify the success-failure condition and estimate the standard error:

$$\hat{p} = \frac{\text{number of "successes"}}{\text{number of cases}} = \frac{\hat{p}_1 n_1 + \hat{p}_2 n_2}{n_1 + n_2}$$

Here $\hat{p}_1 n_1$ represents the number of successes in sample 1 since

$$\hat{p}_1 = \frac{\text{number of successes in sample 1}}{n_1}$$

Similarly, $\hat{p}_2 n_2$ represents the number of successes in sample 2.

In Example 6.13, the pooled proportion was used to check the success-failure condition. In the next example, we see the second place where the pooled proportion comes into play: the standard error calculation.

● **Example 6.14** Compute the point estimate of the difference in breast cancer death rates in the two groups, and use the pooled proportion $\hat{p} = 0.0112$ to calculate the standard error.

The point estimate of the difference in breast cancer death rates is

$$\hat{p}_{mgm} - \hat{p}_{ctrl} = \frac{500}{500 + 44{,}425} - \frac{505}{505 + 44{,}405}$$
$$= 0.01113 - 0.01125$$
$$= -0.00012$$

The breast cancer death rate in the mammogram group was 0.012% less than in the control group. Next, the standard error is calculated *using the pooled proportion, $\hat{p}$*:

$$SE = \sqrt{\frac{\hat{p}(1 - \hat{p})}{n_{mgm}} + \frac{\hat{p}(1 - \hat{p})}{n_{ctrl}}} = 0.00070$$

● **Example 6.15**   Using the point estimate $\hat{p}_{mgm} - \hat{p}_{ctrl} = -0.00012$ and standard error $SE = 0.00070$, calculate a p-value for the hypothesis test and write a conclusion.

Just like in past tests, we first compute a test statistic and draw a picture:

$$Z = \frac{\text{point estimate} - \text{null value}}{SE} = \frac{-0.00012 - 0}{0.00070} = -0.17$$

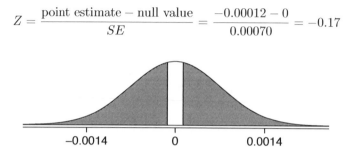

The lower tail area is 0.4325, which we double to get the p-value: 0.8650. Because this p-value is larger than 0.05, we do not reject the null hypothesis. That is, the difference in breast cancer death rates is reasonably explained by chance, and we do not observe benefits or harm from mammograms relative to a regular breast exam.

Can we conclude that mammograms have no benefits or harm? Here are a few important considerations to keep in mind when reviewing the mammogram study as well as any other medical study:

- If mammograms are helpful or harmful, the data suggest the effect isn't very large. So while we do not accept the null hypothesis, we also don't have sufficient evidence to conclude that mammograms reduce or increase breast cancer deaths.

- Are mammograms more or less expensive than a non-mammogram breast exam? If one option is much more expensive than the other and doesn't offer clear benefits, then we should lean towards the less expensive option.

- The study's authors also found that mammograms led to overdiagnosis of breast cancer, which means some breast cancers were found (or thought to be found) but that these cancers would not cause symptoms during patients' lifetimes. That is, something else would kill the patient before breast cancer symptoms appeared. This means some patients may have been treated for breast cancer unnecessarily, and this treatment is another cost to consider. It is also important to recognize that overdiagnosis can cause unnecessary physical or emotional harm to patients.

These considerations highlight the complexity around medical care and treatment recommendations. Experts and medical boards who study medical treatments use considerations like those above to provide their best recommendation based on the current evidence.

🎥 **Calculator videos**

Videos covering confidence intervals and hypothesis tests for the difference of two proportion using TI and Casio graphing calculators are available at openintro.org/videos.

Figure 6.4: A Phantom quadcopter.

6.2.4 More on 2-proportion hypothesis tests (special topic)

When we conduct a 2-proportion hypothesis test, usually H_0 is $p_1 - p_2 = 0$. However, there are rare situations where we want to check for some difference in p_1 and p_2 that is some value other than 0. For example, maybe we care about checking a null hypothesis where $p_1 - p_2 = 0.1$.[11] In contexts like these, we generally use $\hat{p}_1$ and $\hat{p}_2$ to check the success-failure condition and construct the standard error.

⊙ **Guided Practice 6.16** A quadcopter company is considering a new manufacturer for rotor blades. The new manufacturer would be more expensive but their higher-quality blades are more reliable, resulting in happier customers and fewer warranty claims. However, management must be convinced that the more expensive blades are worth the conversion before they approve the switch. If there is strong evidence of a more than 3% improvement in the percent of blades that pass inspection, management says they will switch suppliers, otherwise they will maintain the current supplier. Set up appropriate hypotheses for the test.[12]

● **Example 6.17** The quality control engineer from Guided Practice 6.16 collects a sample of blades, examining 1000 blades from each company and finds that 899 blades pass inspection from the current supplier and 958 pass inspection from the prospective supplier. Using these data, evaluate the hypothesis setup of Guided Practice 6.16 with a significance level of 5%.

First, we check the conditions. The sample is not necessarily random, so to proceed we must assume the blades are all independent; for this sample we will suppose this assumption is reasonable, but the engineer would be more knowledgeable as to whether this assumption is appropriate. The success-failure condition also holds for

---

[11]We can also encounter a similar situation with a difference of two means, though no such example was given in Chapter 5 since the methods remain exactly the same in the context of sample means. On the other hand, the success-failure condition and the calculation of the standard error vary slightly in different proportion contexts.

[12]H_0: The higher-quality blades will pass inspection just 3% more frequently than the standard-quality blades. $p_{highQ} - p_{standard} = 0.03$. H_A: The higher-quality blades will pass inspection >3% more often than the standard-quality blades. $p_{highQ} - p_{standard} > 0.03$.

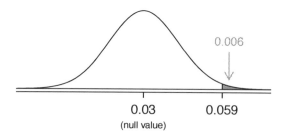

Figure 6.5: Distribution of the test statistic if the null hypothesis was true. The p-value is represented by the shaded area.

each sample. Thus, the difference in sample proportions, $0.958 - 0.899 = 0.059$, can be said to come from a nearly normal distribution.

The standard error is computed using the two sample proportions since we do not use a pooled proportion for this context:

$$SE = \sqrt{\frac{0.958(1 - 0.958)}{1000} + \frac{0.899(1 - 0.899)}{1000}} = 0.0114$$

In this hypothesis test, because the null is that $p_1 - p_2 = 0.03$, the sample proportions were used for the standard error calculation rather than a pooled proportion.

Next, we compute the test statistic and use it to find the p-value, which is depicted in Figure 6.5.

$$Z = \frac{\text{point estimate} - \text{null value}}{SE} = \frac{0.059 - 0.03}{0.0114} = 2.54$$

Using the normal model for this test statistic, we identify the right tail area as 0.006. Since this is a one-sided test, this single tail area is also the p-value, and we reject the null hypothesis because 0.006 is less than 0.05. That is, we have statistically significant evidence that the higher-quality blades actually do pass inspection more than 3% as often as the currently used blades. Based on these results, management will approve the switch to the new supplier.

6.3 Testing for goodness of fit using chi-square (special topic)

In this section, we develop a method for assessing a null model when the data are binned. This technique is commonly used in two circumstances:

- Given a sample of cases that can be classified into several groups, determine if the sample is representative of the general population.

- Evaluate whether data resemble a particular distribution, such as a normal distribution or a geometric distribution.

Each of these scenarios can be addressed using the same statistical test: a chi-square test.

In the first case, we consider data from a random sample of 275 jurors in a small county. Jurors identified their racial group, as shown in Table 6.6, and we would like to determine if these jurors are racially representative of the population. If the jury is representative of the population, then the proportions in the sample should roughly reflect the population of eligible jurors, i.e. registered voters.

Race	White	Black	Hispanic	Other	Total
Representation in juries	205	26	25	19	275
Registered voters	0.72	0.07	0.12	0.09	1.00

Table 6.6: Representation by race in a city's juries and population.

While the proportions in the juries do not precisely represent the population proportions, it is unclear whether these data provide convincing evidence that the sample is not representative. If the jurors really were randomly sampled from the registered voters, we might expect small differences due to chance. However, unusually large differences may provide convincing evidence that the juries were not representative.

A second application, assessing the fit of a distribution, is presented at the end of this section. Daily stock returns from the S&P500 for the years 1990-2011 are used to assess whether stock activity each day is independent of the stock's behavior on previous days.

In these problems, we would like to examine all bins simultaneously, not simply compare one or two bins at a time, which will require us to develop a new test statistic.

6.3.1 Creating a test statistic for one-way tables

● **Example 6.18** Of the people in the city, 275 served on a jury. If the individuals are randomly selected to serve on a jury, about how many of the 275 people would we expect to be white? How many would we expect to be black?

About 72% of the population is white, so we would expect about 72% of the jurors to be white: $0.72 \times 275 = 198$.

Similarly, we would expect about 7% of the jurors to be black, which would correspond to about $0.07 \times 275 = 19.25$ black jurors.

⊙ **Guided Practice 6.19** Twelve percent of the population is Hispanic and 9% represent other races. How many of the 275 jurors would we expect to be Hispanic or from another race? Answers can be found in Table 6.7.

Race	White	Black	Hispanic	Other	Total
Observed data	205	26	25	19	275
Expected counts	198	19.25	33	24.75	275

Table 6.7: Actual and expected make-up of the jurors.

The sample proportion represented from each race among the 275 jurors was not a precise match for any ethnic group. While some sampling variation is expected, we would expect the sample proportions to be fairly similar to the population proportions if there is no bias on juries. We need to test whether the differences are strong enough to provide convincing evidence that the jurors are not a random sample. These ideas can be organized into hypotheses:

H_0: The jurors are a random sample, i.e. there is no racial bias in who serves on a jury, and the observed counts reflect natural sampling fluctuation.

H_A: The jurors are not randomly sampled, i.e. there is racial bias in juror selection.

To evaluate these hypotheses, we quantify how different the observed counts are from the expected counts. Strong evidence for the alternative hypothesis would come in the form of unusually large deviations in the groups from what would be expected based on sampling variation alone.

6.3.2 The chi-square test statistic

In previous hypothesis tests, we constructed a test statistic of the following form:

$$\frac{\text{point estimate} - \text{null value}}{\text{SE of point estimate}}$$

This construction was based on (1) identifying the difference between a point estimate and an expected value if the null hypothesis was true, and (2) standardizing that difference using the standard error of the point estimate. These two ideas will help in the construction of an appropriate test statistic for count data.

Our strategy will be to first compute the difference between the observed counts and the counts we would expect if the null hypothesis was true, then we will standardize the difference:

$$Z_1 = \frac{\text{observed white count} - \text{null white count}}{\text{SE of observed white count}}$$

The standard error for the point estimate of the count in binned data is the square root of the count under the null.[13] Therefore:

$$Z_1 = \frac{205 - 198}{\sqrt{198}} = 0.50$$

The fraction is very similar to previous test statistics: first compute a difference, then standardize it. These computations should also be completed for the black, Hispanic, and other groups:

Black	*Hispanic*	*Other*
$Z_2 = \dfrac{26 - 19.25}{\sqrt{19.25}} = 1.54$	$Z_3 = \dfrac{25 - 33}{\sqrt{33}} = -1.39$	$Z_4 = \dfrac{19 - 24.75}{\sqrt{24.75}} = -1.16$

We would like to use a single test statistic to determine if these four standardized differences are irregularly far from zero. That is, Z_1, Z_2, Z_3, and Z_4 must be combined somehow to help determine if they – as a group – tend to be unusually far from zero. A first thought might be to take the absolute value of these four standardized differences and add them up:

$$|Z_1| + |Z_2| + |Z_3| + |Z_4| = 4.58$$

---

[13]Using some of the rules learned in earlier chapters, we might think that the standard error would be $np(1 - p)$, where n is the sample size and p is the proportion in the population. This would be correct if we were looking only at one count. However, we are computing many standardized differences and adding them together. It can be shown – though not here – that the square root of the count is a better way to standardize the count differences.

Indeed, this does give one number summarizing how far the actual counts are from what was expected. However, it is more common to add the squared values:

$$Z_1^2 + Z_2^2 + Z_3^2 + Z_4^2 = 5.89$$

Squaring each standardized difference before adding them together does two things:

- Any standardized difference that is squared will now be positive.

- Differences that already look unusual – e.g. a standardized difference of 2.5 – will become much larger after being squared.

The test statistic χ^2, which is the sum of the Z^2 values, is generally used for these reasons. We can also write an equation for χ^2 using the observed counts and null counts:

$$\chi^2 = \frac{(\text{observed count}_1 - \text{null count}_1)^2}{\text{null count}_1} + \cdots + \frac{(\text{observed count}_4 - \text{null count}_4)^2}{\text{null count}_4}$$

χ^2
chi-square
test statistic

The final number χ^2 summarizes how strongly the observed counts tend to deviate from the null counts. In Section 6.3.4, we will see that if the null hypothesis is true, then χ^2 follows a new distribution called a *chi-square distribution*. Using this distribution, we will be able to obtain a p-value to evaluate the hypotheses.

6.3.3   The chi-square distribution and finding areas

The **chi-square distribution** is sometimes used to characterize data sets and statistics that are always positive and typically right skewed. Recall the normal distribution had two parameters – mean and standard deviation – that could be used to describe its exact characteristics. The chi-square distribution has just one parameter called **degrees of freedom (df)**, which influences the shape, center, and spread of the distribution.

⊙ **Guided Practice 6.20**   Figure 6.8 shows three chi-square distributions. (a) How does the center of the distribution change when the degrees of freedom is larger? (b) What about the variability (spread)? (c) How does the shape change?[14]

Figure 6.8 and Guided Practice 6.20 demonstrate three general properties of chi-square distributions as the degrees of freedom increases: the distribution becomes more symmetric, the center moves to the right, and the variability inflates.

Our principal interest in the chi-square distribution is the calculation of p-values, which (as we have seen before) is related to finding the relevant area in the tail of a distribution. To do so, a new table is needed: the **chi-square table**, partially shown in Table 6.9. A more complete table is presented in Appendix B.3 on page 432. This table is very similar to the *t*-table: we examine a particular row for distributions with different degrees of freedom, and we identify a range for the area. One important difference from the *t*-table is that the chi-square table only provides upper tail values.

---

[14](a) The center becomes larger. If we look carefully, we can see that the center of each distribution is equal to the distribution's degrees of freedom. (b) The variability increases as the degrees of freedom increases. (c) The distribution is very strongly skewed for $df = 2$, and then the distributions become more symmetric for the larger degrees of freedom $df = 4$ and $df = 9$. We would see this trend continue if we examined distributions with even more larger degrees of freedom.

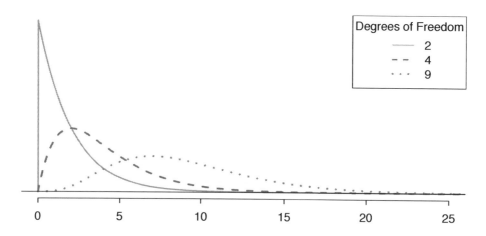

Figure 6.8: Three chi-square distributions with varying degrees of freedom.

Upper tail	0.3	0.2	0.1	0.05	0.02	0.01	0.005	0.001
df 2	2.41	**3.22**	**4.61**	5.99	7.82	9.21	10.60	13.82
3	*3.66*	*4.64*	*6.25*	*7.81*	*9.84*	*11.34*	*12.84*	*16.27*
4	4.88	5.99	7.78	9.49	11.67	13.28	14.86	18.47
5	6.06	7.29	9.24	11.07	13.39	15.09	16.75	20.52
6	7.23	8.56	10.64	12.59	15.03	16.81	18.55	22.46
7	8.38	9.80	12.02	14.07	16.62	18.48	20.28	24.32

Table 6.9: A section of the chi-square table. A complete table is in Appendix B.3 on page 432.

● **Example 6.21** Figure 6.10(a) shows a chi-square distribution with 3 degrees of freedom and an upper shaded tail starting at 6.25. Use Table 6.9 to estimate the shaded area.

This distribution has three degrees of freedom, so only the row with 3 degrees of freedom (df) is relevant. This row has been italicized in the table. Next, we see that the value – 6.25 – falls in the column with upper tail area 0.1. That is, the shaded upper tail of Figure 6.10(a) has area 0.1.

● **Example 6.22** We rarely observe the *exact* value in the table. For instance, Figure 6.10(b) shows the upper tail of a chi-square distribution with 2 degrees of freedom. The bound for this upper tail is at 4.3, which does not fall in Table 6.9. Find the approximate tail area.

The cutoff 4.3 falls between the second and third columns in the 2 degrees of freedom row. Because these columns correspond to tail areas of 0.2 and 0.1, we can be certain that the area shaded in Figure 6.10(b) is between 0.1 and 0.2.

● **Example 6.23** Figure 6.10(c) shows an upper tail for a chi-square distribution with 5 degrees of freedom and a cutoff of 5.1. Find the tail area.

Looking in the row with 5 df, 5.1 falls below the smallest cutoff for this row (6.06). That means we can only say that the area is *greater than 0.3*.

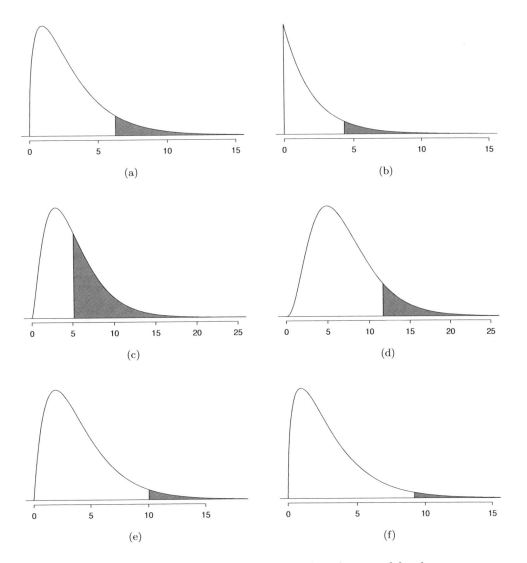

Figure 6.10: **(a)** Chi-square distribution with 3 degrees of freedom, area above 6.25 shaded. **(b)** 2 degrees of freedom, area above 4.3 shaded. **(c)** 5 degrees of freedom, area above 5.1 shaded. **(d)** 7 degrees of freedom, area above 11.7 shaded. **(e)** 4 degrees of freedom, area above 10 shaded. **(f)** 3 degrees of freedom, area above 9.21 shaded.

⊙ **Guided Practice 6.24**   Figure 6.10(d) shows a cutoff of 11.7 on a chi-square distribution with 7 degrees of freedom. Find the area of the upper tail.[15]

⊙ **Guided Practice 6.25**   Figure 6.10(e) shows a cutoff of 10 on a chi-square distribution with 4 degrees of freedom. Find the area of the upper tail.[16]

⊙ **Guided Practice 6.26**   Figure 6.10(f) shows a cutoff of 9.21 with a chi-square distribution with 3 df. Find the area of the upper tail.[17]

6.3.4   Finding a p-value for a chi-square distribution

In Section 6.3.2, we identified a new test statistic (χ^2) within the context of assessing whether there was evidence of racial bias in how jurors were sampled. The null hypothesis represented the claim that jurors were randomly sampled and there was no racial bias. The alternative hypothesis was that there was racial bias in how the jurors were sampled.

We determined that a large χ^2 value would suggest strong evidence favoring the alternative hypothesis: that there was racial bias. However, we could not quantify what the chance was of observing such a large test statistic $(\chi^2 = 5.89)$ if the null hypothesis actually was true. This is where the chi-square distribution becomes useful. If the null hypothesis was true and there was no racial bias, then χ^2 would follow a chi-square distribution, with three degrees of freedom in this case. Under certain conditions, the statistic χ^2 follows a chi-square distribution with $k - 1$ degrees of freedom, where k is the number of bins.

● **Example 6.27**   How many categories were there in the juror example? How many degrees of freedom should be associated with the chi-square distribution used for χ^2?

In the jurors example, there were $k = 4$ categories: white, black, Hispanic, and other. According to the rule above, the test statistic χ^2 should then follow a chi-square distribution with $k - 1 = 3$ degrees of freedom if H_0 is true.

Just like we checked sample size conditions to use the normal model in earlier sections, we must also check a sample size condition to safely apply the chi-square distribution for χ^2. Each expected count must be at least 5. In the juror example, the expected counts were 198, 19.25, 33, and 24.75, all easily above 5, so we can apply the chi-square model to the test statistic, $\chi^2 = 5.89$.

● **Example 6.28**   If the null hypothesis is true, the test statistic $\chi^2 = 5.89$ would be closely associated with a chi-square distribution with three degrees of freedom. Using this distribution and test statistic, identify the p-value.

The chi-square distribution and p-value are shown in Figure 6.11. Because larger chi-square values correspond to stronger evidence against the null hypothesis, we shade the upper tail to represent the p-value. Using the chi-square table in Appendix B.3 or the short table on page 290, we can determine that the area is between 0.1 and 0.2. That is, the p-value is larger than 0.1 but smaller than 0.2. Generally we do not reject the null hypothesis with such a large p-value. In other words, the data do not provide convincing evidence of racial bias in the juror selection.

---

[15]The value 11.7 falls between 9.80 and 12.02 in the 7 df row. Thus, the area is between 0.1 and 0.2.
[16]The area is between 0.02 and 0.05.
[17]Between 0.02 and 0.05.

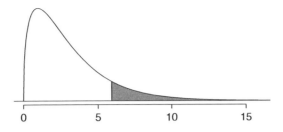

Figure 6.11: The p-value for the juror hypothesis test is shaded in the chi-square distribution with $df = 3$.

Chi-square test for one-way table

Suppose we are to evaluate whether there is convincing evidence that a set of observed counts O_1, O_2, ..., O_k in k categories are unusually different from what might be expected under a null hypothesis. Call the *expected counts* that are based on the null hypothesis E_1, E_2, ..., E_k. If each expected count is at least 5 and the null hypothesis is true, then the test statistic below follows a chi-square distribution with $k - 1$ degrees of freedom:

$$\chi^2 = \frac{(O_1 - E_1)^2}{E_1} + \frac{(O_2 - E_2)^2}{E_2} + \cdots + \frac{(O_k - E_k)^2}{E_k}$$

The p-value for this test statistic is found by looking at the upper tail of this chi-square distribution. We consider the upper tail because larger values of χ^2 would provide greater evidence against the null hypothesis.

TIP: Conditions for the chi-square test

There are two conditions that must be checked before performing a chi-square test:

Independence. Each case that contributes a count to the table must be independent of all the other cases in the table.

Sample size / distribution. Each particular scenario (i.e. cell count) must have at least 5 expected cases.

Failing to check conditions may affect the test's error rates.

When examining a table with just two bins, pick a single bin and use the one-proportion methods introduced in Section 6.1.

6.3.5 Evaluating goodness of fit for a distribution

Section 3.3 would be useful background reading for this example, but it is not a prerequisite.

We can apply our new chi-square testing framework to the second problem in this section: evaluating whether a certain statistical model fits a data set. Daily stock returns from the S&P500 for 1990-2011 can be used to assess whether stock activity each day is independent of the stock's behavior on previous days. This sounds like a very complex question, and it is, but a chi-square test can be used to study the problem. We will label each day as Up or Down (D) depending on whether the market was up or down that day. For example, consider the following changes in price, their new labels of up and down, and then the number of days that must be observed before each Up day:

Change in price	2.52	-1.46	0.51	-4.07	3.36	1.10	-5.46	-1.03	-2.99	1.71
Outcome	Up	D	Up	D	Up	Up	D	D	D	Up
Days to Up	1	-	2	-	2	1	-	-	-	4

If the days really are independent, then the number of days until a positive trading day should follow a geometric distribution. The geometric distribution describes the probability of waiting for the kth trial to observe the first success. Here each up day (Up) represents a success, and down (D) days represent failures. In the data above, it took only one day until the market was up, so the first wait time was 1 day. It took two more days before we observed our next Up trading day, and two more for the third Up day. We would like to determine if these counts (1, 2, 2, 1, 4, and so on) follow the geometric distribution. Table 6.12 shows the number of waiting days for a positive trading day during 1990-2011 for the S&P500.

Days	1	2	3	4	5	6	7+	Total
Observed	1532	760	338	194	74	33	17	2948

Table 6.12: Observed distribution of the waiting time until a positive trading day for the S&P500, 1990-2011.

We consider how many days one must wait until observing an Up day on the S&P500 stock index. If the stock activity was independent from one day to the next and the probability of a positive trading day was constant, then we would expect this waiting time to follow a *geometric distribution*. We can organize this into a hypothesis framework:

H_0: The stock market being up or down on a given day is independent from all other days. We will consider the number of days that pass until an Up day is observed. Under this hypothesis, the number of days until an Up day should follow a geometric distribution.

H_A: The stock market being up or down on a given day is not independent from all other days. Since we know the number of days until an Up day would follow a geometric distribution under the null, we look for deviations from the geometric distribution, which would support the alternative hypothesis.

There are important implications in our result for stock traders: if information from past trading days is useful in telling what will happen today, that information may provide an advantage over other traders.

We consider data for the S&P500 from 1990 to 2011 and summarize the waiting times in Table 6.13 and Figure 6.14. The S&P500 was positive on 53.2% of those days.

Because applying the chi-square framework requires expected counts to be at least 5, we have *binned* together all the cases where the waiting time was at least 7 days to ensure each expected count is well above this minimum. The actual data, shown in the *Observed* row in Table 6.13, can be compared to the expected counts from the *Geometric Model* row. The method for computing expected counts is discussed in Table 6.13. In general, the expected counts are determined by (1) identifying the null proportion associated with each bin, then (2) multiplying each null proportion by the total count to obtain the expected counts. That is, this strategy identifies what proportion of the total count we would expect to be in each bin.

Days	1	2	3	4	5	6	7+	Total
Observed	1532	760	338	194	74	33	17	2948
Geometric Model	1569	734	343	161	75	35	31	2948

Table 6.13: Distribution of the waiting time until a positive trading day. The expected counts based on the geometric model are shown in the last row. To find each expected count, we identify the probability of waiting D days based on the geometric model ($P(D) = (1 - 0.532)^{D-1}(0.532)$) and multiply by the total number of streaks, 2948. For example, waiting for three days occurs under the geometric model about $0.468^2 \times 0.532 = 11.65\%$ of the time, which corresponds to $0.1165 \times 2948 = 343$ streaks.

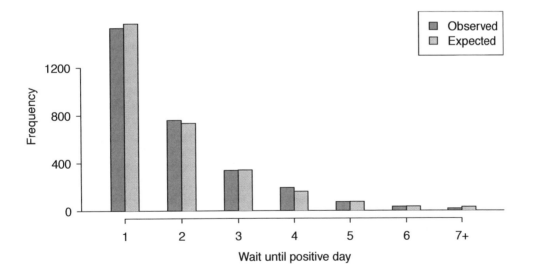

Figure 6.14: Side-by-side bar plot of the observed and expected counts for each waiting time.

● **Example 6.29** Do you notice any unusually large deviations in the graph? Can you tell if these deviations are due to chance just by looking?

It is not obvious whether differences in the observed counts and the expected counts from the geometric distribution are significantly different. That is, it is not clear whether these deviations might be due to chance or whether they are so strong that the data provide convincing evidence against the null hypothesis. However, we can perform a chi-square test using the counts in Table 6.13.

⊙ **Guided Practice 6.30** Table 6.13 provides a set of count data for waiting times ($O_1 = 1532$, $O_2 = 760$, ...) and expected counts under the geometric distribution ($E_1 = 1569$, $E_2 = 734$, ...). Compute the chi-square test statistic, χ^2.[18]

⊙ **Guided Practice 6.31** Because the expected counts are all at least 5, we can safely apply the chi-square distribution to χ^2. However, how many degrees of freedom should we use?[19]

● **Example 6.32** If the observed counts follow the geometric model, then the chi-square test statistic $\chi^2 = 15.08$ would closely follow a chi-square distribution with $df = 6$. Using this information, compute a p-value.

Figure 6.15 shows the chi-square distribution, cutoff, and the shaded p-value. If we look up the statistic $\chi^2 = 15.08$ in Appendix B.3, we find that the p-value is between 0.01 and 0.02. In other words, we have sufficient evidence to reject the notion that the wait times follow a geometric distribution, i.e. trading days are not independent and past days may help predict what the stock market will do today.

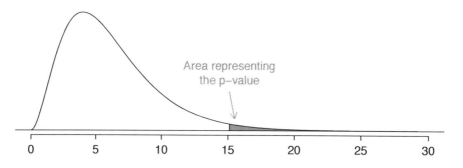

Figure 6.15: Chi-square distribution with 6 degrees of freedom. The p-value for the stock analysis is shaded.

● **Example 6.33** In Example 6.32, we rejected the null hypothesis that the trading days are independent. Why is this so important?

Because the data provided strong evidence that the geometric distribution is not appropriate, we reject the claim that trading days are independent. While it is not obvious how to exploit this information, it suggests there are some hidden patterns in the data that could be interesting and possibly useful to a stock trader.

[18]$\chi^2 = \frac{(1532-1569)^2}{1569} + \frac{(760-734)^2}{734} + \cdots + \frac{(17-31)^2}{31} = 15.08$
[19]There are $k = 7$ groups, so we use $df = k - 1 = 6$.

Calculator videos

Videos covering the chi-square goodness of fit test using TI and Casio graphing calculators are available at openintro.org/videos.

6.4 Testing for independence in two-way tables (special topic)

Google is constantly running experiments to test new search algorithms. For example, Google might test three algorithms using a sample of 10,000 google.com search queries. Table 6.16 shows an example of 10,000 queries split into three algorithm groups.[20] The group sizes were specified before the start of the experiment to be 5000 for the current algorithm and 2500 for each test algorithm.

Search algorithm	current	test 1	test 2	Total
Counts	5000	2500	2500	10000

Table 6.16: Google experiment breakdown of test subjects into three search groups.

Example 6.34 What is the ultimate goal of the Google experiment? What are the null and alternative hypotheses, in regular words?

The ultimate goal is to see whether there is a difference in the performance of the algorithms. The hypotheses can be described as the following:

H_0: The algorithms each perform equally well.

H_A: The algorithms do not perform equally well.

In this experiment, the explanatory variable is the search algorithm. However, an outcome variable is also needed. This outcome variable should somehow reflect whether the search results align with the user's interests. One possible way to quantify this is to determine whether (1) the user clicked one of the links provided and did not try a new search, or (2) the user performed a related search. Under scenario (1), we might think that the user was satisfied with the search results. Under scenario (2), the search results probably were not relevant, so the user tried a second search.

Table 6.17 provides the results from the experiment. These data are very similar to the count data in Section 6.3. However, now the different combinations of two variables are binned in a *two-way* table. In examining these data, we want to evaluate whether there is strong evidence that at least one algorithm is performing better than the others. To do so, we apply a chi-square test to this two-way table. The ideas of this test are similar to those ideas in the one-way table case. However, degrees of freedom and expected counts are computed a little differently than before.

---

[20]Google regularly runs experiments in this manner to help improve their search engine. It is entirely possible that if you perform a search and so does your friend, that you will have different search results. While the data presented in this section resemble what might be encountered in a real experiment, these data are simulated.

Search algorithm	current	test 1	test 2	Total
No new search	3511	1749	1818	7078
New search	1489	751	682	2922
Total	5000	2500	2500	10000

Table 6.17: Results of the Google search algorithm experiment.

What is so different about one-way tables and two-way tables?
A one-way table describes counts for each outcome in a single variable. A two-way table describes counts for *combinations* of outcomes for two variables. When we consider a two-way table, we often would like to know, are these variables related in any way? That is, are they dependent (versus independent)?

The hypothesis test for this Google experiment is really about assessing whether there is statistically significant evidence that the choice of the algorithm affects whether a user performs a second search. In other words, the goal is to check whether the search variable is independent of the algorithm variable.

6.4.1   Expected counts in two-way tables

● **Example 6.35**   From the experiment, we estimate the proportion of users who were satisfied with their initial search (no new search) as $7078/10000 = 0.7078$. If there really is no difference among the algorithms and 70.78% of people are satisfied with the search results, how many of the 5000 people in the "current algorithm" group would be expected to not perform a new search?

About 70.78% of the 5000 would be satisfied with the initial search:

$$0.7078 \times 5000 = 3539 \text{ users}$$

That is, if there was no difference between the three groups, then we would expect 3539 of the current algorithm users not to perform a new search.

⊙ **Guided Practice 6.36**   Using the same rationale described in Example 6.35, about how many users in each test group would not perform a new search if the algorithms were equally helpful?[21]

We can compute the expected number of users who would perform a new search for each group using the same strategy employed in Example 6.35 and Guided Practice 6.36. These expected counts were used to construct Table 6.18, which is the same as Table 6.17, except now the expected counts have been added in parentheses.

The examples and exercises above provided some help in computing expected counts. In general, expected counts for a two-way table may be computed using the row totals, column totals, and the table total. For instance, if there was no difference between the

---

[21]We would expect $0.7078 * 2500 = 1769.5$. It is okay that this is a fraction.

Search algorithm	current		test 1		test 2		Total
No new search	3511	**(3539)**	1749	**(1769.5)**	1818	**(1769.5)**	7078
New search	1489	**(1461)**	751	**(730.5)**	682	**(730.5)**	2922
Total	5000		2500		2500		10000

Table 6.18: The observed counts and the **(expected counts)**.

groups, then about 70.78% of each column should be in the first row:

$$0.7078 \times (\text{column 1 total}) = 3539$$
$$0.7078 \times (\text{column 2 total}) = 1769.5$$
$$0.7078 \times (\text{column 3 total}) = 1769.5$$

Looking back to how the fraction 0.7078 was computed – as the fraction of users who did not perform a new search (7078/10000) – these three expected counts could have been computed as

$$\left(\frac{\text{row 1 total}}{\text{table total}}\right)(\text{column 1 total}) = 3539$$

$$\left(\frac{\text{row 1 total}}{\text{table total}}\right)(\text{column 2 total}) = 1769.5$$

$$\left(\frac{\text{row 1 total}}{\text{table total}}\right)(\text{column 3 total}) = 1769.5$$

This leads us to a general formula for computing expected counts in a two-way table when we would like to test whether there is strong evidence of an association between the column variable and row variable.

Computing expected counts in a two-way table

To identify the expected count for the ith row and j^{th} column, compute

$$\text{Expected Count}_{\text{row } i, \text{ col } j} = \frac{(\text{row } i \text{ total}) \times (\text{column } j \text{ total})}{\text{table total}}$$

6.4.2 The chi-square test for two-way tables

The chi-square test statistic for a two-way table is found the same way it is found for a one-way table. For each table count, compute

General formula	$\dfrac{(\text{observed count} - \text{expected count})^2}{\text{expected count}}$
Row 1, Col 1	$\dfrac{(3511 - 3539)^2}{3539} = 0.222$
Row 1, Col 2	$\dfrac{(1749 - 1769.5)^2}{1769.5} = 0.237$
$\vdots$	$\vdots$
Row 2, Col 3	$\dfrac{(682 - 730.5)^2}{730.5} = 3.220$

Adding the computed value for each cell gives the chi-square test statistic χ^2:

$$\chi^2 = 0.222 + 0.237 + \cdots + 3.220 = 6.120$$

Just like before, this test statistic follows a chi-square distribution. However, the degrees of freedom are computed a little differently for a two-way table.[22] For two way tables, the degrees of freedom is equal to

$$df = (\text{number of rows minus } 1) \times (\text{number of columns minus } 1)$$

In our example, the degrees of freedom parameter is

$$df = (2 - 1) \times (3 - 1) = 2$$

If the null hypothesis is true (i.e. the algorithms are equally useful), then the test statistic $\chi^2 = 6.12$ closely follows a chi-square distribution with 2 degrees of freedom. Using this information, we can compute the p-value for the test, which is depicted in Figure 6.19.

Computing degrees of freedom for a two-way table

When applying the chi-square test to a two-way table, we use

$$df = (R - 1) \times (C - 1)$$

where R is the number of rows in the table and C is the number of columns.

TIP: Use two-proportion methods for 2-by-2 contingency tables

When analyzing 2-by-2 contingency tables, use the two-proportion methods introduced in Section 6.2.

---

[22]Recall: in the one-way table, the degrees of freedom was the number of cells minus 1.

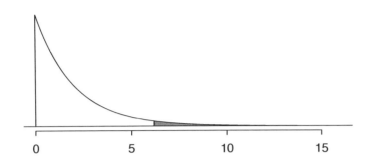

Figure 6.19: Computing the p-value for the Google hypothesis test.

	Obama	Congress Democrats	Republicans	Total
Approve	842	736	541	2119
Disapprove	616	646	842	2104
Total	1458	1382	1383	4223

Table 6.20: Pew Research poll results of a March 2012 poll.

● **Example 6.37** Compute the p-value and draw a conclusion about whether the search algorithms have different performances.

Looking in Appendix B.3 on page 432, we examine the row corresponding to 2 degrees of freedom. The test statistic, $\chi^2 = 6.120$, falls between the fourth and fifth columns, which means the p-value is between 0.02 and 0.05. Because we typically test at a significance level of $\alpha = 0.05$ and the p-value is less than 0.05, the null hypothesis is rejected. That is, the data provide convincing evidence that there is some difference in performance among the algorithms.

● **Example 6.38** Table 6.20 summarizes the results of a Pew Research poll.[23] We would like to determine if there are actually differences in the approval ratings of Barack Obama, Democrats in Congress, and Republicans in Congress. What are appropriate hypotheses for such a test?

H_0: There is no difference in approval ratings between the three groups.

H_A: There is some difference in approval ratings between the three groups, e.g. perhaps Obama's approval differs from Democrats in Congress.

⊙ **Guided Practice 6.39** A chi-square test for a two-way table may be used to test the hypotheses in Example 6.38. As a first step, compute the expected values for each of the six table cells.[24]

---

[23]See the Pew Research website: www.people-press.org/2012/03/14/romney-leads-gop-contest-trails-in-matchup-with-obama. The counts in Table 6.20 are approximate.

[24]The expected count for row one / column one is found by multiplying the row one total (2119) and column one total (1458), then dividing by the table total (4223): $\frac{2119 \times 1458}{4223} = 731.6$. Similarly for the first column and the second row: $\frac{2104 \times 1458}{4223} = 726.4$. Column 2: 693.5 and 688.5. Column 3: 694.0 and 689.0.

⊙ **Guided Practice 6.40**   Compute the chi-square test statistic.[25]

⊙ **Guided Practice 6.41**   Because there are 2 rows and 3 columns, the degrees of freedom for the test is $df = (2 - 1) \times (3 - 1) = 2$. Use $\chi^2 = 106.4$, $df = 2$, and the chi-square table on page 432 to evaluate whether to reject the null hypothesis.[26]

▶ Calculator videos

Videos covering the chi-square test for independence using TI and Casio graphing calculators are available at openintro.org/videos.

6.5   Small sample hypothesis testing for a proportion (special topic)

In this section we develop inferential methods for a single proportion that are appropriate when the sample size is too small to apply the normal model to $\hat{p}$. Just like the methods related to the t-distribution, these methods can also be applied to large samples.

6.5.1   When the success-failure condition is not met

People providing an organ for donation sometimes seek the help of a special "medical consultant". These consultants assist the patient in all aspects of the surgery, with the goal of reducing the possibility of complications during the medical procedure and recovery. Patients might choose a consultant based in part on the historical complication rate of the consultant's clients. One consultant tried to attract patients by noting the average complication rate for liver donor surgeries in the US is about 10%, but her clients have only had 3 complications in the 62 liver donor surgeries she has facilitated. She claims this is strong evidence that her work meaningfully contributes to reducing complications (and therefore she should be hired!).

⊙ **Guided Practice 6.42**   We will let p represent the true complication rate for liver donors working with this consultant. Estimate p using the data, and label this value $\hat{p}$.[27]

● **Example 6.43**   Is it possible to assess the consultant's claim with the data provided?

No. The claim is that there is a causal connection, but the data are observational. Patients who hire this medical consultant may have lower complication rates for other reasons.

While it is not possible to assess this causal claim, it is still possible to test for an association using these data. For this question we ask, could the low complication rate of $\hat{p} = 0.048$ be due to chance?

---

[25]For each cell, compute $\frac{(\text{obs}-\text{exp})^2}{exp}$. For instance, the first row and first column: $\frac{(842-731.6)^2}{731.6} = 16.7$. Adding the results of each cell gives the chi-square test statistic: $\chi^2 = 16.7 + \cdots + 34.0 = 106.4$.

[26]The test statistic is larger than the right-most column of the $df = 2$ row of the chi-square table, meaning the p-value is less than 0.001. That is, we reject the null hypothesis because the p-value is less than 0.05, and we conclude that Americans' approval has differences among Democrats in Congress, Republicans in Congress, and the president.

[27]The sample proportion: $\hat{p} = 3/62 = 0.048$

⊙ **Guided Practice 6.44** Write out hypotheses in both plain and statistical language to test for the association between the consultant's work and the true complication rate, p, for this consultant's clients.[28]

● **Example 6.45** In the examples based on large sample theory, we modeled $\hat{p}$ using the normal distribution. Why is this not appropriate here?

The independence assumption may be reasonable if each of the surgeries is from a different surgical team. However, the success-failure condition is not satisfied. Under the null hypothesis, we would anticipate seeing $62 \times 0.10 = 6.2$ complications, not the 10 required for the normal approximation.

The uncertainty associated with the sample proportion should not be modeled using the normal distribution. However, we would still like to assess the hypotheses from Guided Practice 6.44 in absence of the normal framework. To do so, we need to evaluate the possibility of a sample value ($\hat{p}$) this far below the null value, $p_0 = 0.10$. This possibility is usually measured with a p-value.

The p-value is computed based on the null distribution, which is the distribution of the test statistic if the null hypothesis is true. Supposing the null hypothesis is true, we can compute the p-value by identifying the chance of observing a test statistic that favors the alternative hypothesis at least as strongly as the observed test statistic. This can be done using simulation.

6.5.2 Generating the null distribution and p-value by simulation

We want to identify the sampling distribution of the test statistic ($\hat{p}$) if the null hypothesis was true. In other words, we want to see how the sample proportion changes due to chance alone. Then we plan to use this information to decide whether there is enough evidence to reject the null hypothesis.

Under the null hypothesis, 10% of liver donors have complications during or after surgery. Suppose this rate was really no different for the consultant's clients. If this was the case, we could *simulate* 62 clients to get a sample proportion for the complication rate from the null distribution.

Each client can be simulated using a deck of cards. Take one red card, nine black cards, and mix them up. Then drawing a card is one way of simulating the chance a patient has a complication *if the true complication rate is 10%* for the data. If we do this 62 times and compute the proportion of patients with complications in the simulation, $\hat{p}_{sim}$, then this sample proportion is exactly a sample from the null distribution.

An undergraduate student was paid $2 to complete this simulation. There were 5 simulated cases with a complication and 57 simulated cases without a complication, i.e. $\hat{p}_{sim} = 5/62 = 0.081$.

● **Example 6.46** Is this one simulation enough to determine whether or not we should reject the null hypothesis from Guided Practice 6.44? Explain.

No. To assess the hypotheses, we need to see a distribution of many $\hat{p}_{sim}$, not just a *single* draw from this sampling distribution.

---

[28]H_0: There is no association between the consultant's contributions and the clients' complication rate. In statistical language, $p = 0.10$. H_A: Patients who work with the consultant tend to have a complication rate lower than 10%, i.e. $p < 0.10$.

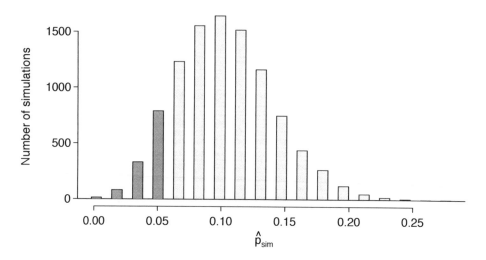

Figure 6.21: The null distribution for $\hat{p}$, created from 10,000 simulated studies. The left tail, representing the p-value for the hypothesis test, contains 12.22% of the simulations.

One simulation isn't enough to get a sense of the null distribution; many simulation studies are needed. Roughly 10,000 seems sufficient. However, paying someone to simulate 10,000 studies by hand is a waste of time and money. Instead, simulations are typically programmed into a computer, which is much more efficient.

Figure 6.21 shows the results of 10,000 simulated studies. The proportions that are equal to or less than $\hat{p} = 0.048$ are shaded. The shaded areas represent sample proportions under the null distribution that provide at least as much evidence as $\hat{p}$ favoring the alternative hypothesis. There were 1222 simulated sample proportions with $\hat{p}_{sim} \leq 0.048$. We use these to construct the null distribution's left-tail area and find the p-value:

$$\text{left tail} = \frac{\text{Number of observed simulations with } \hat{p}_{sim} \leq 0.048}{10000} \tag{6.47}$$

Of the 10,000 simulated $\hat{p}_{sim}$, 1222 were equal to or smaller than $\hat{p}$. Since the hypothesis test is one-sided, the estimated p-value is equal to this tail area: 0.1222.

⊙ **Guided Practice 6.48** Because the estimated p-value is 0.1222, which is larger than the significance level 0.05, we do not reject the null hypothesis. Explain what this means in plain language in the context of the problem.[29]

---

[29]There isn't sufficiently strong evidence to support an association between the consultant's work and fewer surgery complications.

⊙ **Guided Practice 6.49** Does the conclusion in Guided Practice 6.48 imply there is no real association between the surgical consultant's work and the risk of complications? Explain.[30]

One-sided hypothesis test for p with a small sample

The p-value is always derived by analyzing the null distribution of the test statistic. The normal model poorly approximates the null distribution for $\hat{p}$ when the success-failure condition is not satisfied. As a substitute, we can generate the null distribution using simulated sample proportions ($\hat{p}_{sim}$) and use this distribution to compute the tail area, i.e. the p-value.

We continue to use the same rule as before when computing the p-value for a two-sided test: double the single tail area, which remains a reasonable approach even when the sampling distribution is asymmetric. However, this can result in p-values larger than 1 when the point estimate is very near the mean in the null distribution; in such cases, we write that the p-value is 1. Also, very large p-values computed in this way (e.g. 0.85), may also be slightly inflated.

Guided Practice 6.48 said the p-value is *estimated*. It is not exact because the simulated null distribution itself is not exact, only a close approximation. However, we can generate an exact null distribution and p-value using the binomial model from Section 3.4.

6.5.3 Generating the exact null distribution and p-value

The number of successes in n independent cases can be described using the binomial model, which was introduced in Section 3.4. Recall that the probability of observing exactly k successes is given by

$$P(k \text{ successes}) = \binom{n}{k} p^k (1-p)^{n-k} = \frac{n!}{k!(n-k)!} p^k (1-p)^{n-k} \qquad (6.50)$$

where p is the true probability of success. The expression $\binom{n}{k}$ is read as n *choose* k, and the exclamation points represent factorials. For instance, 3! is equal to $3 \times 2 \times 1 = 6$, 4! is equal to $4 \times 3 \times 2 \times 1 = 24$, and so on (see Section 3.4).

The tail area of the null distribution is computed by adding up the probability in Equation (6.50) for each k that provides at least as strong of evidence favoring the alternative hypothesis as the data. If the hypothesis test is one-sided, then the p-value is represented by a single tail area. If the test is two-sided, compute the single tail area and double it to get the p-value, just as we have done in the past.

---

[30]No. It might be that the consultant's work is associated with a reduction but that there isn't enough data to convincingly show this connection.

● **Example 6.51** Compute the exact p-value to check the consultant's claim that her clients' complication rate is below 10%.

Exactly $k = 3$ complications were observed in the $n = 62$ cases cited by the consultant. Since we are testing against the 10% national average, our null hypothesis is $p = 0.10$. We can compute the p-value by adding up the cases where there are 3 or fewer complications:

$$
\begin{aligned}
\text{p-value} &= \sum_{j=0}^{3} \binom{n}{j} p^j (1-p)^{n-j} \\
&= \sum_{j=0}^{3} \binom{62}{j} 0.1^j (1-0.1)^{62-j} \\
&= \binom{62}{0} 0.1^0 (1-0.1)^{62-0} + \binom{62}{1} 0.1^1 (1-0.1)^{62-1} \\
&\quad + \binom{62}{2} 0.1^2 (1-0.1)^{62-2} + \binom{62}{3} 0.1^3 (1-0.1)^{62-3} \\
&= 0.0015 + 0.0100 + 0.0340 + 0.0755 \\
&= 0.1210
\end{aligned}
$$

This exact p-value is very close to the p-value based on the simulations (0.1222), and we come to the same conclusion. We do not reject the null hypothesis, and there is not statistically significant evidence to support the association.

If it were plotted, the exact null distribution would look almost identical to the simulated null distribution shown in Figure 6.21 on page 304.

6.5.4 Using simulation for goodness of fit tests

Simulation methods may also be used to test goodness of fit. In short, we simulate a new sample based on the purported bin probabilities, then compute a chi-square test statistic X_{sim}^2. We do this many times (e.g. 10,000 times), and then examine the distribution of these simulated chi-square test statistics. This distribution will be a very precise null distribution for the test statistic χ^2 if the probabilities are accurate, and we can find the upper tail of this null distribution, using a cutoff of the observed test statistic, to calculate the p-value.

● **Example 6.52** Section 6.3 introduced an example where we considered whether jurors were racially representative of the population. Would our findings differ if we used a simulation technique?

Since the minimum bin count condition was satisfied, the chi-square distribution is an excellent approximation of the null distribution, meaning the results should be very similar. Figure 6.22 shows the simulated null distribution using 100,000 simulated X_{sim}^2 values with an overlaid curve of the chi-square distribution. The distributions are almost identical, and the p-values are essentially indistinguishable: 0.115 for the simulated null distribution and 0.117 for the theoretical null distribution.

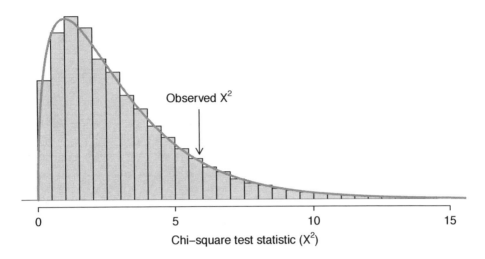

Figure 6.22: The precise null distribution for the juror example from Section 6.3 is shown as a histogram of simulated X^2_{sim} statistics, and the theoretical chi-square distribution is also shown.

6.6 Randomization test (special topic)

Cardiopulmonary resuscitation (CPR) is a procedure commonly used on individuals suffering a heart attack when other emergency resources are not available. This procedure is helpful in maintaining some blood circulation, but the chest compressions involved can also cause internal injuries. Internal bleeding and other injuries complicate additional treatment efforts following arrival at a hospital. For example, blood thinners may be used to release a clot responsible for a heart attack. However, the blood thinner would negatively affect internal bleeding.

We consider an experiment for patients who underwent CPR for a heart attack and were subsequently admitted to a hospital.[31] These patients were randomly divided into a treatment group where they received a blood thinner or the control group where they did not receive a blood thinner. The outcome variable of interest was whether the patients survived for at least 24 hours.

● **Example 6.53** What is an appropriate set of hypotheses for this study? Let p_c represent the true survival rate of people who do not receive a blood thinner (corresponding to the control group) and p_t represent the survival rate for people receiving a blood thinner (corresponding to the treatment group).

We are interested in whether the blood thinners are helpful or harmful, so a two-sided test is appropriate.

H_0: Blood thinners do not have an overall survival effect, i.e. the survival proportions are the same in each group. $p_t - p_c = 0$.

H_A: Blood thinners do have an impact on survival. $p_t - p_c \neq 0$.

[31]*Efficacy and safety of thrombolytic therapy after initially unsuccessful cardiopulmonary resuscitation: a prospective clinical trial*, by Böttiger et al., The Lancet, 2001.

6.6.1   Large sample framework for a difference in two proportions

There were 50 patients in the experiment who did not receive the blood thinner and 40 patients who did. The study results are shown in Table 6.23.

	Survived	Died	Total
Control	11	39	50
Treatment	14	26	40
Total	25	65	90

Table 6.23: Results for the CPR study. Patients in the treatment group were given a blood thinner, and patients in the control group were not.

⊙ **Guided Practice 6.54**   What is the observed survival rate in the control group? And in the treatment group? Also, provide a point estimate of the difference in survival proportions of the two groups: $\hat{p}_t - \hat{p}_c$.[32]

According to the point estimate, for patients who have undergone CPR outside of the hospital, an additional 13% survive when they are treated with blood thinners. However, this difference might be explainable by chance. We'd like to investigate this using a large sample framework, but we first need to check the conditions for such an approach.

● **Example 6.55**   Can the point estimate of the difference in survival proportions be adequately modeled using a normal distribution?

We will assume the patients are independent, which is probably reasonable. The success-failure condition is also satisfied. Since the proportions are equal under the null, we can compute the pooled proportion, $\hat{p} = (11 + 14)/(50 + 40) = 0.278$, for checking conditions. We find the expected number of successes (13.9, 11.1) and failures (36.1, 28.9) are above 10. The normal model is reasonable.

While we can apply a normal framework as an approximation to find a p-value, we might keep in mind that the expected number of successes is only 13.9 in one group and 11.1 in the other. Below we conduct an analysis relying on the large sample normal theory. We will follow up with a small sample analysis and compare the results.

● **Example 6.56**   Assess the hypotheses presented in Example 6.53 using a large sample framework. Use a significance level of $\alpha = 0.05$.

We suppose the null distribution of the sample difference follows a normal distribution with mean 0 (the null value) and a standard deviation equal to the standard error of the estimate. The null hypothesis in this case would be that the two proportions are the same, so we compute the standard error using the pooled proportion:

$$SE = \sqrt{\frac{p(1-p)}{n_t} + \frac{p(1-p)}{n_c}} \approx \sqrt{\frac{0.278(1 - 0.278)}{40} + \frac{0.278(1 - 0.278)}{50}} = 0.095$$

where we have used the pooled estimate $\left(\hat{p} = \frac{11+14}{50+40} = 0.278\right)$ in place of the true proportion, p.

---

[32]Observed control survival rate: $p_c = \frac{11}{50} = 0.22$. Treatment survival rate: $p_t = \frac{14}{40} = 0.35$. Observed difference: $\hat{p}_t - \hat{p}_c = 0.35 - 0.22 = 0.13$.

The null distribution with mean zero and standard deviation 0.095 is shown in Figure 6.24. We compute the tail areas to identify the p-value. To do so, we use the Z-score of the point estimate:

$$Z = \frac{(\hat{p}_t - \hat{p}_c) - \text{null value}}{SE} = \frac{0.13 - 0}{0.095} = 1.37$$

If we look this Z-score up in Appendix B.1, we see that the right tail has area 0.0853. The p-value is twice the single tail area: 0.176. This p-value does not provide convincing evidence that the blood thinner helps. Thus, there is insufficient evidence to conclude whether or not the blood thinner helps or hurts. (Remember, we never "accept" the null hypothesis – we can only reject or fail to reject.)

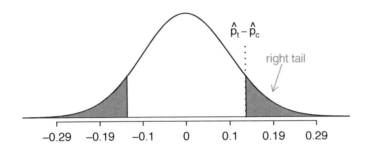

Figure 6.24: The null distribution of the point estimate $\hat{p}_t - \hat{p}_c$ under the large sample framework is a normal distribution with mean 0 and standard deviation equal to the standard error, in this case $SE = 0.095$. The p-value is represented by the shaded areas.

The p-value 0.176 relies on the normal approximation. We know that when the samples sizes are large, this approximation is quite good. However, when the sample sizes are relatively small as in this example, the approximation may only be adequate. Next we develop a simulation technique, apply it to these data, and compare our results. In general, the small sample method we develop may be used for any size sample, small or large, and should be considered as more accurate than the corresponding large sample technique.

6.6.2 Simulating a difference under the null distribution

The ideas in this section were first introduced in the optional Section 1.8 on page 50. For the interested reader, this earlier section provides a more in-depth discussion.

Suppose the null hypothesis is true. Then the blood thinner has no impact on survival and the 13% difference was due to chance. In this case, we can simulate *null* differences that are due to chance using a *randomization technique*.[33] By randomly assigning "fake treatment" and "fake control" stickers to the patients' files, we could get a new grouping – one that is completely due to chance. The expected difference between the two proportions under this simulation is zero.

We run this simulation by taking 40 `treatment_fake` and 50 `control_fake` labels and randomly assigning them to the patients. The label counts of 40 and 50 correspond to the number of treatment and control assignments in the actual study. We use a computer program to randomly assign these labels to the patients, and we organize the simulation results into Table 6.25.

---

[33]The test procedure we employ in this section is formally called a **permutation test**.

	Survived	Died	Total
control_fake	15	35	50
treatment_fake	10	30	40
Total	25	65	90

Table 6.25: Simulated results for the CPR study under the null hypothesis. The labels were randomly assigned and are independent of the outcome of the patient.

⊙ **Guided Practice 6.57**   What is the difference in survival rates between the two fake groups in Table 6.25? How does this compare to the observed 13% in the real groups?[34]

The difference computed in Guided Practice 6.57 represents a draw from the null distribution of the sample differences. Next we generate many more simulated experiments to build up the null distribution, much like we did in Section 6.5.2 to build a null distribution for a one sample proportion.

> **Caution: Simulation in the two proportion case requires that the null difference is zero**
>
> The technique described here to simulate a difference from the null distribution relies on an important condition in the null hypothesis: there is no connection between the two variables considered. In some special cases, the null difference might not be zero, and more advanced methods (or a large sample approximation, if appropriate) would be necessary.

6.6.3   Null distribution for the difference in two proportions

We build up an approximation to the null distribution by repeatedly creating tables like the one shown in Table 6.25 and computing the sample differences. The null distribution from 10,000 simulations is shown in Figure 6.26.

● **Example 6.58**   Compare Figures 6.24 and 6.26. How are they similar? How are they different?

The shapes are similar, but the simulated results show that the continuous approximation of the normal distribution is not very good. We might wonder, how close are the p-values?

⊙ **Guided Practice 6.59**   The right tail area is about 0.13. (It is only a coincidence that we also have $\hat{p}_t - \hat{p}_c = 0.13$.) The p-value is computed by doubling the right tail area: 0.26. How does this value compare with the large sample approximation for the p-value?[35]

---

[34]The difference is $\hat{p}_{t,fake} - \hat{p}_{c,fake} = \frac{10}{40} - \frac{15}{50} = -0.05$, which is closer to the null value $p_0 = 0$ than what we observed.

[35]The approximation in this case is fairly poor (p-values: 0.174 vs. 0.26), though we come to the same conclusion. The data do not provide convincing evidence showing the blood thinner helps or hurts patients.

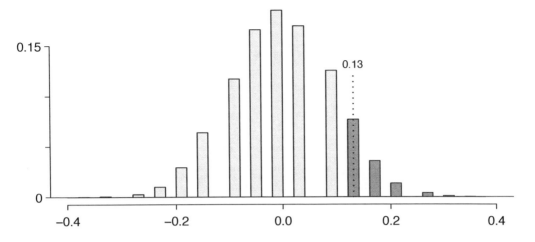

Figure 6.26: An approximation of the null distribution of the point estimate, $\hat{p}_t - \hat{p}_c$. The p-value is twice the right tail area.

In general, small sample methods produce more accurate results since they rely on fewer assumptions. However, they often require some extra work or simulations. For this reason, many statisticians use small sample methods only when conditions for large sample methods are not satisfied.

6.6.4 Randomization for two-way tables and chi-square

Randomization methods may also be used for the contingency tables. In short, we create a randomized contingency table, then compute a chi-square test statistic X^2_{sim}. We repeat this many times using a computer, and then we examine the distribution of these simulated test statistics. This randomization approach is valid for any sized sample, and it will be more accurate for cases where one or more expected bin counts do not meet the minimum threshold of 5. When the minimum threshold is met, the simulated null distribution will very closely resemble the chi-square distribution. As before, we use the upper tail of the null distribution to calculate the p-value.

6.7   Exercises

6.7.1   Inference for a single proportion

6.1   Vegetarian college students. Suppose that 8% of college students are vegetarians. Determine if the following statements are true or false, and explain your reasoning.

(a) The distribution of the sample proportions of vegetarians in random samples of size 60 is approximately normal since $n \geq 30$.

(b) The distribution of the sample proportions of vegetarian college students in random samples of size 50 is right skewed.

(c) A random sample of 125 college students where 12% are vegetarians would be considered unusual.

(d) A random sample of 250 college students where 12% are vegetarians would be considered unusual.

(e) The standard error would be reduced by one-half if we increased the sample size from 125 to 250.

6.2   Young Americans, Part I. About 77% of young adults think they can achieve the American dream. Determine if the following statements are true or false, and explain your reasoning.[36]

(a) The distribution of sample proportions of young Americans who think they can achieve the American dream in samples of size 20 is left skewed.

(b) The distribution of sample proportions of young Americans who think they can achieve the American dream in random samples of size 40 is approximately normal since $n \geq 30$.

(c) A random sample of 60 young Americans where 85% think they can achieve the American dream would be considered unusual.

(d) A random sample of 120 young Americans where 85% think they can achieve the American dream would be considered unusual.

6.3   Orange tabbies. Suppose that 90% of orange tabby cats are male. Determine if the following statements are true or false, and explain your reasoning.

(a) The distribution of sample proportions of random samples of size 30 is left skewed.

(b) Using a sample size that is 4 times as large will reduce the standard error of the sample proportion by one-half.

(c) The distribution of sample proportions of random samples of size 140 is approximately normal.

(d) The distribution of sample proportions of random samples of size 280 is approximately normal.

6.4   Young Americans, Part II. About 25% of young Americans have delayed starting a family due to the continued economic slump. Determine if the following statements are true or false, and explain your reasoning.[37]

(a) The distribution of sample proportions of young Americans who have delayed starting a family due to the continued economic slump in random samples of size 12 is right skewed.

(b) In order for the distribution of sample proportions of young Americans who have delayed starting a family due to the continued economic slump to be approximately normal, we need random samples where the sample size is at least 40.

(c) A random sample of 50 young Americans where 20% have delayed starting a family due to the continued economic slump would be considered unusual.

(d) A random sample of 150 young Americans where 20% have delayed starting a family due to the continued economic slump would be considered unusual.

(e) Tripling the sample size will reduce the standard error of the sample proportion by one-third.

[36]A. Vaughn. "Poll finds young adults optimistic, but not about money". In: *Los Angeles Times* (2011).
[37]Demos.org. "The State of Young America: The Poll". In: (2011).

6.5 Prop 19 in California. In a 2010 Survey USA poll, 70% of the 119 respondents between the ages of 18 and 34 said they would vote in the 2010 general election for Prop 19, which would change California law to legalize marijuana and allow it to be regulated and taxed. At a 95% confidence level, this sample has an 8% margin of error. Based on this information, determine if the following statements are true or false, and explain your reasoning.[38]

(a) We are 95% confident that between 62% and 78% of the California voters in this sample support Prop 19.

(b) We are 95% confident that between 62% and 78% of all California voters between the ages of 18 and 34 support Prop 19.

(c) If we considered many random samples of 119 California voters between the ages of 18 and 34, and we calculated 95% confidence intervals for each, 95% of them will include the true population proportion of 18-34 year old Californians who support Prop 19.

(d) In order to decrease the margin of error to 4%, we would need to quadruple (multiply by 4) the sample size.

(e) Based on this confidence interval, there is sufficient evidence to conclude that a majority of California voters between the ages of 18 and 34 support Prop 19.

6.6 2010 Healthcare Law. On June 28, 2012 the U.S. Supreme Court upheld the much debated 2010 healthcare law, declaring it constitutional. A Gallup poll released the day after this decision indicates that 46% of 1,012 Americans agree with this decision. At a 95% confidence level, this sample has a 3% margin of error. Based on this information, determine if the following statements are true or false, and explain your reasoning.[39]

(a) We are 95% confident that between 43% and 49% of Americans in this sample support the decision of the U.S. Supreme Court on the 2010 healthcare law.

(b) We are 95% confident that between 43% and 49% of Americans support the decision of the U.S. Supreme Court on the 2010 healthcare law.

(c) If we considered many random samples of 1,012 Americans, and we calculated the sample proportions of those who support the decision of the U.S. Supreme Court, 95% of those sample proportions will be between 43% and 49%.

(d) The margin of error at a 90% confidence level would be higher than 3%.

6.7 Fireworks on July 4th. In late June 2012, Survey USA published results of a survey stating that 56% of the 600 randomly sampled Kansas residents planned to set off fireworks on July 4th. Determine the margin of error for the 56% point estimate using a 95% confidence level.[40]

6.8 Elderly drivers. In January 2011, The Marist Poll published a report stating that 66% of adults nationally think licensed drivers should be required to retake their road test once they reach 65 years of age. It was also reported that interviews were conducted on 1,018 American adults, and that the margin of error was 3% using a 95% confidence level.[41]

(a) Verify the margin of error reported by The Marist Poll.

(b) Based on a 95% confidence interval, does the poll provide convincing evidence that *more than* 70% of the population think that licensed drivers should be required to retake their road test once they turn 65?

---

[38]Survey USA, Election Poll #16804, data collected July 8-11, 2010.
[39]Gallup, Americans Issue Split Decision on Healthcare Ruling, data collected June 28, 2012.
[40]Survey USA, News Poll #19333, data collected on June 27, 2012.
[41]Marist Poll, Road Rules: Re-Testing Drivers at Age 65?, March 4, 2011.

6.9 Life after college. We are interested in estimating the proportion of graduates at a mid-sized university who found a job within one year of completing their undergraduate degree. Suppose we conduct a survey and find out that 348 of the 400 randomly sampled graduates found jobs. The graduating class under consideration included over 4500 students.

(a) Describe the population parameter of interest. What is the value of the point estimate of this parameter?

(b) Check if the conditions for constructing a confidence interval based on these data are met.

(c) Calculate a 95% confidence interval for the proportion of graduates who found a job within one year of completing their undergraduate degree at this university, and interpret it in the context of the data.

(d) What does "95% confidence" mean?

(e) Now calculate a 99% confidence interval for the same parameter and interpret it in the context of the data.

(f) Compare the widths of the 95% and 99% confidence intervals. Which one is wider? Explain.

6.10 Life rating in Greece. Greece has faced a severe economic crisis since the end of 2009. A Gallup poll surveyed 1,000 randomly sampled Greeks in 2011 and found that 25% of them said they would rate their lives poorly enough to be considered "suffering".[42]

(a) Describe the population parameter of interest. What is the value of the point estimate of this parameter?

(b) Check if the conditions required for constructing a confidence interval based on these data are met.

(c) Construct a 95% confidence interval for the proportion of Greeks who are "suffering".

(d) Without doing any calculations, describe what would happen to the confidence interval if we decided to use a higher confidence level.

(e) Without doing any calculations, describe what would happen to the confidence interval if we used a larger sample.

6.11 Study abroad. A survey on 1,509 high school seniors who took the SAT and who completed an optional web survey between April 25 and April 30, 2007 shows that 55% of high school seniors are fairly certain that they will participate in a study abroad program in college.[43]

(a) Is this sample a representative sample from the population of all high school seniors in the US? Explain your reasoning.

(b) Let's suppose the conditions for inference are met. Even if your answer to part (a) indicated that this approach would not be reliable, this analysis may still be interesting to carry out (though not report). Construct a 90% confidence interval for the proportion of high school seniors (of those who took the SAT) who are fairly certain they will participate in a study abroad program in college, and interpret this interval in context.

(c) What does "90% confidence" mean?

(d) Based on this interval, would it be appropriate to claim that the majority of high school seniors are fairly certain that they will participate in a study abroad program in college?

---

[42]Gallup World, More Than One in 10 "Suffering" Worldwide, data collected throughout 2011.

[43]studentPOLL, College-Bound Students' Interests in Study Abroad and Other International Learning Activities, January 2008.

6.12 Legalization of marijuana, Part I. The 2010 General Social Survey asked 1,259 US residents: "Do you think the use of marijuana should be made legal, or not?" 48% of the respondents said it should be made legal.[44]

(a) Is 48% a sample statistic or a population parameter? Explain.

(b) Construct a 95% confidence interval for the proportion of US residents who think marijuana should be made legal, and interpret it in the context of the data.

(c) A critic points out that this 95% confidence interval is only accurate if the statistic follows a normal distribution, or if the normal model is a good approximation. Is this true for these data? Explain.

(d) A news piece on this survey's findings states, "Majority of Americans think marijuana should be legalized." Based on your confidence interval, is this news piece's statement justified?

6.13 Public option, Part I. A *Washington Post* article from 2009 reported that "support for a government-run health-care plan to compete with private insurers has rebounded from its summertime lows and wins clear majority support from the public." More specifically, the article says "seven in 10 Democrats back the plan, while almost nine in 10 Republicans oppose it. Independents divide 52 percent against, 42 percent in favor of the legislation." (6% responded with "other".) There were 819 Democrats, 566 Republicans and 783 Independents surveyed.[45]

(a) A political pundit on TV claims that a majority of Independents oppose the health care public option plan. Do these data provide strong evidence to support this statement?

(b) Would you expect a confidence interval for the proportion of Independents who oppose the public option plan to include 0.5? Explain.

6.14 The Civil War. A national survey conducted in 2011 among a simple random sample of 1,507 adults shows that 56% of Americans think the Civil War is still relevant to American politics and political life.[46]

(a) Conduct a hypothesis test to determine if these data provide strong evidence that the majority of the Americans think the Civil War is still relevant.

(b) Interpret the p-value in this context.

(c) Calculate a 90% confidence interval for the proportion of Americans who think the Civil War is still relevant. Interpret the interval in this context, and comment on whether or not the confidence interval agrees with the conclusion of the hypothesis test.

6.15 Browsing on the mobile device. A 2012 survey of 2,254 American adults indicates that 17% of cell phone owners do their browsing on their phone rather than a computer or other device.[47]

(a) According to an online article, a report from a mobile research company indicates that 38 percent of Chinese mobile web users only access the internet through their cell phones.[48] Conduct a hypothesis test to determine if these data provide strong evidence that the proportion of Americans who only use their cell phones to access the internet is different than the Chinese proportion of 38%.

(b) Interpret the p-value in this context.

(c) Calculate a 95% confidence interval for the proportion of Americans who access the internet on their cell phones, and interpret the interval in this context.

---

[44]National Opinion Research Center, General Social Survey, 2010.

[45]D. Balz and J. Cohen. "Most support public option for health insurance, poll finds". In: *The Washington Post* (2009).

[46]Pew Research Center Publications, Civil War at 150: Still Relevant, Still Divisive, data collected between March 30 - April 3, 2011.

[47]Pew Internet, Cell Internet Use 2012, data collected between March 15 - April 13, 2012.

[48]S. Chang. "The Chinese Love to Use Feature Phone to Access the Internet". In: *M.I.C Gadget* (2012).

6.16 Is college worth it? Part I. Among a simple random sample of 331 American adults who do not have a four-year college degree and are not currently enrolled in school, 48% said they decided not to go to college because they could not afford school.[49]

(a) A newspaper article states that only a minority of the Americans who decide not to go to college do so because they cannot afford it and uses the point estimate from this survey as evidence. Conduct a hypothesis test to determine if these data provide strong evidence supporting this statement.

(b) Would you expect a confidence interval for the proportion of American adults who decide not to go to college because they cannot afford it to include 0.5? Explain.

6.17 Taste test. Some people claim that they can tell the difference between a diet soda and a regular soda in the first sip. A researcher wanting to test this claim randomly sampled 80 such people. He then filled 80 plain white cups with soda, half diet and half regular through random assignment, and asked each person to take one sip from their cup and identify the soda as diet or regular. 53 participants correctly identified the soda.

(a) Do these data provide strong evidence that these people are able to detect the difference between diet and regular soda, in other words, are the results significantly better than just random guessing?

(b) Interpret the p-value in this context.

6.18 Is college worth it? Part II. Exercise 6.16 presents the results of a poll where 48% of 331 Americans who decide to not go to college do so because they cannot afford it.

(a) Calculate a 90% confidence interval for the proportion of Americans who decide to not go to college because they cannot afford it, and interpret the interval in context.

(b) Suppose we wanted the margin of error for the 90% confidence level to be about 1.5%. How large of a survey would you recommend?

6.19 College smokers. We are interested in estimating the proportion of students at a university who smoke. Out of a random sample of 200 students from this university, 40 students smoke.

(a) Calculate a 95% confidence interval for the proportion of students at this university who smoke, and interpret this interval in context. (Reminder: Check conditions.)

(b) If we wanted the margin of error to be no larger than 2% at a 95% confidence level for the proportion of students who smoke, how big of a sample would we need?

6.20 Legalize Marijuana, Part II. As discussed in Exercise 6.12, the 2010 General Social Survey reported a sample where about 48% of US residents thought marijuana should be made legal. If we wanted to limit the margin of error of a 95% confidence interval to 2%, about how many Americans would we need to survey ?

6.21 Public option, Part II. Exercise 6.13 presents the results of a poll evaluating support for the health care public option in 2009, reporting that 52% of Independents in the sample opposed the public option. If we wanted to estimate this number to within 1% with 90% confidence, what would be an appropriate sample size?

---

[49]Pew Research Center Publications, Is College Worth It?, data collected between March 15-29, 2011.

6.22 Acetaminophen and liver damage. It is believed that large doses of acetaminophen (the active ingredient in over the counter pain relievers like Tylenol) may cause damage to the liver. A researcher wants to conduct a study to estimate the proportion of acetaminophen users who have liver damage. For participating in this study, he will pay each subject $20 and provide a free medical consultation if the patient has liver damage.

(a) If he wants to limit the margin of error of his 98% confidence interval to 2%, what is the minimum amount of money he needs to set aside to pay his subjects?

(b) The amount you calculated in part (a) is substantially over his budget so he decides to use fewer subjects. How will this affect the width of his confidence interval?

6.7.2 Difference of two proportions

6.23 Social experiment, Part I. A "social experiment" conducted by a TV program questioned what people do when they see a very obviously bruised woman getting picked on by her boyfriend. On two different occasions at the same restaurant, the same couple was depicted. In one scenario the woman was dressed "provocatively" and in the other scenario the woman was dressed "conservatively". The table below shows how many restaurant diners were present under each scenario, and whether or not they intervened.

		Scenario		
		Provocative	Conservative	Total
Intervene	Yes	5	15	20
	No	15	10	25
	Total	20	25	45

Explain why the sampling distribution of the difference between the proportions of interventions under provocative and conservative scenarios does not follow an approximately normal distribution.

6.24 Heart transplant success. The Stanford University Heart Transplant Study was conducted to determine whether an experimental heart transplant program increased lifespan. Each patient entering the program was officially designated a heart transplant candidate, meaning that he was gravely ill and might benefit from a new heart. Patients were randomly assigned into treatment and control groups. Patients in the treatment group received a transplant, and those in the control group did not. The table below displays how many patients survived and died in each group.[50]

	control	treatment
alive	4	24
dead	30	45

A hypothesis test would reject the conclusion that the survival rate is the same in each group, and so we might like to calculate a confidence interval. Explain why we cannot construct such an interval using the normal approximation. What might go wrong if we constructed the confidence interval despite this problem?

---

[50]B. Turnbull et al. "Survivorship of Heart Transplant Data". In: *Journal of the American Statistical Association* 69 (1974), pp. 74–80.

6.25 Gender and color preference. A 2001 study asked 1,924 male and 3,666 female undergraduate college students their favorite color. A 95% confidence interval for the difference between the proportions of males and females whose favorite color is black ($p_{male} - p_{female}$) was calculated to be (0.02, 0.06). Based on this information, determine if the following statements are true or false, and explain your reasoning for each statement you identify as false.[51]

(a) We are 95% confident that the true proportion of males whose favorite color is black is 2% lower to 6% higher than the true proportion of females whose favorite color is black.

(b) We are 95% confident that the true proportion of males whose favorite color is black is 2% to 6% higher than the true proportion of females whose favorite color is black.

(c) 95% of random samples will produce 95% confidence intervals that include the true difference between the population proportions of males and females whose favorite color is black.

(d) We can conclude that there is a significant difference between the proportions of males and females whose favorite color is black and that the difference between the two sample proportions is too large to plausibly be due to chance.

(e) The 95% confidence interval for ($p_{female} - p_{male}$) cannot be calculated with only the information given in this exercise.

6.26 The Daily Show. A 2010 Pew Research foundation poll indicates that among 1,099 college graduates, 33% watch The Daily Show. Meanwhile, 22% of the 1,110 people with a high school degree but no college degree in the poll watch The Daily Show. A 95% confidence interval for ($p_{college\ grad} - p_{HS\ or\ less}$), where p is the proportion of those who watch The Daily Show, is (0.07, 0.15). Based on this information, determine if the following statements are true or false, and explain your reasoning if you identify the statement as false.[52]

(a) At the 5% significance level, the data provide convincing evidence of a difference between the proportions of college graduates and those with a high school degree or less who watch The Daily Show.

(b) We are 95% confident that 7% less to 15% more college graduates watch The Daily Show than those with a high school degree or less.

(c) 95% of random samples of 1,099 college graduates and 1,110 people with a high school degree or less will yield differences in sample proportions between 7% and 15%.

(d) A 90% confidence interval for ($p_{college\ grad} - p_{HS\ or\ less}$) would be wider.

(e) A 95% confidence interval for ($p_{HS\ or\ less} - p_{college\ grad}$) is (-0.15,-0.07).

6.27 Public Option, Part III. Exercise 6.13 presents the results of a poll evaluating support for the health care public option plan in 2009. 70% of 819 Democrats and 42% of 783 Independents support the public option.

(a) Calculate a 95% confidence interval for the difference between ($p_D - p_I$) and interpret it in this context. We have already checked conditions for you.

(b) True or false: If we had picked a random Democrat and a random Independent at the time of this poll, it is more likely that the Democrat would support the public option than the Independent.

[51]L Ellis and C Ficek. "Color preferences according to gender and sexual orientation". In: *Personality and Individual Differences* 31.8 (2001), pp. 1375–1379.
[52]The Pew Research Center, Americans Spending More Time Following the News, data collected June 8-28, 2010.

6.28 Sleep deprivation, CA vs. OR, Part I. According to a report on sleep deprivation by the Centers for Disease Control and Prevention, the proportion of California residents who reported insufficient rest or sleep during each of the preceding 30 days is 8.0%, while this proportion is 8.8% for Oregon residents. These data are based on simple random samples of 11,545 California and 4,691 Oregon residents. Calculate a 95% confidence interval for the difference between the proportions of Californians and Oregonians who are sleep deprived and interpret it in context of the data.[53]

6.29 Offshore drilling, Part I. A 2010 survey asked 827 randomly sampled registered voters in California "Do you support? Or do you oppose? Drilling for oil and natural gas off the Coast of California? Or do you not know enough to say?" Below is the distribution of responses, separated based on whether or not the respondent graduated from college.[54]

(a) What percent of college graduates and what percent of the non-college graduates in this sample do not know enough to have an opinion on drilling for oil and natural gas off the Coast of California?

(b) Conduct a hypothesis test to determine if the data provide strong evidence that the proportion of college graduates who do not have an opinion on this issue is different than that of non-college graduates.

	College Grad	
	Yes	No
Support	154	132
Oppose	180	126
Do not know	104	131
Total	438	389

6.30 Sleep deprivation, CA vs. OR, Part II. Exercise 6.28 provides data on sleep deprivation rates of Californians and Oregonians. The proportion of California residents who reported insufficient rest or sleep during each of the preceding 30 days is 8.0%, while this proportion is 8.8% for Oregon residents. These data are based on simple random samples of 11,545 California and 4,691 Oregon residents.

(a) Conduct a hypothesis test to determine if these data provide strong evidence the rate of sleep deprivation is different for the two states. (Reminder: Check conditions.)

(b) It is possible the conclusion of the test in part (a) is incorrect. If this is the case, what type of error was made?

6.31 Offshore drilling, Part II. Results of a poll evaluating support for drilling for oil and natural gas off the coast of California were introduced in Exercise 6.29.

	College Grad	
	Yes	No
Support	154	132
Oppose	180	126
Do not know	104	131
Total	438	389

(a) What percent of college graduates and what percent of the non-college graduates in this sample support drilling for oil and natural gas off the Coast of California?

(b) Conduct a hypothesis test to determine if the data provide strong evidence that the proportion of college graduates who support off-shore drilling in California is different than that of non-college graduates.

---

[53]CDC, Perceived Insufficient Rest or Sleep Among Adults — United States, 2008.
[54]Survey USA, Election Poll #16804, data collected July 8-11, 2010.

6.32 Full body scan, Part I. A news article reports that "Americans have differing views on two potentially inconvenient and invasive practices that airports could implement to uncover potential terrorist attacks." This news piece was based on a survey conducted among a random sample of 1,137 adults nationwide, interviewed by telephone November 7-10, 2010, where one of the questions on the survey was "Some airports are now using 'full-body' digital x-ray machines to electronically screen passengers in airport security lines. Do you think these new x-ray machines should or should not be used at airports?" Below is a summary of responses based on party affiliation.[55]

		Party Affiliation		
		Republican	Democrat	Independent
	Should	264	299	351
Answer	Should not	38	55	77
	Don't know/No answer	16	15	22
	Total	318	369	450

(a) Conduct an appropriate hypothesis test evaluating whether there is a difference in the proportion of Republicans and Democrats who think the full-body scans should be applied in airports. Assume that all relevant conditions are met.

(b) The conclusion of the test in part (a) may be incorrect, meaning a testing error was made. If an error was made, was it a Type 1 or a Type 2 Error? Explain.

6.33 Sleep deprived transportation workers. The National Sleep Foundation conducted a survey on the sleep habits of randomly sampled transportation workers and a control sample of non-transportation workers. The results of the survey are shown below.[56]

		Transportation Professionals		
	Control	Pilots	Truck Drivers	Train Operators	Bus/Taxi/Limo Drivers
Less than 6 hours of sleep	35	19	35	29	21
6 to 8 hours of sleep	193	132	117	119	131
More than 8 hours	64	51	51	32	58
Total	292	202	203	180	210

Conduct a hypothesis test to evaluate if these data provide evidence of a difference between the proportions of truck drivers and non-transportation workers (the control group) who get less than 6 hours of sleep per day, i.e. are considered sleep deprived.

---

[55]S. Condon. "Poll: 4 in 5 Support Full-Body Airport Scanners". In: *CBS News* (2010).
[56]National Sleep Foundation, 2012 Sleep in America Poll: Transportation Workers' Sleep, 2012.

6.34 Prenatal vitamins and Autism. Researchers studying the link between prenatal vitamin use and autism surveyed the mothers of a random sample of children aged 24 - 60 months with autism and conducted another separate random sample for children with typical development. The table below shows the number of mothers in each group who did and did not use prenatal vitamins during the three months before pregnancy (periconceptional period).[57]

		Autism		
		Autism	Typical development	Total
Periconceptional	No vitamin	111	70	181
prenatal vitamin	Vitamin	143	159	302
	Total	254	229	483

(a) State appropriate hypotheses to test for independence of use of prenatal vitamins during the three months before pregnancy and autism.

(b) Complete the hypothesis test and state an appropriate conclusion. (Reminder: Verify any necessary conditions for the test.)

(c) A New York Times article reporting on this study was titled "Prenatal Vitamins May Ward Off Autism". Do you find the title of this article to be appropriate? Explain your answer. Additionally, propose an alternative title.[58]

6.35 HIV in sub-Saharan Africa. In July 2008 the US National Institutes of Health announced that it was stopping a clinical study early because of unexpected results. The study population consisted of HIV-infected women in sub-Saharan Africa who had been given single dose Nevaripine (a treatment for HIV) while giving birth, to prevent transmission of HIV to the infant. The study was a randomized comparison of continued treatment of a woman (after successful childbirth) with Nevaripine vs. Lopinavir, a second drug used to treat HIV. 240 women participated in the study; 120 were randomized to each of the two treatments. Twenty-four weeks after starting the study treatment, each woman was tested to determine if the HIV infection was becoming worse (an outcome called *virologic failure*). Twenty-six of the 120 women treated with Nevaripine experienced virologic failure, while 10 of the 120 women treated with the other drug experienced virologic failure.[59]

(a) Create a two-way table presenting the results of this study.

(b) State appropriate hypotheses to test for independence of treatment and virologic failure.

(c) Complete the hypothesis test and state an appropriate conclusion. (Reminder: Verify any necessary conditions for the test.)

6.36 Diabetes and unemployment. A 2012 Gallup poll surveyed Americans about their employment status and whether or not they have diabetes. The survey results indicate that 1.5% of the 47,774 employed (full or part time) and 2.5% of the 5,855 unemployed 18-29 year olds have diabetes.[60]

(a) Create a two-way table presenting the results of this study.

(b) State appropriate hypotheses to test for independence of incidence of diabetes and employment status.

(c) The sample difference is about 1%. If we completed the hypothesis test, we would find that the p-value is very small (about 0), meaning the difference is statistically significant. Use this result to explain the difference between statistically significant and practically significant findings.

[57]R.J. Schmidt et al. "Prenatal vitamins, one-carbon metabolism gene variants, and risk for autism". In: *Epidemiology* 22.4 (2011), p. 476.

[58]R.C. Rabin. "Patterns: Prenatal Vitamins May Ward Off Autism". In: *New York Times* (2011).

[59]S. Lockman et al. "Response to antiretroviral therapy after a single, peripartum dose of nevirapine". In: *Obstetrical & gynecological survey* 62.6 (2007), p. 361.

[60]Gallup Wellbeing, Employed Americans in Better Health Than the Unemployed, data collected Jan. 2, 2011 - May 21, 2012.

6.37 Active learning. A teacher wanting to increase the active learning component of her course is concerned about student reactions to changes she is planning to make. She conducts a survey in her class, asking students whether they believe more active learning in the classroom (hands on exercises) instead of traditional lecture will helps improve their learning. She does this at the beginning and end of the semester and wants to evaluate whether students' opinions have changed over the semester. Can she used the methods we learned in this chapter for this analysis? Explain your reasoning.

6.38 An apple a day keeps the doctor away. A physical education teacher at a high school wanting to increase awareness on issues of nutrition and health asked her students at the beginning of the semester whether they believed the expression "an apple a day keeps the doctor away", and 40% of the students responded yes. Throughout the semester she started each class with a brief discussion of a study highlighting positive effects of eating more fruits and vegetables. She conducted the same apple-a-day survey at the end of the semester, and this time 60% of the students responded yes. Can she used the methods we learned in this chapter for this analysis? Explain your reasoning.

6.7.3 Testing for goodness of fit using chi-square

6.39 True or false, Part I. Determine if the statements below are true or false. For each false statement, suggest an alternative wording to make it a true statement.

(a) The chi-square distribution, just like the normal distribution, has two parameters, mean and standard deviation.

(b) The chi-square distribution is always right skewed, regardless of the value of the degrees of freedom parameter.

(c) The chi-square statistic is always positive.

(d) As the degrees of freedom increases, the shape of the chi-square distribution becomes more skewed.

6.40 True or false, Part II. Determine if the statements below are true or false. For each false statement, suggest an alternative wording to make it a true statement.

(a) As the degrees of freedom increases, the mean of the chi-square distribution increases.

(b) If you found $\chi^2 = 10$ with $df = 5$ you would fail to reject H_0 at the 5% significance level.

(c) When finding the p-value of a chi-square test, we always shade the tail areas in both tails.

(d) As the degrees of freedom increases, the variability of the chi-square distribution decreases.

6.41 Open source textbook. A professor using an open source introductory statistics book predicts that 60% of the students will purchase a hard copy of the book, 25% will print it out from the web, and 15% will read it online. At the end of the semester he asks his students to complete a survey where they indicate what format of the book they used. Of the 126 students, 71 said they bought a hard copy of the book, 30 said they printed it out from the web, and 25 said they read it online.

(a) State the hypotheses for testing if the professor's predictions were inaccurate.

(b) How many students did the professor expect to buy the book, print the book, and read the book exclusively online?

(c) This is an appropriate setting for a chi-square test. List the conditions required for a test and verify they are satisfied.

(d) Calculate the chi-squared statistic, the degrees of freedom associated with it, and the p-value.

(e) Based on the p-value calculated in part (d), what is the conclusion of the hypothesis test? Interpret your conclusion in this context.

6.42 Evolution vs. creationism. A Gallup Poll released in December 2010 asked 1019 adults living in the Continental U.S. about their belief in the origin of humans. These results, along with results from a more comprehensive poll from 2001 (that we will assume to be exactly accurate), are summarized in the table below:[61]

Response	2010	2001
Humans evolved, with God guiding (1)	38%	37%
Humans evolved, but God had no part in process (2)	16%	12%
God created humans in present form (3)	40%	45%
Other / No opinion (4)	6%	6%

Year columns: 2010, 2001

(a) Calculate the actual number of respondents in 2010 that fall in each response category.

(b) State hypotheses for the following research question: have beliefs on the origin of human life changed since 2001?

(c) Calculate the expected number of respondents in each category under the condition that the null hypothesis from part (b) is true.

(d) Conduct a chi-square test and state your conclusion. (Reminder: Verify conditions.)

6.43 Rock-paper-scissors. Rock-paper-scissors is a hand game played by two or more people where players choose to sign either rock, paper, or scissors with their hands. For your statistics class project, you want to evaluate whether players choose between these three options randomly, or if certain options are favored above others. You ask two friends to play rock-paper-scissors and count the times each option is played. The following table summarizes the data:

Rock	Paper	Scissors
43	21	35

Use these data to evaluate whether players choose between these three options randomly, or if certain options are favored above others. Make sure to clearly outline each step of your analysis, and interpret your results in context of the data and the research question.

6.44 Barking deer. Microhabitat factors associated with forage and bed sites of barking deer in Hainan Island, China were examined from 2001 to 2002. In this region woods make up 4.8% of the land, cultivated grass plot makes up 14.7%, and deciduous forests makes up 39.6%. Of the 426 sites where the deer forage, 4 were categorized as woods, 16 as cultivated grassplot, and 61 as deciduous forests. The table below summarizes these data.[62]

Woods	Cultivated grassplot	Deciduous forests	Other	Total
4	16	61	345	426

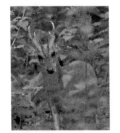

(a) Write the hypotheses for testing if barking deer prefer to forage in certain habitats over others.

(b) What type of test can we use to answer this research question?

(c) Check if the assumptions and conditions required for this test are satisfied.

(d) Do these data provide convincing evidence that barking deer prefer to forage in certain habitats over others? Conduct an appropriate hypothesis test to answer this research question.

Photo by Shrikant Rao
(http://flic.kr/p/4Xjdkk)
CC BY 2.0 license

---

[61] Four in 10 Americans Believe in Strict Creationism, December 17, 2010, www.gallup.com/poll/145286/Four-Americans-Believe-Strict-Creationism.aspx.

[62] Liwei Teng et al. "Forage and bed sites characteristics of Indian muntjac (Muntiacus muntjak) in Hainan Island, China". In: *Ecological Research* 19.6 (2004), pp. 675–681.

6.7.4   Testing for independence in two-way tables

6.45 Quitters. Does being part of a support group affect the ability of people to quit smoking? A county health department enrolled 300 smokers in a randomized experiment. 150 participants were assigned to a group that used a nicotine patch and met weekly with a support group; the other 150 received the patch and did not meet with a support group. At the end of the study, 40 of the participants in the patch plus support group had quit smoking while only 30 smokers had quit in the other group.

(a) Create a two-way table presenting the results of this study.

(b) Answer each of the following questions under the null hypothesis that being part of a support group does not affect the ability of people to quit smoking, and indicate whether the expected values are higher or lower than the observed values.

    i. How many subjects in the "patch + support" group would you expect to quit?

    ii. How many subjects in the "patch only" group would you expect to not quit?

6.46 Full body scan, Part II. The table below summarizes a data set we first encountered in Exercise 6.32 regarding views on full-body scans and political affiliation. The differences in each political group may be due to chance. Complete the following computations under the null hypothesis of independence between an individual's party affiliation and his support of full-body scans. It may be useful to first add on an extra column for row totals before proceeding with the computations.

		Party Affiliation		
		Republican	Democrat	Independent
	Should	264	299	351
Answer	Should not	38	55	77
	Don't know/No answer	16	15	22
	Total	318	369	450

(a) How many Republicans would you expect to not support the use of full-body scans?

(b) How many Democrats would you expect to support the use of full-body scans?

(c) How many Independents would you expect to not know or not answer?

6.47 Offshore drilling, Part III. The table below summarizes a data set we first encountered in Exercise 6.29 that examines the responses of a random sample of college graduates and non-graduates on the topic of oil drilling. Complete a chi-square test for these data to check whether there is a statistically significant difference in responses from college graduates and non-graduates.

	College Grad	
	Yes	No
Support	154	132
Oppose	180	126
Do not know	104	131
Total	438	389

6.48 Coffee and Depression. Researchers conducted a study investigating the relationship between caffeinated coffee consumption and risk of depression in women. They collected data on 50,739 women free of depression symptoms at the start of the study in the year 1996, and these women were followed through 2006. The researchers used questionnaires to collect data on caffeinated coffee consumption, asked each individual about physician-diagnosed depression, and also asked about the use of antidepressants. The table below shows the distribution of incidences of depression by amount of caffeinated coffee consumption.[63]

		Caffeinated coffee consumption					
		≤ 1 cup/week	2-6 cups/week	1 cup/day	2-3 cups/day	≥ 4 cups/day	Total
Clinical	Yes	670	373	905	564	95	2,607
depression	No	11,545	6,244	16,329	11,726	2,288	48,132
	Total	12,215	6,617	17,234	12,290	2,383	50,739

(a) What type of test is appropriate for evaluating if there is an association between coffee intake and depression?

(b) Write the hypotheses for the test you identified in part (a).

(c) Calculate the overall proportion of women who do and do not suffer from depression.

(d) Identify the expected count for the highlighted cell, and calculate the contribution of this cell to the test statistic, i.e. $(Observed - Expected)^2/Expected$.

(e) The test statistic is $\chi^2 = 20.93$. What is the p-value?

(f) What is the conclusion of the hypothesis test?

(g) One of the authors of this study was quoted on the NYTimes as saying it was "too early to recommend that women load up on extra coffee" based on just this study.[64] Do you agree with this statement? Explain your reasoning.

6.49 Shipping holiday gifts. A December 2010 survey asked 500 randomly sampled Los Angeles residents which shipping carrier they prefer to use for shipping holiday gifts. The table below shows the distribution of responses by age group as well as the expected counts for each cell (shown in parentheses).

		Age						Total
		18-34		35-54		55+		
	USPS	72	(81)	97	(102)	76	(62)	245
	UPS	52	(53)	76	(68)	34	(41)	162
Shipping Method	FedEx	31	(21)	24	(27)	9	(16)	64
	Something else	7	(5)	6	(7)	3	(4)	16
	Not sure	3	(5)	6	(5)	4	(3)	13
	Total	165		209		126		500

(a) State the null and alternative hypotheses for testing for independence of age and preferred shipping method for holiday gifts among Los Angeles residents.

(b) Are the conditions for inference using a chi-square test satisfied?

---

[63]M. Lucas et al. "Coffee, caffeine, and risk of depression among women". In: *Archives of internal medicine* 171.17 (2011), p. 1571.

[64]A. O'Connor. "Coffee Drinking Linked to Less Depression in Women". In: *New York Times* (2011).

6.50 How's it going? The American National Election Studies (ANES) collects data on voter attitudes and intentions as well as demographic information. In this question we will focus on two variables from the 2012 ANES dataset:[65]

- region (levels: Northeast, North Central, South, and West), and

- whether the respondent feels things in this country are generally going in the right direction or things have pretty seriously gotten off on the wrong track.

To keep calculations simple we will work with a random sample of 500 respondents from the ANES dataset. The distribution of responses are as follows:

	Right Direction	Wrong Track	Total
Northeast	29	54	83
North Central	44	77	121
South	62	131	193
West	36	67	103
Total	171	329	500

(a) Region: According to the 2010 Census, 18% of US residents live in the Northeast, 22% live in the North Central region, 37% live in the South, and 23% live in the West. Evaluate whether the ANES sample is representative of the population distribution of US residents. Make sure to clearly state the hypotheses, check conditions, calculate the appropriate test statistic and the p-value, and make your conclusion in context of the data. Also comment on what your conclusion says about whether or not this sample can be considered to be representative.

(b) Region and direction:

   (i) We would like to evaluate the relationship between region and feeling about the country's direction. What is the response variable and what is the explanatory variable?

  (ii) What are the hypotheses for evaluating this relationship?

 (iii) Complete the hypothesis test and interpret your results in context of the data and the research question.

---

[65]The American National Election Studies (ANES). The ANES 2012 Time Series Study [dataset]. Stanford University and the University of Michigan [producers].

6.7.5 Small sample hypothesis testing for a proportion

6.51 Bullying in schools. A 2012 Survey USA poll asked Florida residents how big of a problem they thought bullying was in local schools. 9 out of 191 18-34 year olds responded that bullying is no problem at all. Using these data, is it appropriate to construct a confidence interval using the formula $\hat{p} \pm z^\star \sqrt{\hat{p}(1-\hat{p})/n}$ for the true proportion of 18-34 year old Floridians who think bullying is no problem at all? If it is appropriate, construct the confidence interval. If it is not, explain why.

6.52 Choose a test. We would like to test the following hypotheses:

$H_0 : p = 0.1$

$H_A : p \neq 0.1$

The sample size is 120 and the sample proportion is 8.5%. Determine which of the below test(s) is/are appropriate for this situation and explain your reasoning.

I. Z-test for a proportion, i.e. proportion test using normal model

II. Z-test for comparing two proportions

III. χ^2 test of independence

IV. Simulation test for a proportion

V. t-test for a mean

VI. ANOVA

6.53 The Egyptian Revolution. A popular uprising that started on January 25, 2011 in Egypt led to the 2011 Egyptian Revolution. Polls show that about 69% of American adults followed the news about the political crisis and demonstrations in Egypt closely during the first couple weeks following the start of the uprising. Among a random sample of 30 high school students, it was found that only 17 of them followed the news about Egypt closely during this time.[66]

(a) Write the hypotheses for testing if the proportion of high school students who followed the news about Egypt is different than the proportion of American adults who did.

(b) Calculate the proportion of high schoolers in this sample who followed the news about Egypt closely during this time.

(c) Based on large sample theory, we modeled $\hat{p}$ using the normal distribution. Why should we be cautious about this approach for these data?

(d) The normal approximation will not be as reliable as a simulation, especially for a sample of this size. Describe how to perform such a simulation and, once you had results, how to estimate the p-value.

(e) Below is a histogram showing the distribution of $\hat{p}_{sim}$ in 10,000 simulations under the null hypothesis. Estimate the p-value using the plot and determine the conclusion of the hypothesis test.

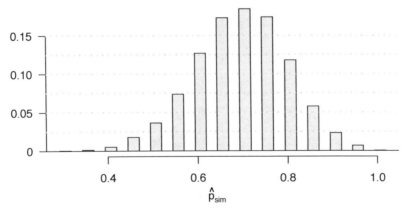

---

[66]Gallup Politics, Americans' Views of Egypt Sharply More Negative, data collected February 2-5, 2011.

6.54 Assisted Reproduction. Assisted Reproductive Technology (ART) is a collection of techniques that help facilitate pregnancy (e.g. in vitro fertilization). A 2008 report by the Centers for Disease Control and Prevention estimated that ART has been successful in leading to a live birth in 31% of cases[67]. A new fertility clinic claims that their success rate is higher than average. A random sample of 30 of their patients yielded a success rate of 40%. A consumer watchdog group would like to determine if this provides strong evidence to support the company's claim.

(a) Write the hypotheses to test if the success rate for ART at this clinic is significantly higher than the success rate reported by the CDC.

(b) Based on large sample theory, we modeled $\hat{p}$ using the normal distribution. Why is this not appropriate here?

(c) The normal approximation would be less reliable here, so we should use a simulation strategy. Describe a setup for a simulation that would be appropriate in this situation and how the p-value can be calculated using the simulation results.

(d) Below is a histogram showing the distribution of $\hat{p}_{sim}$ in 10,000 simulations under the null hypothesis. Estimate the p-value using the plot and use it to evaluate the hypotheses.

(e) After performing this analysis, the consumer group releases the following news headline: "Infertility clinic falsely advertises better success rates". Comment on the appropriateness of this statement.

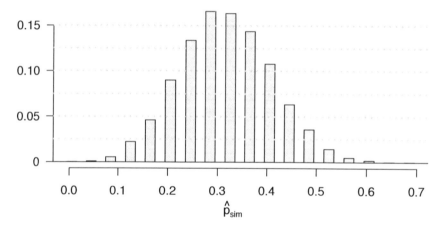

---

[67]CDC. *2008 Assisted Reproductive Technology Report.*

6.7.6 Randomization test

6.55 Social experiment, Part II. Exercise 6.23 introduces a "social experiment" conducted by a TV program that questioned what people do when they see a very obviously bruised woman getting picked on by her boyfriend. On two different occasions at the same restaurant, the same couple was depicted. In one scenario the woman was dressed "provocatively" and in the other scenario the woman was dressed "conservatively". The table below shows how many restaurant diners were present under each scenario, and whether or not they intervened.

		Scenario		
		Provocative	Conservative	Total
Intervene	Yes	5	15	20
	No	15	10	25
	Total	20	25	45

A simulation was conducted to test if people react differently under the two scenarios. 10,000 simulated differences were generated to construct the null distribution shown. The value $\hat{p}_{pr,sim}$ represents the proportion of diners who intervened in the simulation for the provocatively dressed woman, and $\hat{p}_{con,sim}$ is the proportion for the conservatively dressed woman.

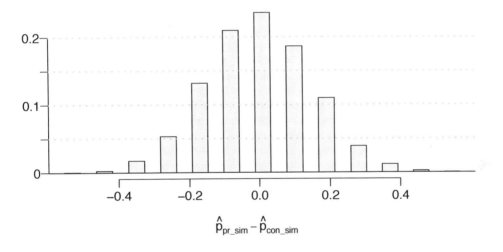

(a) What are the hypotheses? For the purposes of this exercise, you may assume that each observed person at the restaurant behaved independently, though we would want to evaluate this assumption more rigorously if we were reporting these results.

(b) Calculate the observed difference between the rates of intervention under the provocative and conservative scenarios: $\hat{p}_{pr} - \hat{p}_{con}$.

(c) Estimate the p-value using the figure above and determine the conclusion of the hypothesis test.

6.56 Is yawning contagious? An experiment conducted by the *MythBusters*, a science entertainment TV program on the Discovery Channel, tested if a person can be subconsciously influenced into yawning if another person near them yawns. 50 people were randomly assigned to two groups: 34 to a group where a person near them yawned (treatment) and 16 to a group where there wasn't a person yawning near them (control). The following table shows the results of this experiment.[68]

		Group		
		Treatment	Control	Total
Result	Yawn	10	4	14
	Not Yawn	24	12	36
	Total	34	16	50

A simulation was conducted to understand the distribution of the test statistic under the assumption of independence: having someone yawn near another person has no influence on if the other person will yawn. In order to conduct the simulation, a researcher wrote yawn on 14 index cards and not yawn on 36 index cards to indicate whether or not a person yawned. Then he shuffled the cards and dealt them into two groups of size 34 and 16 for treatment and control, respectively. He counted how many participants in each simulated group yawned in an apparent response to a nearby yawning person, and calculated the difference between the simulated proportions of yawning as $\hat{p}_{trtmt,sim} - \hat{p}_{ctrl,sim}$. This simulation was repeated 10,000 times using software to obtain 10,000 differences that are due to chance alone. The histogram shows the distribution of the simulated differences.

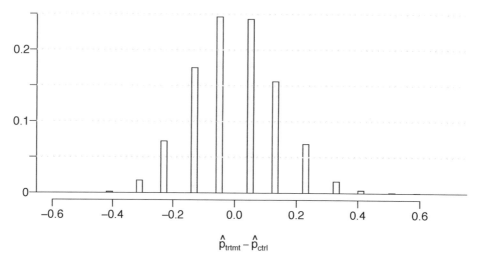

(a) What are the hypotheses for testing if yawning is contagious, i.e. whether it is more likely for someone to yawn if they see someone else yawning?

(b) Calculate the observed difference between the yawning rates under the two scenarios.

(c) Estimate the p-value using the figure above and determine the conclusion of the hypothesis test.

---

[68]MythBusters, Season 3, Episode 28.

Chapter 7

Introduction to linear regression

Linear regression is a very powerful statistical technique. Many people have some familiarity with regression just from reading the news, where graphs with straight lines are overlaid on scatterplots. Linear models can be used for prediction or to evaluate whether there is a linear relationship between two numerical variables.

Figure 7.1 shows two variables whose relationship can be modeled perfectly with a straight line. The equation for the line is

$$y = 5 + 57.49x$$

Imagine what a perfect linear relationship would mean: you would know the exact value of y just by knowing the value of x. This is unrealistic in almost any natural process. For example, if we took family income x, this value would provide some useful information about how much financial support y a college may offer a prospective student. However, there would still be variability in financial support, even when comparing students whose families have similar financial backgrounds.

Linear regression assumes that the relationship between two variables, x and y, can be modeled by a straight line:

$$y = \beta_0 + \beta_1 x \qquad (7.1)$$

β_0, β_1
Linear model parameters

where β_0 and β_1 represent two model parameters (β is the Greek letter *beta*). These parameters are estimated using data, and we write their point estimates as b_0 and b_1. When we use x to predict y, we usually call x the explanatory or **predictor** variable, and we call y the response.

It is rare for all of the data to fall on a straight line, as seen in the three scatterplots in Figure 7.2. In each case, the data fall around a straight line, even if none of the observations fall exactly on the line. The first plot shows a relatively strong downward linear trend, where the remaining variability in the data around the line is minor relative to the strength of the relationship between x and y. The second plot shows an upward trend that, while evident, is not as strong as the first. The last plot shows a very weak downward trend in the data, so slight we can hardly notice it. In each of these examples, we will have some uncertainty regarding our estimates of the model parameters, β_0 and β_1. For instance, we might wonder, should we move the line up or down a little, or should we tilt it more or less?

331

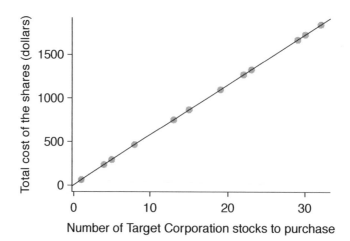

Figure 7.1: Requests from twelve separate buyers were simultaneously placed with a trading company to purchase Target Corporation stock (ticker TGT, April 26th, 2012), and the total cost of the shares were reported. Because the cost is computed using a linear formula, the linear fit is perfect.

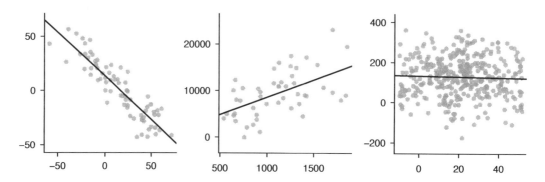

Figure 7.2: Three data sets where a linear model may be useful even though the data do not all fall exactly on the line.

As we move forward in this chapter, we will learn different criteria for line-fitting, and we will also learn about the uncertainty associated with estimates of model parameters.

We will also see examples in this chapter where fitting a straight line to the data, even if there is a clear relationship between the variables, is not helpful. One such case is shown in Figure 7.3 where there is a very strong relationship between the variables even though the trend is not linear. We will discuss nonlinear trends in this chapter and the next, but the details of fitting nonlinear models are saved for a later course.

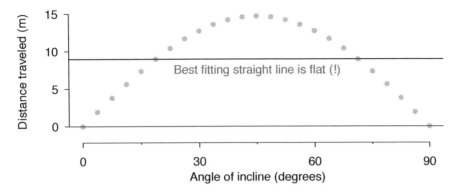

Figure 7.3: A linear model is not useful in this nonlinear case. These data are from an introductory physics experiment.

7.1   Line fitting, residuals, and correlation

It is helpful to think deeply about the line fitting process. In this section, we examine criteria for identifying a linear model and introduce a new statistic, *correlation.*

7.1.1   Beginning with straight lines

Scatterplots were introduced in Chapter 1 as a graphical technique to present two numerical variables simultaneously. Such plots permit the relationship between the variables to be examined with ease. Figure 7.4 shows a scatterplot for the head length and total length of 104 brushtail possums from Australia. Each point represents a single possum from the data.

The head and total length variables are associated. Possums with an above average total length also tend to have above average head lengths. While the relationship is not perfectly linear, it could be helpful to partially explain the connection between these variables with a straight line.

Straight lines should only be used when the data appear to have a linear relationship, such as the case shown in the left panel of Figure 7.6. The right panel of Figure 7.6 shows a case where a curved line would be more useful in understanding the relationship between the two variables.

Caution: Watch out for curved trends

We only consider models based on straight lines in this chapter. If data show a nonlinear trend, like that in the right panel of Figure 7.6, more advanced techniques should be used.

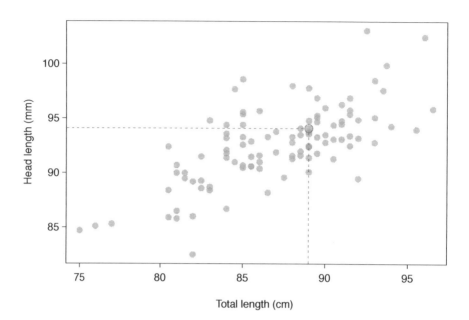

Figure 7.4: A scatterplot showing head length against total length for 104 brushtail possums. A point representing a possum with head length 94.1mm and total length 89cm is highlighted.

Figure 7.5: The common brushtail possum of Australia.

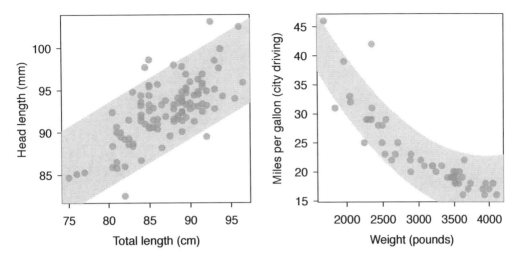

Figure 7.6: The figure on the left shows head length versus total length, and reveals that many of the points could be captured by a straight band. On the right, we see that a curved band is more appropriate in the scatterplot for weight and mpgCity from the cars data set.

7.1.2 Fitting a line by eye

We want to describe the relationship between the head length and total length variables in the possum data set using a line. In this example, we will use the total length as the predictor variable, x, to predict a possum's head length, y. We could fit the linear relationship by eye, as in Figure 7.7. The equation for this line is

$$\hat{y} = 41 + 0.59x \tag{7.2}$$

We can use this line to discuss properties of possums. For instance, the equation predicts a possum with a total length of 80 cm will have a head length of

$$\hat{y} = 41 + 0.59 \times 80$$
$$= 88.2$$

A "hat" on y is used to signify that this is an estimate. This estimate may be viewed as an average: the equation predicts that possums with a total length of 80 cm will have an average head length of 88.2 mm. Absent further information about an 80 cm possum, the prediction for head length that uses the average is a reasonable estimate.

7.1.3 Residuals

Residuals are the leftover variation in the data after accounting for the model fit:

$$\text{Data} = \text{Fit} + \text{Residual}$$

Each observation will have a residual. If an observation is above the regression line, then its residual, the vertical distance from the observation to the line, is positive. Observations below the line have negative residuals. One goal in picking the right linear model is for these residuals to be as small as possible.

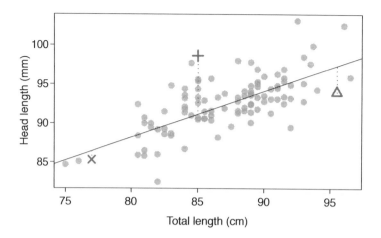

Figure 7.7: A reasonable linear model was fit to represent the relationship between head length and total length.

Three observations are noted specially in Figure 7.7. The observation marked by an "×" has a small, negative residual of about -1; the observation marked by "+" has a large residual of about +7; and the observation marked by "△" has a moderate residual of about -4. The size of a residual is usually discussed in terms of its absolute value. For example, the residual for "△" is larger than that of "×" because $|-4|$ is larger than $|-1|$.

Residual: difference between observed and expected

The residual of the ith observation (x_i, y_i) is the difference of the observed response (y_i) and the response we would predict based on the model fit $(\hat{y}_i)$:

$$e_i = y_i - \hat{y}_i$$

We typically identify $\hat{y}_i$ by plugging x_i into the model.

● **Example 7.3** The linear fit shown in Figure 7.7 is given as $\hat{y} = 41 + 0.59x$. Based on this line, formally compute the residual of the observation $(77.0, 85.3)$. This observation is denoted by "×" on the plot. Check it against the earlier visual estimate, -1.

We first compute the predicted value of point "×" based on the model:

$$\hat{y}_\times = 41 + 0.59x_\times = 41 + 0.59 \times 77.0 = 86.4$$

Next we compute the difference of the actual head length and the predicted head length:

$$e_\times = y_\times - \hat{y}_\times = 85.3 - 86.4 = -1.1$$

This is very close to the visual estimate of -1.

⊙ **Guided Practice 7.4** If a model underestimates an observation, will the residual be positive or negative? What about if it overestimates the observation?[1]

---

[1]If a model underestimates an observation, then the model estimate is below the actual. The residual, which is the actual observation value minus the model estimate, must then be positive. The opposite is true when the model overestimates the observation: the residual is negative.

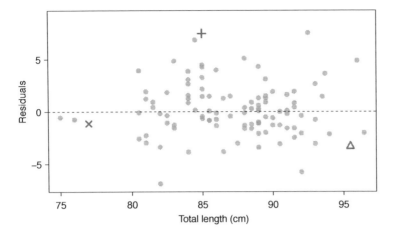

Figure 7.8: Residual plot for the model in Figure 7.7.

 Guided Practice 7.5 Compute the residuals for the observations $(85.0, 98.6)$ ("+" in the figure) and $(95.5, 94.0)$ ("$\triangle$") using the linear relationship $\hat{y} = 41 + 0.59x$. [2]

Residuals are helpful in evaluating how well a linear model fits a data set. We often display them in a **residual plot** such as the one shown in Figure 7.8 for the regression line in Figure 7.7. The residuals are plotted at their original horizontal locations but with the vertical coordinate as the residual. For instance, the point $(85.0, 98.6)_+$ had a residual of 7.45, so in the residual plot it is placed at $(85.0, 7.45)$. Creating a residual plot is sort of like tipping the scatterplot over so the regression line is horizontal.

● **Example 7.6** One purpose of residual plots is to identify characteristics or patterns still apparent in data after fitting a model. Figure 7.9 shows three scatterplots with linear models in the first row and residual plots in the second row. Can you identify any patterns remaining in the residuals?

In the first data set (first column), the residuals show no obvious patterns. The residuals appear to be scattered randomly around the dashed line that represents 0.

The second data set shows a pattern in the residuals. There is some curvature in the scatterplot, which is more obvious in the residual plot. We should not use a straight line to model these data. Instead, a more advanced technique should be used.

The last plot shows very little upwards trend, and the residuals also show no obvious patterns. It is reasonable to try to fit a linear model to the data. However, it is unclear whether there is statistically significant evidence that the slope parameter is different from zero. The point estimate of the slope parameter, labeled b_1, is not zero, but we might wonder if this could just be due to chance. We will address this sort of scenario in Section 7.4.

---

[2](+) First compute the predicted value based on the model:

$$\hat{y}_+ = 41 + 0.59x_+ = 41 + 0.59 \times 85.0 = 91.15$$

Then the residual is given by

$$e_+ = y_+ - \hat{y}_+ = 98.6 - 91.15 = 7.45$$

This was close to the earlier estimate of 7.
($\triangle$) $\hat{y}_\triangle = 41 + 0.59x_\triangle = 97.3$. $e_\triangle = y_\triangle - \hat{y}_\triangle = -3.3$, close to the estimate of -4.

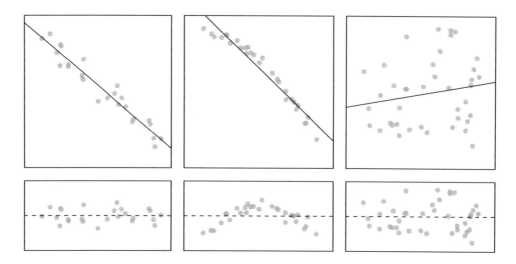

Figure 7.9: Sample data with their best fitting lines (top row) and their corresponding residual plots (bottom row).

7.1.4   Describing linear relationships with correlation

R
correlation

> **Correlation: strength of a linear relationship**
>
> **Correlation**, which always takes values between -1 and 1, describes the strength of the linear relationship between two variables. We denote the correlation by R.

We can compute the correlation using a formula, just as we did with the sample mean and standard deviation. However, this formula is rather complex,[3] so we generally perform the calculations on a computer or calculator. Figure 7.10 shows eight plots and their corresponding correlations. Only when the relationship is perfectly linear is the correlation either -1 or 1. If the relationship is strong and positive, the correlation will be near +1. If it is strong and negative, it will be near -1. If there is no apparent linear relationship between the variables, then the correlation will be near zero.

The correlation is intended to quantify the strength of a linear trend. Nonlinear trends, even when strong, sometimes produce correlations that do not reflect the strength of the relationship; see three such examples in Figure 7.11.

⊙ **Guided Practice 7.7**   It appears no straight line would fit any of the datasets represented in Figure 7.11. Try drawing nonlinear curves on each plot. Once you create a curve for each, describe what is important in your fit.[4]

---

[3]Formally, we can compute the correlation for observations (x_1, y_1), (x_2, y_2), ..., (x_n, y_n) using the formula

$$R = \frac{1}{n-1} \sum_{i=1}^{n} \frac{x_i - \bar{x}}{s_x} \frac{y_i - \bar{y}}{s_y}$$

where $\bar{x}$, $\bar{y}$, s_x, and s_y are the sample means and standard deviations for each variable.

[4]We'll leave it to you to draw the lines. In general, the lines you draw should be close to most points and reflect overall trends in the data.

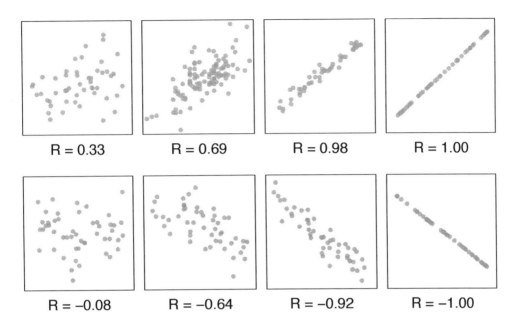

Figure 7.10: Sample scatterplots and their correlations. The first row shows variables with a positive relationship, represented by the trend up and to the right. The second row shows variables with a negative trend, where a large value in one variable is associated with a low value in the other.

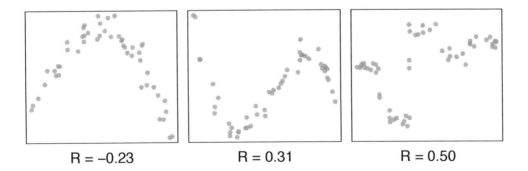

Figure 7.11: Sample scatterplots and their correlations. In each case, there is a strong relationship between the variables. However, the correlation is not very strong, and the relationship is not linear.

7.2 Fitting a line by least squares regression 📹

Fitting linear models by eye is open to criticism since it is based on an individual preference. In this section, we use *least squares regression* as a more rigorous approach.

This section considers family income and gift aid data from a random sample of fifty students in the 2011 freshman class of Elmhurst College in Illinois.[5] Gift aid is financial aid that does not need to be paid back, as opposed to a loan. A scatterplot of the data is shown in Figure 7.12 along with two linear fits. The lines follow a negative trend in the data; students who have higher family incomes tended to have lower gift aid from the university.

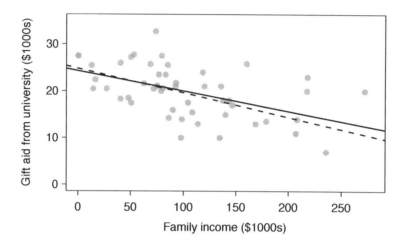

Figure 7.12: Gift aid and family income for a random sample of 50 freshman students from Elmhurst College. Two lines are fit to the data, the solid line being the *least squares line*.

⊙ **Guided Practice 7.8**   Is the correlation positive or negative in Figure 7.12?[6]

7.2.1 An objective measure for finding the best line

We begin by thinking about what we mean by "best". Mathematically, we want a line that has small residuals. Perhaps our criterion could minimize the sum of the residual magnitudes:

$$|e_1| + |e_2| + \cdots + |e_n| \tag{7.9}$$

which we could accomplish with a computer program. The resulting dashed line shown in Figure 7.12 demonstrates this fit can be quite reasonable. However, a more common practice is to choose the line that minimizes the sum of the squared residuals:

$$e_1^2 + e_2^2 + \cdots + e_n^2 \tag{7.10}$$

---

[5]These data were sampled from a table of data for all freshman from the 2011 class at Elmhurst College that accompanied an article titled *What Students Really Pay to Go to College* published online by *The Chronicle of Higher Education*: chronicle.com/article/What-Students-Really-Pay-to-Go/131435

[6]Larger family incomes are associated with lower amounts of aid, so the correlation will be negative. Using a computer, the correlation can be computed: -0.499.

The line that minimizes this **least squares criterion** is represented as the solid line in Figure 7.12. This is commonly called the **least squares line**. The following are three possible reasons to choose Criterion (7.10) over Criterion (7.9):

1. It is the most commonly used method.

2. Computing the line based on Criterion (7.10) is much easier by hand and in most statistical software.

3. In many applications, a residual twice as large as another residual is more than twice as bad. For example, being off by 4 is usually more than twice as bad as being off by 2. Squaring the residuals accounts for this discrepancy.

The first two reasons are largely for tradition and convenience; the last reason explains why Criterion (7.10) is typically most helpful.[7]

7.2.2 Conditions for the least squares line

When fitting a least squares line, we generally require

Linearity. The data should show a linear trend. If there is a nonlinear trend (e.g. left panel of Figure 7.13), an advanced regression method from another book or later course should be applied.

Nearly normal residuals. Generally the residuals must be nearly normal. When this condition is found to be unreasonable, it is usually because of outliers or concerns about influential points, which we will discuss in greater depth in Section 7.3. An example of non-normal residuals is shown in the second panel of Figure 7.13.

Constant variability. The variability of points around the least squares line remains roughly constant. An example of non-constant variability is shown in the third panel of Figure 7.13.

Independent observations. Be cautious about applying regression to **time series** data, which are sequential observations in time such as a stock price each day. Such data may have an underlying structure that should be considered in a model and analysis. An example of a data set where successive observations are not independent is shown in the fourth panel of Figure 7.13. There are also other instances where correlations within the data are important, which is further discussed in Chapter 8.

⊙ **Guided Practice 7.11** Should we have concerns about applying least squares regression to the Elmhurst data in Figure 7.12?[8]

---

[7]There are applications where Criterion (7.9) may be more useful, and there are plenty of other criteria we might consider. However, this book only applies the least squares criterion.

[8]The trend appears to be linear, the data fall around the line with no obvious outliers, the variance is roughly constant. These are also not time series observations. Least squares regression can be applied to these data.

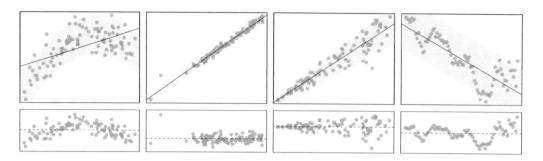

Figure 7.13: Four examples showing when the methods in this chapter are insufficient to apply to the data. In the left panel, a straight line does not fit the data. In the second panel, there are outliers; two points on the left are relatively distant from the rest of the data, and one of these points is very far away from the line. In the third panel, the variability of the data around the line increases with larger values of x. In the last panel, a time series data set is shown, where successive observations are highly correlated.

7.2.3 Finding the least squares line

For the Elmhurst data, we could write the equation of the least squares regression line as

$$\widehat{aid} = \beta_0 + \beta_1 \times family\_income$$

Here the equation is set up to predict gift aid based on a student's family income, which would be useful to students considering Elmhurst. These two values, β_0 and β_1, are the *parameters* of the regression line.

As in Chapters 4-6, the parameters are estimated using observed data. In practice, this estimation is done using a computer in the same way that other estimates, like a sample mean, can be estimated using a computer or calculator. However, we can also find the parameter estimates by applying two properties of the least squares line:

- The slope of the least squares line can be estimated by

$$b_1 = \frac{s_y}{s_x} R \tag{7.12}$$

  where R is the correlation between the two variables, and s_x and s_y are the sample standard deviations of the explanatory variable and response, respectively.

- If $\bar{x}$ is the mean of the horizontal variable (from the data) and $\bar{y}$ is the mean of the vertical variable, then the point $(\bar{x}, \bar{y})$ is on the least squares line.

b_0, b_1
Sample estimates of β_0, β_1

We use b_0 and b_1 to represent the point estimates of the parameters β_0 and β_1.

⊙ **Guided Practice 7.13** Table 7.14 shows the sample means for the family income and gift aid as \$101,800 and \$19,940, respectively. Plot the point $(101.8, 19.94)$ on Figure 7.12 on page 340 to verify it falls on the least squares line (the solid line).[9]

---

[9]If you need help finding this location, draw a straight line up from the x-value of 100 (or thereabout). Then draw a horizontal line at 20 (or thereabout). These lines should intersect on the least squares line.

	family income, in $1000s ("$x$")	gift aid, in $1000s ("$y$")
mean	$\bar{x} = 101.8$	$\bar{y} = 19.94$
sd	$s_x = 63.2$	$s_y = 5.46$
		$R = -0.499$

Table 7.14: Summary statistics for family income and gift aid.

⊙ **Guided Practice 7.14**   Using the summary statistics in Table 7.14, compute the slope for the regression line of gift aid against family income.[10]

You might recall the **point-slope** form of a line from math class (another common form is *slope-intercept*). Given the slope of a line and a point on the line, (x_0, y_0), the equation for the line can be written as

$$y - y_0 = slope \times (x - x_0) \tag{7.15}$$

A common exercise to become more familiar with foundations of least squares regression is to use basic summary statistics and point-slope form to produce the least squares line.

TIP: Identifying the least squares line from summary statistics

To identify the least squares line from summary statistics:

- Estimate the slope parameter, b_1, using Equation (7.12).

- Noting that the point $(\bar{x}, \bar{y})$ is on the least squares line, use $x_0 = \bar{x}$ and $y_0 = \bar{y}$ along with the slope b_1 in the point-slope equation:

$$y - \bar{y} = b_1(x - \bar{x})$$

- Simplify the equation.

● **Example 7.16**   Using the point $(101.8, 19.94)$ from the sample means and the slope estimate $b_1 = -0.0431$ from Guided Practice 7.14, find the least-squares line for predicting aid based on family income.

Apply the point-slope equation using $(101.8, 19.94)$ and the slope $b_1 = -0.0431$:

$$y - y_0 = b_1(x - x_0)$$
$$y - 19.94 = -0.0431(x - 101.8)$$

Expanding the right side and then adding 19.94 to each side, the equation simplifies:

$$\widehat{aid} = 24.3 - 0.0431 \times family\_income$$

Here we have replaced y with $\widehat{aid}$ and x with $family\_income$ to put the equation in context.

---

[10]Apply Equation (7.12) with the summary statistics from Table 7.14 to compute the slope:

$$b_1 = \frac{s_y}{s_x} R = \frac{5.46}{63.2}(-0.499) = -0.0431$$

We mentioned earlier that a computer is usually used to compute the least squares line. A summary table based on computer output is shown in Table 7.15 for the Elmhurst data. The first column of numbers provides estimates for b_0 and b_1, respectively. Compare these to the result from Example 7.16.

	Estimate	Std. Error	t value	Pr($>$|t|)
(Intercept)	24.3193	1.2915	18.83	0.0000
family_income	-0.0431	0.0108	-3.98	0.0002

Table 7.15: Summary of least squares fit for the Elmhurst data. Compare the parameter estimates in the first column to the results of Example 7.16.

● **Example 7.17** Examine the second, third, and fourth columns in Table 7.15. Can you guess what they represent?

We'll describe the meaning of the columns using the second row, which corresponds to β_1. The first column provides the point estimate for β_1, as we calculated in an earlier example: -0.0431. The second column is a standard error for this point estimate: 0.0108. The third column is a t-test statistic for the null hypothesis that $\beta_1 = 0$: $T = -3.98$. The last column is the p-value for the t-test statistic for the null hypothesis $\beta_1 = 0$ and a two-sided alternative hypothesis: 0.0002. We will get into more of these details in Section 7.4.

● **Example 7.18** Suppose a high school senior is considering Elmhurst College. Can she simply use the linear equation that we have estimated to calculate her financial aid from the university?

She may use it as an estimate, though some qualifiers on this approach are important. First, the data all come from one freshman class, and the way aid is determined by the university may change from year to year. Second, the equation will provide an imperfect estimate. While the linear equation is good at capturing the trend in the data, no individual student's aid will be perfectly predicted.

7.2.4   Interpreting regression line parameter estimates

Interpreting parameters in a regression model is often one of the most important steps in the analysis.

● **Example 7.19**   The slope and intercept estimates for the Elmhurst data are -0.0431 and 24.3. What do these numbers really mean?

Interpreting the slope parameter is helpful in almost any application. For each additional \$1,000 of family income, we would expect a student to receive a net difference of $\$1,000 \times (-0.0431) = -\$43.10$ in aid on average, i.e. \$43.10 *less*. Note that a higher family income corresponds to less aid because the coefficient of family income is negative in the model. We must be cautious in this interpretation: while there is a real association, we cannot interpret a causal connection between the variables because these data are observational. That is, increasing a student's family income may not cause the student's aid to drop. (It would be reasonable to contact the college and ask if the relationship is causal, i.e. if Elmhurst College's aid decisions are partially based on students' family income.)

The estimated intercept $b_0 = 24.3$ (in \$1000s) describes the average aid if a student's family had no income. The meaning of the intercept is relevant to this application since the family income for some students at Elmhurst is \$0. In other applications, the intercept may have little or no practical value if there are no observations where x is near zero.

Interpreting parameters estimated by least squares

The slope describes the estimated difference in the y variable if the explanatory variable x for a case happened to be one unit larger. The intercept describes the average outcome of y if $x = 0$ *and* the linear model is valid all the way to $x = 0$, which in many applications is not the case.

7.2.5   Extrapolation is treacherous

When those blizzards hit the East Coast this winter, it proved to my satisfaction that global warming was a fraud. That snow was freezing cold. But in an alarming trend, temperatures this spring have risen. Consider this: On February 6th it was 10 degrees. Today it hit almost 80. At this rate, by August it will be 220 degrees. So clearly folks the climate debate rages on.

Stephen Colbert
April 6th, 2010[11]

Linear models can be used to approximate the relationship between two variables. However, these models have real limitations. Linear regression is simply a modeling framework. The truth is almost always much more complex than our simple line. For example, we do not know how the data outside of our limited window will behave.

---

[11]www.colbertnation.com/the-colbert-report-videos/269929

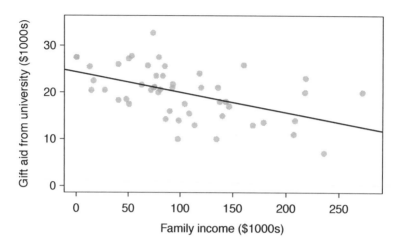

Figure 7.16: Gift aid and family income for a random sample of 50 freshman students from Elmhurst College, shown with the least squares regression line.

● **Example 7.20** Use the model $\widehat{aid} = 24.3 - 0.0431 \times family\_income$ to estimate the aid of another freshman student whose family had income of \$1 million.

Recall that the units of family income are in \$1000s, so we want to calculate the aid for $family\_income = 1000$:

$$24.3 - 0.0431 \times family\_income = 24.3 - 0.0431 \times 1000 = -18.8$$

The model predicts this student will have -\$18,800 in aid (!). Elmhurst College cannot (or at least does not) require any students to pay extra on top of tuition to attend.

Applying a model estimate to values outside of the realm of the original data is called **extrapolation**. Generally, a linear model is only an approximation of the real relationship between two variables. If we extrapolate, we are making an unreliable bet that the approximate linear relationship will be valid in places where it has not been analyzed.

7.2.6 Using R^2 to describe the strength of a fit

We evaluated the strength of the linear relationship between two variables earlier using the correlation, R. However, it is more common to explain the strength of a linear fit using R^2, called **R-squared**. If provided with a linear model, we might like to describe how closely the data cluster around the linear fit.

The R^2 of a linear model describes the amount of variation in the response that is explained by the least squares line. For example, consider the Elmhurst data, shown in Figure 7.16. The variance of the response variable, aid received, is $s^2_{aid} = 29.8$. However, if we apply our least squares line, then this model reduces our uncertainty in predicting aid using a student's family income. The variability in the residuals describes how much variation remains after using the model: $s^2_{_{RES}} = 22.4$. In short, there was a reduction of

$$\frac{s^2_{aid} - s^2_{_{RES}}}{s^2_{aid}} = \frac{29.8 - 22.4}{29.8} = \frac{7.5}{29.8} = 0.25$$

or about 25% in the data's variation by using information about family income for predicting aid using a linear model. This corresponds exactly to the R-squared value:

$$R = -0.499 \qquad\qquad R^2 = 0.25$$

⊙ **Guided Practice 7.21** If a linear model has a very strong negative relationship with a correlation of -0.97, how much of the variation in the response is explained by the explanatory variable?[12]

▶ Calculator videos

Videos covering how to find regression coefficients using TI and Casio graphing calculators are available at openintro.org/videos.

7.2.7 Categorical predictors with two levels

Categorical variables are also useful in predicting outcomes. Here we consider a categorical predictor with two levels (recall that a *level* is the same as a *category*). We'll consider Ebay auctions for a video game, *Mario Kart* for the Nintendo Wii, where both the total price of the auction and the condition of the game were recorded.[13] Here we want to predict total price based on game condition, which takes values used and new. A plot of the auction data is shown in Figure 7.17.

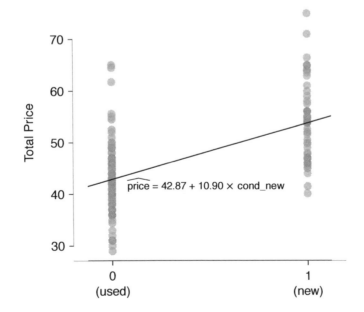

Figure 7.17: Total auction prices for the video game *Mario Kart*, divided into used ($x = 0$) and new ($x = 1$) condition games. The least squares regression line is also shown.

---

[12]About $R^2 = (-0.97)^2 = 0.94$ or 94% of the variation is explained by the linear model.

[13]These data were collected in Fall 2009 and may be found at openintro.org.

	Estimate	Std. Error	t value	Pr(>|t|)
(Intercept)	42.87	0.81	52.67	0.0000
cond_new	10.90	1.26	8.66	0.0000

Table 7.18: Least squares regression summary for the final auction price against the condition of the game.

To incorporate the game condition variable into a regression equation, we must convert the categories into a numerical form. We will do so using an **indicator variable** called cond_new, which takes value 1 when the game is new and 0 when the game is used. Using this indicator variable, the linear model may be written as

$$\widehat{price} = \beta_0 + \beta_1 \times \texttt{cond\_new}$$

The fitted model is summarized in Table 7.18, and the model with its parameter estimates is given as

$$\widehat{price} = 42.87 + 10.90 \times \texttt{cond\_new}$$

For categorical predictors with just two levels, the linearity assumption will always be satisfied. However, we must evaluate whether the residuals in each group are approximately normal and have approximately equal variance. As can be seen in Figure 7.17, both of these conditions are reasonably satisfied by the auction data.

● **Example 7.22** Interpret the two parameters estimated in the model for the price of *Mario Kart* in eBay auctions.

The intercept is the estimated price when cond_new takes value 0, i.e. when the game is in used condition. That is, the average selling price of a used version of the game is $42.87.

The slope indicates that, on average, new games sell for about $10.90 more than used games.

TIP: Interpreting model estimates for categorical predictors.
The estimated intercept is the value of the response variable for the first category (i.e. the category corresponding to an indicator value of 0). The estimated slope is the average change in the response variable between the two categories.

We'll elaborate further on this Ebay auction data in Chapter 8, where we examine the influence of many predictor variables simultaneously using multiple regression. In multiple regression, we will consider the association of auction price with regard to each variable while controlling for the influence of other variables. This is especially important since some of the predictors are associated. For example, auctions with games in new condition also often came with more accessories.

7.3 Types of outliers in linear regression

In this section, we identify criteria for determining which outliers are important and influential. Outliers in regression are observations that fall far from the "cloud" of points. These points are especially important because they can have a strong influence on the least squares line.

● **Example 7.23** There are six plots shown in Figure 7.19 along with the least squares line and residual plots. For each scatterplot and residual plot pair, identify the outliers and note how they influence the least squares line. Recall that an outlier is any point that doesn't appear to belong with the vast majority of the other points.

(1) There is one outlier far from the other points, though it only appears to slightly influence the line.

(2) There is one outlier on the right, though it is quite close to the least squares line, which suggests it wasn't very influential.

(3) There is one point far away from the cloud, and this outlier appears to pull the least squares line up on the right; examine how the line around the primary cloud doesn't appear to fit very well.

(4) There is a primary cloud and then a small secondary cloud of four outliers. The secondary cloud appears to be influencing the line somewhat strongly, making the least square line fit poorly almost everywhere. There might be an interesting explanation for the dual clouds, which is something that could be investigated.

(5) There is no obvious trend in the main cloud of points and the outlier on the right appears to largely control the slope of the least squares line.

(6) There is one outlier far from the cloud, however, it falls quite close to the least squares line and does not appear to be very influential.

Examine the residual plots in Figure 7.19. You will probably find that there is some trend in the main clouds of (3) and (4). In these cases, the outliers influenced the slope of the least squares lines. In (5), data with no clear trend were assigned a line with a large trend simply due to one outlier (!).

Leverage

Points that fall horizontally away from the center of the cloud tend to pull harder on the line, so we call them points with **high leverage**.

Points that fall horizontally far from the line are points of high leverage; these points can strongly influence the slope of the least squares line. If one of these high leverage points does appear to actually invoke its influence on the slope of the line – as in cases (3), (4), and (5) of Example 7.23 – then we call it an **influential point**. Usually we can say a point is influential if, had we fitted the line without it, the influential point would have been unusually far from the least squares line.

It is tempting to remove outliers. Don't do this without a very good reason. Models that ignore exceptional (and interesting) cases often perform poorly. For instance, if a financial firm ignored the largest market swings – the "outliers" – they would soon go bankrupt by making poorly thought-out investments.

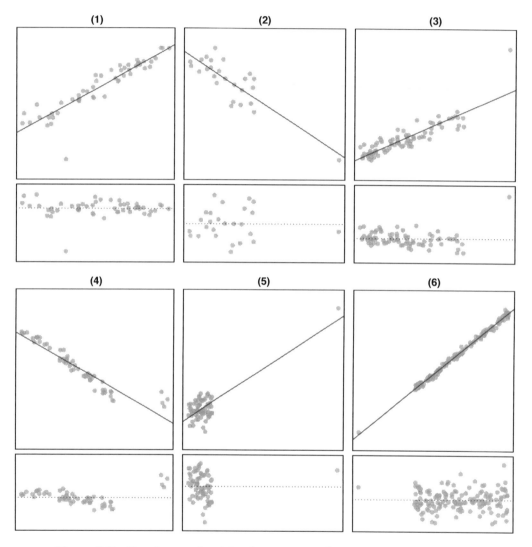

Figure 7.19: Six plots, each with a least squares line and residual plot. All data sets have at least one outlier.

Caution: Don't ignore outliers when fitting a final model

If there are outliers in the data, they should not be removed or ignored without a good reason. Whatever final model is fit to the data would not be very helpful if it ignores the most exceptional cases.

Caution: Outliers for a categorical predictor with two levels

Be cautious about using a categorical predictor when one of the levels has very few observations. When this happens, those few observations become influential points.

7.4 Inference for linear regression

In this section we discuss uncertainty in the estimates of the slope and y-intercept for a regression line. Just as we identified standard errors for point estimates in previous chapters, we first discuss standard errors for these new estimates. However, in the case of regression, we will identify standard errors using statistical software.

7.4.1 Midterm elections and unemployment

Elections for members of the United States House of Representatives occur every two years, coinciding every four years with the U.S. Presidential election. The set of House elections occurring during the middle of a Presidential term are called midterm elections. In America's two-party system, one political theory suggests the higher the unemployment rate, the worse the President's party will do in the midterm elections.

To assess the validity of this claim, we can compile historical data and look for a connection. We consider every midterm election from 1898 to 2010, with the exception of those elections during the Great Depression. Figure 7.20 shows these data and the least-squares regression line:

$$\% \text{ change in House seats for President's party}$$
$$= -6.71 - 1.00 \times (\text{unemployment rate})$$

We consider the percent change in the number of seats of the President's party (e.g. percent change in the number of seats for Democrats in 2010) against the unemployment rate.

Examining the data, there are no clear deviations from linearity, the constant variance condition, or in the normality of residuals (though we don't examine a normal probability plot here). While the data are collected sequentially, a separate analysis was used to check for any apparent correlation between successive observations; no such correlation was found.

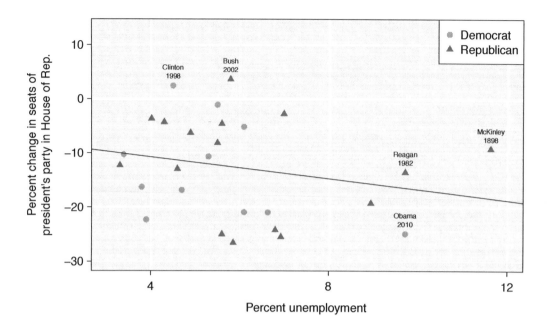

Figure 7.20: The percent change in House seats for the President's party in each election from 1898 to 2010 plotted against the unemployment rate. The two points for the Great Depression have been removed, and a least squares regression line has been fit to the data.

⊙ **Guided Practice 7.24**   The data for the Great Depression (1934 and 1938) were removed because the unemployment rate was 21% and 18%, respectively. Do you agree that they should be removed for this investigation? Why or why not?[14]

There is a negative slope in the line shown in Figure 7.20. However, this slope (and the y-intercept) are only estimates of the parameter values. We might wonder, is this convincing evidence that the "true" linear model has a negative slope? That is, do the data provide strong evidence that the political theory is accurate? We can frame this investigation into a one-sided statistical hypothesis test:

H_0: $\beta_1 = 0$. The true linear model has slope zero.

H_A: $\beta_1 < 0$. The true linear model has a slope less than zero. The higher the unemployment, the greater the loss for the President's party in the House of Representatives.

We would reject H_0 in favor of H_A if the data provide strong evidence that the true slope parameter is less than zero. To assess the hypotheses, we identify a standard error for the estimate, compute an appropriate test statistic, and identify the p-value.

---

[14]We will provide two considerations. Each of these points would have very high leverage on any least-squares regression line, and years with such high unemployment may not help us understand what would happen in other years where the unemployment is only modestly high. On the other hand, these are exceptional cases, and we would be discarding important information if we exclude them from a final analysis.

7.4.2 Understanding regression output from software

Just like other point estimates we have seen before, we can compute a standard error and test statistic for b_1. We will generally label the test statistic using a T, since it follows the t-distribution.

We will rely on statistical software to compute the standard error and leave the explanation of how this standard error is determined to a second or third statistics course. Table 7.21 shows software output for the least squares regression line in Figure 7.20. The row labeled *unemp* represents the information for the slope, which is the coefficient of the unemployment variable.

	Estimate	Std. Error	t value	Pr($>$|t|)
(Intercept)	-6.7142	5.4567	-1.23	0.2300
unemp	-1.0010	0.8717	-1.15	0.2617
				$df = 25$

Table 7.21: Output from statistical software for the regression line modeling the midterm election losses for the President's party as a response to unemployment.

● **Example 7.25** What do the first and second columns of Table 7.21 represent?

The entries in the first column represent the least squares estimates, b_0 and b_1, and the values in the second column correspond to the standard errors of each estimate.

We previously used a t-test statistic for hypothesis testing in the context of numerical data. Regression is very similar. In the hypotheses we consider, the null value for the slope is 0, so we can compute the test statistic using the T (or Z) score formula:

$$T = \frac{\text{estimate} - \text{null value}}{\text{SE}} = \frac{-1.0010 - 0}{0.8717} = -1.15$$

We can look for the one-sided p-value – shown in Figure 7.22 – using the probability table for the t-distribution in Appendix B.2 on page 430.

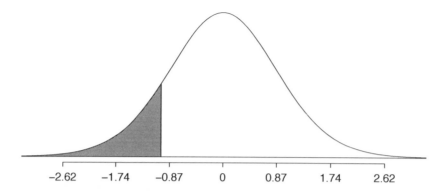

Figure 7.22: The distribution shown here is the sampling distribution for b_1, if the null hypothesis was true. The shaded tail represents the p-value for the hypothesis test evaluating whether there is convincing evidence that higher unemployment corresponds to a greater loss of House seats for the President's party during a midterm election.

● **Example 7.26** Table 7.21 offers the degrees of freedom for the test statistic T: $df = 25$. Identify the p-value for the hypothesis test.

Looking in the 25 degrees of freedom row in Appendix B.2, we see that the absolute value of the test statistic is smaller than any value listed, which means the tail area and therefore also the p-value is larger than 0.100 (one tail!). Because the p-value is so large, we fail to reject the null hypothesis. That is, the data do not provide convincing evidence that a higher unemployment rate has any correspondence with smaller or larger losses for the President's party in the House of Representatives in midterm elections.

We could have identified the t-test statistic from the software output in Table 7.21, shown in the second row (unemp) and third column (t value). The entry in the second row and last column in Table 7.21 represents the p-value for the two-sided hypothesis test where the null value is zero. The corresponding one-sided test would have a p-value half of the listed value.

Inference for regression

We usually rely on statistical software to identify point estimates and standard errors for parameters of a regression line. After verifying conditions hold for fitting a line, we can use the methods learned in Section 5.1 for the t-distribution to create confidence intervals for regression parameters or to evaluate hypothesis tests.

Caution: Don't carelessly use the p-value from regression output

The last column in regression output often lists p-values for one particular hypothesis: a two-sided test where the null value is zero. If your test is one-sided and the point estimate is in the direction of H_A, then you can halve the software's p-value to get the one-tail area. If neither of these scenarios match your hypothesis test, be cautious about using the software output to obtain the p-value.

● **Example 7.27** Examine Figure 7.16 on page 346, which relates the Elmhurst College aid and student family income. How sure are you that the slope is statistically significantly different from zero? That is, do you think a formal hypothesis test would reject the claim that the true slope of the line should be zero?

While the relationship between the variables is not perfect, there is an evident decreasing trend in the data. This suggests the hypothesis test will reject the null claim that the slope is zero.

⊙ **Guided Practice 7.28** Table 7.23 shows statistical software output from fitting the least squares regression line shown in Figure 7.16. Use this output to formally evaluate the following hypotheses. H_0: The true coefficient for family income is zero. H_A: The true coefficient for family income is not zero.[15]

	Estimate	Std. Error	t value	$Pr(>|t|)$
(Intercept)	24.3193	1.2915	18.83	0.0000
family_income	-0.0431	0.0108	-3.98	0.0002
				$df = 48$

Table 7.23: Summary of least squares fit for the Elmhurst College data.

TIP: Always check assumptions
If conditions for fitting the regression line do not hold, then the methods presented here should not be applied. The standard error or distribution assumption of the point estimate – assumed to be normal when applying the t-test statistic – may not be valid.

▶ **Calculator videos**
Videos covering hypothesis testing for a regression coefficient using TI and Casio graphing calculators are available at openintro.org/videos.

---

[15]We look in the second row corresponding to the family income variable. We see the point estimate of the slope of the line is -0.0431, the standard error of this estimate is 0.0108, and the t-test statistic is -3.98. The p-value corresponds exactly to the two-sided test we are interested in: 0.0002. The p-value is so small that we reject the null hypothesis and conclude that family income and financial aid at Elmhurst College for freshman entering in the year 2011 are negatively correlated and the true slope parameter is indeed less than 0, just as we believed in Example 7.27.

7.5   Exercises

7.5.1   Line fitting, residuals, and correlation

7.1   Visualize the residuals. The scatterplots shown below each have a superimposed regression line. If we were to construct a residual plot (residuals versus x) for each, describe what those plots would look like.

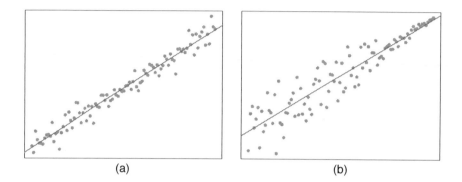

(a)   (b)

7.2   Trends in the residuals. Shown below are two plots of residuals remaining after fitting a linear model to two different sets of data. Describe important features and determine if a linear model would be appropriate for these data. Explain your reasoning.

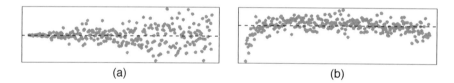

(a)   (b)

7.3   Identify relationships, Part I. For each of the six plots, identify the strength of the relationship (e.g. weak, moderate, or strong) in the data and whether fitting a linear model would be reasonable.

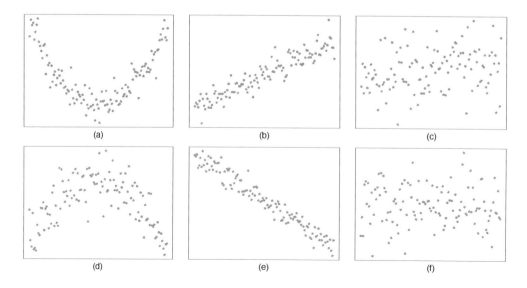

(a)   (b)   (c)

(d)   (e)   (f)

7.4 Identify relationships, Part II. For each of the six plots, identify the strength of the relationship (e.g. weak, moderate, or strong) in the data and whether fitting a linear model would be reasonable.

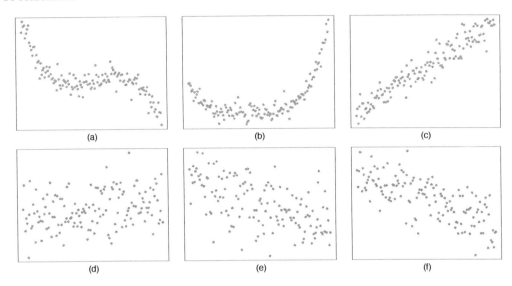

7.5 Exams and grades. The two scatterplots below show the relationship between final and mid-semester exam grades recorded during several years for a Statistics course at a university.

(a) Based on these graphs, which of the two exams has the strongest correlation with the final exam grade? Explain.

(b) Can you think of a reason why the correlation between the exam you chose in part (a) and the final exam is higher?

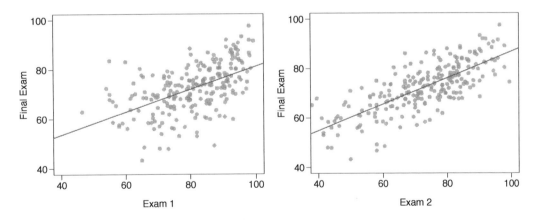

7.6 Husbands and wives, Part I. The Great Britain Office of Population Census and Surveys once collected data on a random sample of 170 married couples in Britain, recording the age (in years) and heights (converted here to inches) of the husbands and wives.[16] The scatterplot on the left shows the wife's age plotted against her husband's age, and the plot on the right shows wife's height plotted against husband's height.

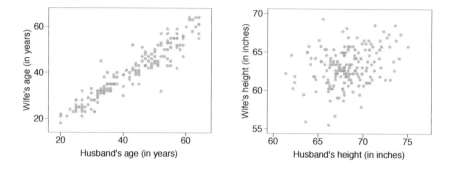

(a) Describe the relationship between husbands' and wives' ages.

(b) Describe the relationship between husbands' and wives' heights.

(c) Which plot shows a stronger correlation? Explain your reasoning.

(d) Data on heights were originally collected in centimeters, and then converted to inches. Does this conversion affect the correlation between husbands' and wives' heights?

7.7 Match the correlation, Part I. Match the calculated correlations to the corresponding scatterplot.

(a) $r = -0.7$

(b) $r = 0.45$

(c) $r = 0.06$

(d) $r = 0.92$

7.8 Match the correlation, Part II. Match the calculated correlations to the corresponding scatterplot.

(a) $r = 0.49$

(b) $r = -0.48$

(c) $r = -0.03$

(d) $r = -0.85$

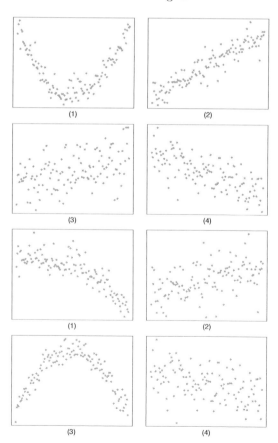

[16]D.J. Hand. *A handbook of small data sets.* Chapman & Hall/CRC, 1994.

7.9 True / False. Determine if the following statements are true or false. If false, explain why.

(a) A correlation coefficient of -0.90 indicates a stronger linear relationship than a correlation coefficient of 0.5.

(b) Correlation is a measure of the association between any two variables.

7.10 Guess the correlation. Eduardo and Rosie are both collecting data on number of rainy days in a year and the total rainfall for the year. Eduardo records rainfall in inches and Rosie in centimeters. How will their correlation coefficients compare?

7.11 Speed and height. 1,302 UCLA students were asked to fill out a survey where they were asked about their height, fastest speed they have ever driven, and gender. The scatterplot on the left displays the relationship between height and fastest speed, and the scatterplot on the right displays the breakdown by gender in this relationship.

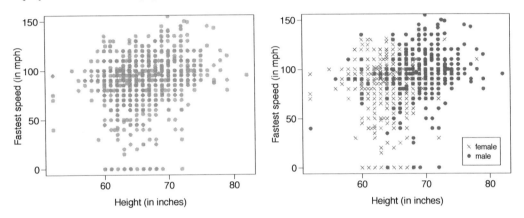

(a) Describe the relationship between height and fastest speed.

(b) Why do you think these variables are positively associated?

(c) What role does gender play in the relationship between height and fastest driving speed?

7.12   Trees. The scatterplots below show the relationship between height, diameter, and volume of timber in 31 felled black cherry trees. The diameter of the tree is measured 4.5 feet above the ground.[17]

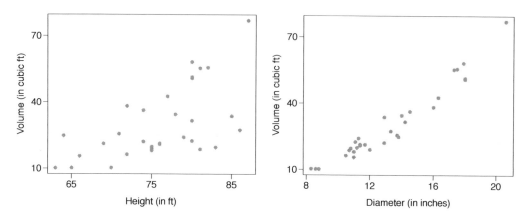

(a) Describe the relationship between volume and height of these trees.

(b) Describe the relationship between volume and diameter of these trees.

(c) Suppose you have height and diameter measurements for another black cherry tree. Which of these variables would be preferable to use to predict the volume of timber in this tree using a simple linear regression model? Explain your reasoning.

7.13   The Coast Starlight, Part I. The Coast Starlight Amtrak train runs from Seattle to Los Angeles. The scatterplot below displays the distance between each stop (in miles) and the amount of time it takes to travel from one stop to another (in minutes).

(a) Describe the relationship between distance and travel time.

(b) How would the relationship change if travel time was instead measured in hours, and distance was instead measured in kilometers?

(c) Correlation between travel time (in miles) and distance (in minutes) is $r = 0.636$. What is the correlation between travel time (in kilometers) and distance (in hours)?

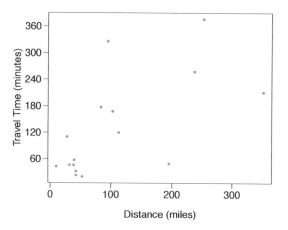

[17]Source: R Dataset, stat.ethz.ch/R-manual/R-patched/library/datasets/html/trees.html.

7.14 Crawling babies, Part I. A study conducted at the University of Denver investigated whether babies take longer to learn to crawl in cold months, when they are often bundled in clothes that restrict their movement, than in warmer months.[18] Infants born during the study year were split into twelve groups, one for each birth month. We consider the average crawling age of babies in each group against the average temperature when the babies are six months old (that's when babies often begin trying to crawl). Temperature is measured in degrees Fahrenheit (°F) and age is measured in weeks.

(a) Describe the relationship between temperature and crawling age.

(b) How would the relationship change if temperature was measured in degrees Celsius (°C) and age was measured in months?

(c) The correlation between temperature in °F and age in weeks was $r = -0.70$. If we converted the temperature to °C and age to months, what would the correlation be?

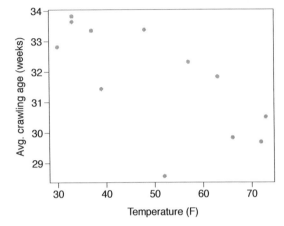

7.15 Body measurements, Part I. Researchers studying anthropometry collected body girth measurements and skeletal diameter measurements, as well as age, weight, height and gender for 507 physically active individuals.[19] The scatterplot below shows the relationship between height and shoulder girth (over deltoid muscles), both measured in centimeters.

(a) Describe the relationship between shoulder girth and height.

(b) How would the relationship change if shoulder girth was measured in inches while the units of height remained in centimeters?

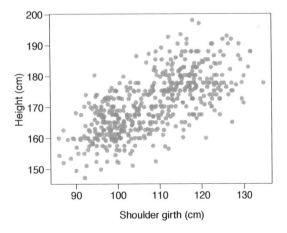

[18] J.B. Benson. "Season of birth and onset of locomotion: Theoretical and methodological implications". In: *Infant behavior and development* 16.1 (1993), pp. 69–81. ISSN: 0163-6383.

[19] G. Heinz et al. "Exploring relationships in body dimensions". In: *Journal of Statistics Education* 11.2 (2003).

7.16 Body measurements, Part II. The scatterplot below shows the relationship between weight measured in kilograms and hip girth measured in centimeters from the data described in Exercise 7.15.

(a) Describe the relationship between hip girth and weight.

(b) How would the relationship change if weight was measured in pounds while the units for hip girth remained in centimeters?

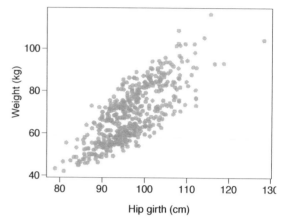

7.17 Correlation, Part I. What would be the correlation between the ages of husbands and wives if men always married woman who were

(a) 3 years younger than themselves?

(b) 2 years older than themselves?

(c) half as old as themselves?

7.18 Correlation, Part II. What would be the correlation between the annual salaries of males and females at a company if for a certain type of position men always made

(a) \$5,000 more than women?

(b) 25% more than women?

(c) 15% less than women?

7.5.2   Fitting a line by least squares regression

7.19 Units of regression. Consider a regression predicting weight (kg) from height (cm) for a sample of adult males. What are the units of the correlation coefficient, the intercept, and the slope?

7.20 Which is higher? Determine if I or II is higher or if they are equal. Explain your reasoning.

For a regression line, the uncertainty associated with the slope estimate, b_1, is higher when

   I. there is a lot of scatter around the regression line or

   II. there is very little scatter around the regression line

7.21 Over-under, Part I. Suppose we fit a regression line to predict the shelf life of an apple based on its weight. For a particular apple, we predict the shelf life to be 4.6 days. The apple's residual is -0.6 days. Did we over or under estimate the shelf-life of the apple? Explain your reasoning.

7.22 Over-under, Part II. Suppose we fit a regression line to predict the number of incidents of skin cancer per 1,000 people from the number of sunny days in a year. For a particular year, we predict the incidence of skin cancer to be 1.5 per 1,000 people, and the residual for this year is 0.5. Did we over or under estimate the incidence of skin cancer? Explain your reasoning.

7.23 Tourism spending. The Association of Turkish Travel Agencies reports the number of foreign tourists visiting Turkey and tourist spending by year.[20] Three plots are provided: scatterplot showing the relationship between these two variables along with the least squares fit, residuals plot, and histogram of residuals.

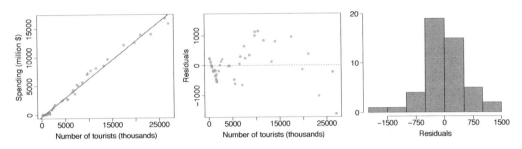

(a) Describe the relationship between number of tourists and spending.

(b) What are the explanatory and response variables?

(c) Why might we want to fit a regression line to these data?

(d) Do the data meet the conditions required for fitting a least squares line? In addition to the scatterplot, use the residual plot and histogram to answer this question.

7.24 Nutrition at Starbucks, Part I. The scatterplot below shows the relationship between the number of calories and amount of carbohydrates (in grams) Starbucks food menu items contain.[21] Since Starbucks only lists the number of calories on the display items, we are interested in predicting the amount of carbs a menu item has based on its calorie content.

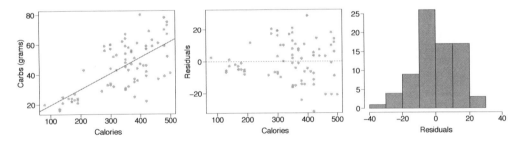

(a) Describe the relationship between number of calories and amount of carbohydrates (in grams) that Starbucks food menu items contain.

(b) In this scenario, what are the explanatory and response variables?

(c) Why might we want to fit a regression line to these data?

(d) Do these data meet the conditions required for fitting a least squares line?

---

[20]Association of Turkish Travel Agencies, Foreign Visitors Figure & Tourist Spendings By Years.

[21]Source: Starbucks.com, collected on March 10, 2011,
www.starbucks.com/menu/nutrition.

7.25   The Coast Starlight, Part II. Exercise 7.13 introduces data on the Coast Starlight Amtrak train that runs from Seattle to Los Angeles. The mean travel time from one stop to the next on the Coast Starlight is 129 mins, with a standard deviation of 113 minutes. The mean distance traveled from one stop to the next is 108 miles with a standard deviation of 99 miles. The correlation between travel time and distance is 0.636.

(a) Write the equation of the regression line for predicting travel time.

(b) Interpret the slope and the intercept in this context.

(c) Calculate R^2 of the regression line for predicting travel time from distance traveled for the Coast Starlight, and interpret R^2 in the context of the application.

(d) The distance between Santa Barbara and Los Angeles is 103 miles. Use the model to estimate the time it takes for the Starlight to travel between these two cities.

(e) It actually takes the Coast Starlight about 168 mins to travel from Santa Barbara to Los Angeles. Calculate the residual and explain the meaning of this residual value.

(f) Suppose Amtrak is considering adding a stop to the Coast Starlight 500 miles away from Los Angeles. Would it be appropriate to use this linear model to predict the travel time from Los Angeles to this point?

7.26   Body measurements, Part III. Exercise 7.15 introduces data on shoulder girth and height of a group of individuals. The mean shoulder girth is 107.20 cm with a standard deviation of 10.37 cm. The mean height is 171.14 cm with a standard deviation of 9.41 cm. The correlation between height and shoulder girth is 0.67.

(a) Write the equation of the regression line for predicting height.

(b) Interpret the slope and the intercept in this context.

(c) Calculate R^2 of the regression line for predicting height from shoulder girth, and interpret it in the context of the application.

(d) A randomly selected student from your class has a shoulder girth of 100 cm. Predict the height of this student using the model.

(e) The student from part (d) is 160 cm tall. Calculate the residual, and explain what this residual means.

(f) A one year old has a shoulder girth of 56 cm. Would it be appropriate to use this linear model to predict the height of this child?

7.27   Nutrition at Starbucks, Part II. Exercise 7.24 introduced a data set on nutrition information on Starbucks food menu items. Based on the scatterplot and the residual plot provided, describe the relationship between the protein content and calories of these menu items, and determine if a simple linear model is appropriate to predict amount of protein from the number of calories.

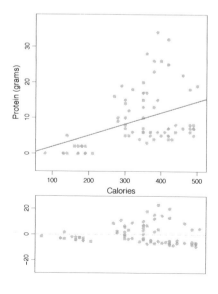

7.28 Helmets and lunches. The scatterplot shows the relationship between socioeconomic status measured as the percentage of children in a neighborhood receiving reduced-fee lunches at school (`lunch`) and the percentage of bike riders in the neighborhood wearing helmets (`helmet`). The average percentage of children receiving reduced-fee lunches is 30.8% with a standard deviation of 26.7% and the average percentage of bike riders wearing helmets is 38.8% with a standard deviation of 16.9%.

(a) If the R^2 for the least-squares regression line for these data is 72%, what is the correlation between `lunch` and `helmet`?

(b) Calculate the slope and intercept for the least-squares regression line for these data.

(c) Interpret the intercept of the least-squares regression line in the context of the application.

(d) Interpret the slope of the least-squares regression line in the context of the application.

(e) What would the value of the residual be for a neighborhood where 40% of the children receive reduced-fee lunches and 40% of the bike riders wear helmets? Interpret the meaning of this residual in the context of the application.

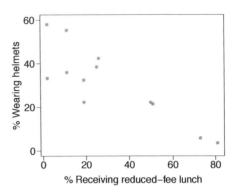

7.29 Murders and poverty, Part I. The following regression output is for predicting annual murders per million from percentage living in poverty in a random sample of 20 metropolitan areas.

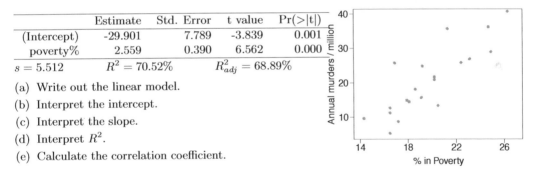

	Estimate	Std. Error	t value	Pr(>|t|)
(Intercept)	-29.901	7.789	-3.839	0.001
poverty%	2.559	0.390	6.562	0.000

$s = 5.512 \qquad R^2 = 70.52\% \qquad R^2_{adj} = 68.89\%$

(a) Write out the linear model.

(b) Interpret the intercept.

(c) Interpret the slope.

(d) Interpret R^2.

(e) Calculate the correlation coefficient.

7.30 Cats, Part I. The following regression output is for predicting the heart weight (in g) of cats from their body weight (in kg). The coefficients are estimated using a dataset of 144 domestic cats.

	Estimate	Std. Error	t value	Pr(>|t|)
(Intercept)	-0.357	0.692	-0.515	0.607
body wt	4.034	0.250	16.119	0.000

$s = 1.452 \qquad R^2 = 64.66\% \qquad R^2_{adj} = 64.41\%$

(a) Write out the linear model.

(b) Interpret the intercept.

(c) Interpret the slope.

(d) Interpret R^2.

(e) Calculate the correlation coefficient.

7.5.3   Types of outliers in linear regression

7.31   Outliers, Part I. Identify the outliers in the scatterplots shown below, and determine what type of outliers they are. Explain your reasoning.

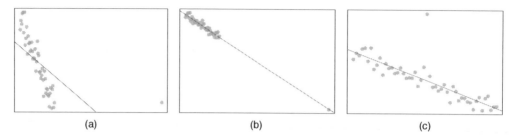

| (a) | (b) | (c) |

7.32   Outliers, Part II. Identify the outliers in the scatterplots shown below and determine what type of outliers they are. Explain your reasoning.

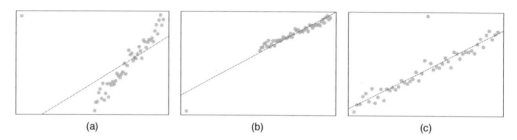

| (a) | (b) | (c) |

7.33   Urban homeowners, Part I. The scatterplot below shows the percent of families who own their home vs. the percent of the population living in urban areas in 2010.[22] There are 52 observations, each corresponding to a state in the US. Puerto Rico and District of Columbia are also included.

(a) Describe the relationship between the percent of families who own their home and the percent of the population living in urban areas in 2010.

(b) The outlier at the bottom right corner is District of Columbia, where 100% of the population is considered urban. What type of outlier is this observation?

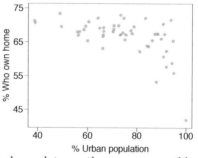

7.34   Crawling babies, Part II. Exercise 7.14 introduces data on the average monthly temperature during the month babies first try to crawl (about 6 months after birth) and the average first crawling age for babies born in a given month. A scatterplot of these two variables reveals a potential outlying month when the average temperature is about 53°F and average crawling age is about 28.5 weeks. Does this point have high leverage? Is it an influential point?

---

[22]United States Census Bureau, 2010 Census Urban and Rural Classification and Urban Area Criteria and Housing Characteristics: 2010.

7.5.4 Inference for linear regression

In the following exercises, visually check the conditions for fitting a least squares regression line, but you do not need to report these conditions in your solutions.

7.35 Body measurements, Part IV. The scatterplot and least squares summary below show the relationship between weight measured in kilograms and height measured in centimeters of 507 physically active individuals.

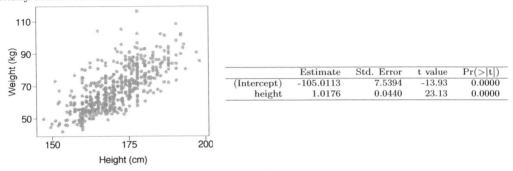

	Estimate	Std. Error	t value	Pr(>|t|)
(Intercept)	-105.0113	7.5394	-13.93	0.0000
height	1.0176	0.0440	23.13	0.0000

(a) Describe the relationship between height and weight.

(b) Write the equation of the regression line. Interpret the slope and intercept in context.

(c) Do the data provide strong evidence that an increase in height is associated with an increase in weight? State the null and alternative hypotheses, report the p-value, and state your conclusion.

(d) The correlation coefficient for height and weight is 0.72. Calculate R^2 and interpret it in context.

7.36 Beer and blood alcohol content. Many people believe that gender, weight, drinking habits, and many other factors are much more important in predicting blood alcohol content (BAC) than simply considering the number of drinks a person consumed. Here we examine data from sixteen student volunteers at Ohio State University who each drank a randomly assigned number of cans of beer. These students were evenly divided between men and women, and they differed in weight and drinking habits. Thirty minutes later, a police officer measured their blood alcohol content (BAC) in grams of alcohol per deciliter of blood.[23] The scatterplot and regression table summarize the findings.

	Estimate	Std. Error	t value	Pr(>|t|)
(Intercept)	-0.0127	0.0126	-1.00	0.3320
beers	0.0180	0.0024	7.48	0.0000

(a) Describe the relationship between the number of cans of beer and BAC.

(b) Write the equation of the regression line. Interpret the slope and intercept in context.

(c) Do the data provide strong evidence that drinking more cans of beer is associated with an increase in blood alcohol? State the null and alternative hypotheses, report the p-value, and state your conclusion.

(d) The correlation coefficient for number of cans of beer and BAC is 0.89. Calculate R^2 and interpret it in context.

(e) Suppose we visit a bar, ask people how many drinks they have had, and also take their BAC. Do you think the relationship between number of drinks and BAC would be as strong as the relationship found in the Ohio State study?

7.37 Husbands and wives, Part II. The scatterplot below summarizes husbands' and wives' heights in a random sample of 170 married couples in Britain, where both partners' ages are below 65 years. Summary output of the least squares fit for predicting wife's height from husband's height is also provided in the table.

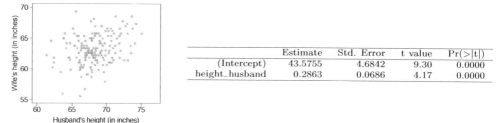

	Estimate	Std. Error	t value	Pr(>|t|)
(Intercept)	43.5755	4.6842	9.30	0.0000
height_husband	0.2863	0.0686	4.17	0.0000

(a) Is there strong evidence that taller men marry taller women? State the hypotheses and include any information used to conduct the test.

(b) Write the equation of the regression line for predicting wife's height from husband's height.

(c) Interpret the slope and intercept in the context of the application.

(d) Given that $R^2 = 0.09$, what is the correlation of heights in this data set?

(e) You meet a married man from Britain who is 5'9" (69 inches). What would you predict his wife's height to be? How reliable is this prediction?

(f) You meet another married man from Britain who is 6'7" (79 inches). Would it be wise to use the same linear model to predict his wife's height? Why or why not?

---

[23] J. Malkevitch and L.M. Lesser. *For All Practical Purposes: Mathematical Literacy in Today's World.* WH Freeman & Co, 2008.

7.38 Husbands and wives, Part III. Exercise 7.37 presents a scatterplot displaying the relationship between husbands' and wives' ages in a random sample of 170 married couples in Britain, where both partners' ages are below 65 years. Given below is summary output of the least squares fit for predicting wife's age from husband's age.

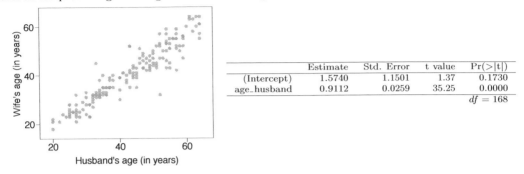

	Estimate	Std. Error	t value	Pr($>$|t|)
(Intercept)	1.5740	1.1501	1.37	0.1730
age_husband	0.9112	0.0259	35.25	0.0000
				$df = 168$

(a) We might wonder, is the age difference between husbands and wives consistent across ages? If this were the case, then the slope parameter would be $\beta_1 = 1$. Use the information above to evaluate if there is strong evidence that the difference in husband and wife ages differs for different ages.

(b) Write the equation of the regression line for predicting wife's age from husband's age.

(c) Interpret the slope and intercept in context.

(d) Given that $R^2 = 0.88$, what is the correlation of ages in this data set?

(e) You meet a married man from Britain who is 55 years old. What would you predict his wife's age to be? How reliable is this prediction?

(f) You meet another married man from Britain who is 85 years old. Would it be wise to use the same linear model to predict his wife's age? Explain.

7.39 Urban homeowners, Part II.

Exercise 7.33 gives a scatterplot displaying the relationship between the percent of families that own their home and the percent of the population living in urban areas. Below is a similar scatterplot, excluding District of Columbia, as well as the residuals plot. There were 51 cases.

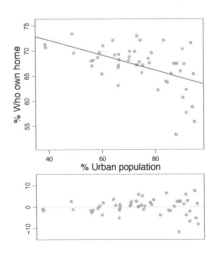

(a) For these data, $R^2 = 0.28$. What is the correlation? How can you tell if it is positive or negative?

(b) Examine the residual plot. What do you observe? Is a simple least squares fit appropriate for these data?

7.40 Rate my professor. Many college courses conclude by giving students the opportunity to evaluate the course and the instructor anonymously. However, the use of these student evaluations as an indicator of course quality and teaching effectiveness is often criticized because these measures may reflect the influence of non-teaching related characteristics, such as the physical appearance of the instructor. Researchers at University of Texas, Austin collected data on teaching evaluation score (higher score means better) and standardized beauty score (a score of 0 means average, negative score means below average, and a positive score means above average) for a sample of 463 professors.[24] The scatterplot below shows the relationship between these variables, and also provided is a regression output for predicting teaching evaluation score from beauty score.

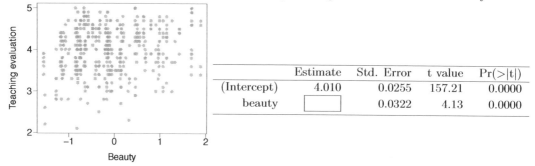

	Estimate	Std. Error	t value	Pr($>$|t|)
(Intercept)	4.010	0.0255	157.21	0.0000
beauty		0.0322	4.13	0.0000

(a) Given that the average standardized beauty score is -0.0883 and average teaching evaluation score is 3.9983, calculate the slope. Alternatively, the slope may be computed using just the information provided in the model summary table.

(b) Do these data provide convincing evidence that the slope of the relationship between teaching evaluation and beauty is positive? Explain your reasoning.

(c) List the conditions required for linear regression and check if each one is satisfied for this model based on the following diagnostic plots.

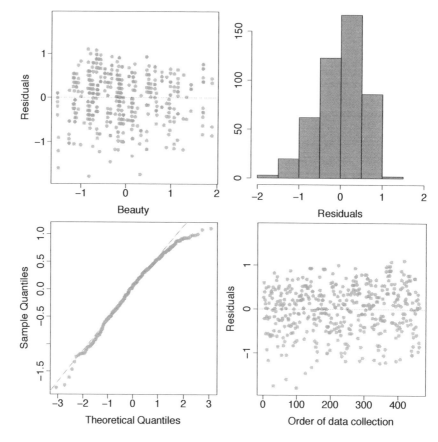

[24]Daniel S Hamermesh and Amy Parker. "Beauty in the classroom: Instructors pulchritude and putative pedagogical productivity". In: *Economics of Education Review* 24.4 (2005), pp. 369–376.

7.41 Murders and poverty, Part II. Exercise 7.29 presents regression output from a model for predicting annual murders per million from percentage living in poverty based on a random sample of 20 metropolitan areas. The model output is also provided below.

	Estimate	Std. Error	t value	Pr(>|t|)
(Intercept)	-29.901	7.789	-3.839	0.001
poverty%	2.559	0.390	6.562	0.000

$$s = 5.512 \qquad R^2 = 70.52\% \qquad R^2_{adj} = 68.89\%$$

(a) What are the hypotheses for evaluating whether poverty percentage is a significant predictor of murder rate?

(b) State the conclusion of the hypothesis test from part (a) in context of the data.

(c) Calculate a 95% confidence interval for the slope of poverty percentage, and interpret it in context of the data.

(d) Do your results from the hypothesis test and the confidence interval agree? Explain.

7.42 Babies. Is the gestational age (time between conception and birth) of a low birth-weight baby useful in predicting head circumference at birth? Twenty-five low birth-weight babies were studied at a Harvard teaching hospital; the investigators calculated the regression of head circumference (measured in centimeters) against gestational age (measured in weeks). The estimated regression line is

$$\widehat{head\_circumference} = 3.91 + 0.78 \times gestational\_age$$

(a) What is the predicted head circumference for a baby whose gestational age is 28 weeks?

(b) The standard error for the coefficient of gestational age is 0.35, which is associated with $df = 23$. Does the model provide strong evidence that gestational age is significantly associated with head circumference?

7.43 Murders and poverty, Part III. In Exercises 7.41 you evaluated whether poverty percentage is a significant predictor of murder rate. How, if at all, would your answer change if we wanted to find out whether poverty percentage is positively associated with murder rate. Make sure to include the appropriate p-value for this hypothesis test in your answer.

7.44 Cats, Part II. Exercise 7.30 presents regression output from a model for predicting the heart weight (in g) of cats from their body weight (in kg). The coefficients are estimated using a dataset of 144 domestic cats. The model output is also provided below.

	Estimate	Std. Error	t value	Pr(>|t|)
(Intercept)	-0.357	0.692	-0.515	0.607
body wt	4.034	0.250	16.119	0.000

$$s = 1.452 \qquad R^2 = 64.66\% \qquad R^2_{adj} = 64.41\%$$

(a) What are the hypotheses for evaluating whether body weight is positively associated with heart weight in cats?

(b) State the conclusion of the hypothesis test from part (a) in context of the data.

(c) Calculate a 95% confidence interval for the slope of body weight, and interpret it in context of the data.

(d) Do your results from the hypothesis test and the confidence interval agree? Explain.

Chapter 8

Multiple and logistic regression

The principles of simple linear regression lay the foundation for more sophisticated regression methods used in a wide range of challenging settings. In Chapter 8, we explore multiple regression, which introduces the possibility of more than one predictor, and logistic regression, a technique for predicting categorical outcomes with two possible categories.

8.1 Introduction to multiple regression

Multiple regression extends simple two-variable regression to the case that still has one response but many predictors (denoted x_1, x_2, x_3, ...). The method is motivated by scenarios where many variables may be simultaneously connected to an output.

We will consider Ebay auctions of a video game called *Mario Kart* for the Nintendo Wii. The outcome variable of interest is the total price of an auction, which is the highest bid plus the shipping cost. We will try to determine how total price is related to each characteristic in an auction while simultaneously controlling for other variables. For instance, all other characteristics held constant, are longer auctions associated with higher or lower prices? And, on average, how much more do buyers tend to pay for additional Wii wheels (plastic steering wheels that attach to the Wii controller) in auctions? Multiple regression will help us answer these and other questions.

The data set `mario_kart` includes results from 141 auctions.[1] Four observations from this data set are shown in Table 8.1, and descriptions for each variable are shown in Table 8.2. Notice that the condition and stock photo variables are indicator variables. For instance, the `cond_new` variable takes value 1 if the game up for auction is new and 0 if it is used. Using indicator variables in place of category names allows for these variables to be directly used in regression. See Section 7.2.7 for additional details. Multiple regression also allows for categorical variables with many levels, though we do not have any such variables in this analysis, and we save these details for a second or third course.

---

[1] Diez DM, Barr CD, Çetinkaya-Rundel M. 2015. `openintro`: OpenIntro data sets and supplement functions. github.com/OpenIntroOrg/openintro-r-package.

	price	cond_new	stock_photo	duration	wheels
1	51.55	1	1	3	1
2	37.04	0	1	7	1
⋮	⋮	⋮	⋮	⋮	⋮
140	38.76	0	0	7	0
141	54.51	1	1	1	2

Table 8.1: Four observations from the `mario_kart` data set.

variable	description
price	final auction price plus shipping costs, in US dollars
cond_new	a coded two-level categorical variable, which takes value 1 when the game is new and 0 if the game is used
stock_photo	a coded two-level categorical variable, which takes value 1 if the primary photo used in the auction was a stock photo and 0 if the photo was unique to that auction
duration	the length of the auction, in days, taking values from 1 to 10
wheels	the number of Wii wheels included with the auction (a *Wii wheel* is a plastic racing wheel that holds the Wii controller and is an optional but helpful accessory for playing Mario Kart)

Table 8.2: Variables and their descriptions for the `mario_kart` data set.

8.1.1 A single-variable model for the Mario Kart data

Let's fit a linear regression model with the game's condition as a predictor of auction price. The model may be written as

$$\widehat{price} = 42.87 + 10.90 \times cond\_new$$

Results of this model are shown in Table 8.3 and a scatterplot for price versus game condition is shown in Figure 8.4.

	Estimate	Std. Error	t value	Pr(>|t|)
(Intercept)	42.8711	0.8140	52.67	0.0000
cond_new	10.8996	1.2583	8.66	0.0000
				$df = 139$

Table 8.3: Summary of a linear model for predicting auction price based on game condition.

⊙ **Guided Practice 8.1** Examine Figure 8.4. Does the linear model seem reasonable?[2]

---

[2]Yes. Constant variability, nearly normal residuals, and linearity all appear reasonable.

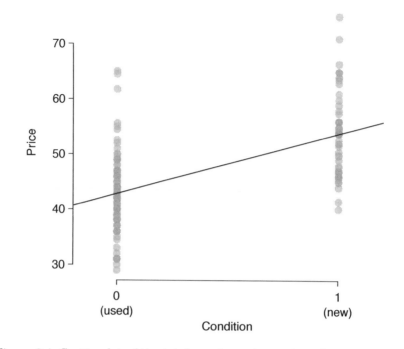

Figure 8.4: Scatterplot of the total auction price against the game's condition. The least squares line is also shown.

● **Example 8.2**   Interpret the coefficient for the game's condition in the model. Is this coefficient significantly different from 0?

Note that `cond_new` is a two-level categorical variable that takes value 1 when the game is new and value 0 when the game is used. So 10.90 means that the model predicts an extra \$10.90 for those games that are new versus those that are used. (See Section 7.2.7 for a review of the interpretation for two-level categorical predictor variables.) Examining the regression output in Table 8.3, we can see that the p-value for `cond_new` is very close to zero, indicating there is strong evidence that the coefficient is different from zero when using this simple one-variable model.

8.1.2   Including and assessing many variables in a model

Sometimes there are underlying structures or relationships between predictor variables. For instance, new games sold on Ebay tend to come with more Wii wheels, which may have led to higher prices for those auctions. We would like to fit a model that includes all potentially important variables simultaneously. This would help us evaluate the relationship between a predictor variable and the outcome while controlling for the potential influence of other variables. This is the strategy used in **multiple regression**. While we remain cautious about making any causal interpretations using multiple regression, such models are a common first step in providing evidence of a causal connection.

We want to construct a model that accounts for not only the game condition, as in Section 8.1.1, but simultaneously accounts for three other variables: stock_photo, duration, and wheels.

$$\widehat{\text{price}} = \beta_0 + \beta_1 \times \text{cond\_new} + \beta_2 \times \text{stock\_photo}$$
$$+ \beta_3 \times \text{duration} + \beta_4 \times \text{wheels}$$
$$\hat{y} = \beta_0 + \beta_1 x_1 + \beta_2 x_2 + \beta_3 x_3 + \beta_4 x_4 \tag{8.3}$$

In this equation, y represents the total price, x_1 indicates whether the game is new, x_2 indicates whether a stock photo was used, x_3 is the duration of the auction, and x_4 is the number of Wii wheels included with the game. Just as with the single predictor case, a multiple regression model may be missing important components or it might not precisely represent the relationship between the outcome and the available explanatory variables. While no model is perfect, we wish to explore the possibility that this one may fit the data reasonably well.

We estimate the parameters β_0, β_1, ..., β_4 in the same way as we did in the case of a single predictor. We select b_0, b_1, ..., b_4 that minimize the sum of the squared residuals:

$$SSE = e_1^2 + e_2^2 + \cdots + e_{141}^2 = \sum_{i=1}^{141} e_i^2 = \sum_{i=1}^{141} (y_i - \hat{y}_i)^2 \tag{8.4}$$

Here there are 141 residuals, one for each observation. We typically use a computer to minimize the sum in Equation (8.4) and compute point estimates, as shown in the sample output in Table 8.5. Using this output, we identify the point estimates b_i of each β_i, just as we did in the one-predictor case.

	Estimate	Std. Error	t value	Pr(>|t|)
(Intercept)	36.2110	1.5140	23.92	0.0000
cond_new	5.1306	1.0511	4.88	0.0000
stock_photo	1.0803	1.0568	1.02	0.3085
duration	-0.0268	0.1904	-0.14	0.8882
wheels	7.2852	0.5547	13.13	0.0000
				df = 136

Table 8.5: Output for the regression model where price is the outcome and cond_new, stock_photo, duration, and wheels are the predictors.

Multiple regression model

A multiple regression model is a linear model with many predictors. In general, we write the model as

$$\hat{y} = \beta_0 + \beta_1 x_1 + \beta_2 x_2 + \cdots + \beta_k x_k$$

when there are k predictors. We often estimate the β_i parameters using a computer.

⊙ **Guided Practice 8.5**   Write out the model in Equation (8.3) using the point estimates from Table 8.5. How many predictors are there in this model?[3]

⊙ **Guided Practice 8.6**   What does β_4, the coefficient of variable x_4 (Wii wheels), represent? What is the point estimate of β_4?[4]

⊙ **Guided Practice 8.7**   Compute the residual of the first observation in Table 8.1 on page 373 using the equation identified in Guided Practice 8.5.[5]

● **Example 8.8**   We estimated a coefficient for cond_new in Section 8.1.1 of $b_1 = 10.90$ with a standard error of $SE_{b_1} = 1.26$ when using simple linear regression. Why might there be a difference between that estimate and the one in the multiple regression setting?

If we examined the data carefully, we would see that some predictors are correlated. For instance, when we estimated the connection of the outcome price and predictor cond_new using simple linear regression, we were unable to control for other variables like the number of Wii wheels included in the auction. That model was biased by the confounding variable wheels. When we use both variables, this particular underlying and unintentional bias is reduced or eliminated (though bias from other confounding variables may still remain).

Example 8.8 describes a common issue in multiple regression: correlation among predictor variables. We say the two predictor variables are **collinear** (pronounced as *co-linear*) when they are correlated, and this collinearity complicates model estimation. While it is impossible to prevent collinearity from arising in observational data, experiments are usually designed to prevent predictors from being collinear.

⊙ **Guided Practice 8.9**   The estimated value of the intercept is 36.21, and one might be tempted to make some interpretation of this coefficient, such as, it is the model's predicted price when each of the variables take value zero: the game is used, the primary image is not a stock photo, the auction duration is zero days, and there are no wheels included. Is there any value gained by making this interpretation?[6]

8.1.3   Adjusted R^2 as a better estimate of explained variance

We first used R^2 in Section 7.2 to determine the amount of variability in the response that was explained by the model:

$$R^2 = 1 - \frac{\text{variability in residuals}}{\text{variability in the outcome}} = 1 - \frac{Var(e_i)}{Var(y_i)}$$

where e_i represents the residuals of the model and y_i the outcomes. This equation remains valid in the multiple regression framework, but a small enhancement can often be even more informative.

---

[3]$\hat{y} = 36.21 + 5.13x_1 + 1.08x_2 - 0.03x_3 + 7.29x_4$, and there are $k = 4$ predictor variables.

[4]It is the average difference in auction price for each additional Wii wheel included when holding the other variables constant. The point estimate is $b_4 = 7.29$.

[5]$e_i = y_i - \hat{y}_i = 51.55 - 49.62 = 1.93$, where 49.62 was computed using the variables values from the observation and the equation identified in Guided Practice 8.5.

[6]Three of the variables (cond_new, stock_photo, and wheels) do take value 0, but the auction duration is always one or more days. If the auction is not up for any days, then no one can bid on it! That means the total auction price would always be zero for such an auction; the interpretation of the intercept in this setting is not insightful.

⊙ **Guided Practice 8.10** The variance of the residuals for the model given in Guided Practice 8.7 is 23.34, and the variance of the total price in all the auctions is 83.06. Calculate R^2 for this model.[7]

This strategy for estimating R^2 is acceptable when there is just a single variable. However, it becomes less helpful when there are many variables. The regular R^2 is a less estimate of the amount of variability explained by the model. To get a better estimate, we use the adjusted R^2.

Adjusted R^2 as a tool for model assessment

The **adjusted R^2** is computed as

$$R^2_{adj} = 1 - \frac{Var(e_i)/(n-k-1)}{Var(y_i)/(n-1)} = 1 - \frac{Var(e_i)}{Var(y_i)} \times \frac{n-1}{n-k-1}$$

where n is the number of cases used to fit the model and k is the number of predictor variables in the model.

Because k is never negative, the adjusted R^2 will be smaller – often times just a little smaller – than the unadjusted R^2. The reasoning behind the adjusted R^2 lies in the **degrees of freedom** associated with each variance.[8]

⊙ **Guided Practice 8.11** There were $n = 141$ auctions in the `mario_kart` data set and $k = 4$ predictor variables in the model. Use n, k, and the variances from Guided Practice 8.10 to calculate R^2_{adj} for the Mario Kart model.[9]

⊙ **Guided Practice 8.12** Suppose you added another predictor to the model, but the variance of the errors $Var(e_i)$ didn't go down. What would happen to the R^2? What would happen to the adjusted R^2?[10]

Adjusted R^2 could have been used in Chapter 7. However, when there is only $k = 1$ predictors, adjusted R^2 is very close to regular R^2, so this nuance isn't typically important when considering only one predictor.

---

[7]$R^2 = 1 - \frac{23.34}{83.06} = 0.719$.

[8]In multiple regression, the degrees of freedom associated with the variance of the estimate of the residuals is $n - k - 1$, not $n - 1$. For instance, if we were to make predictions for new data using our current model, we would find that the unadjusted R^2 is an overly optimistic estimate of the reduction in variance in the response, and using the degrees of freedom in the adjusted R^2 formula helps correct this bias.

[9]$R^2_{adj} = 1 - \frac{23.34}{83.06} \times \frac{141-1}{141-4-1} = 0.711$.

[10]The unadjusted R^2 would stay the same and the adjusted R^2 would go down.

8.2   Model selection 🎥

The best model is not always the most complicated. Sometimes including variables that are not evidently important can actually reduce the accuracy of predictions. In this section we discuss model selection strategies, which will help us eliminate variables from the model that are found to be less important.

In practice, the model that includes all available explanatory variables is often referred to as the **full model**. The full model may not be the best model, and if it isn't, we want to identify a smaller model that is preferable.

8.2.1   Identifying variables in the model that may not be helpful

Adjusted R^2 describes the strength of a model fit, and it is a useful tool for evaluating which predictors are adding value to the model, where *adding value* means they are (likely) improving the accuracy in predicting future outcomes.

Let's consider two models, which are shown in Tables 8.6 and 8.7. The first table summarizes the full model since it includes all predictors, while the second does not include the duration variable.

	Estimate	Std. Error	t value	Pr(>|t|)
(Intercept)	36.2110	1.5140	23.92	0.0000
cond_new	5.1306	1.0511	4.88	0.0000
stock_photo	1.0803	1.0568	1.02	0.3085
duration	-0.0268	0.1904	-0.14	0.8882
wheels	7.2852	0.5547	13.13	0.0000
$R^2_{adj} = 0.7108$				$df = 136$

Table 8.6: The fit for the full regression model, including the adjusted R^2.

	Estimate	Std. Error	t value	Pr(>|t|)
(Intercept)	36.0483	0.9745	36.99	0.0000
cond_new	5.1763	0.9961	5.20	0.0000
stock_photo	1.1177	1.0192	1.10	0.2747
wheels	7.2984	0.5448	13.40	0.0000
$R^2_{adj} = 0.7128$				$df = 137$

Table 8.7: The fit for the regression model for predictors cond_new, stock_photo, and wheels.

● **Example 8.13**   Which of the two models is better?

We compare the adjusted R^2 of each model to determine which to choose. Since the first model has an R^2_{adj} smaller than the R^2_{adj} of the second model, we prefer the second model to the first.

Will the model without duration be better than the model with duration? We cannot know for sure, but based on the adjusted R^2, this is our best assessment.

8.2.2 Two model selection strategies

Two common strategies for adding or removing variables in a multiple regression model are called *backward elimination* and *forward selection*. These techniques are often referred to as **stepwise** model selection strategies, because they add or delete one variable at a time as they "step" through the candidate predictors.

Backward elimination starts with the model that includes all potential predictor variables. Variables are eliminated one-at-a-time from the model until we cannot improve the adjusted R^2. The strategy within each elimination step is to eliminate the variable that leads to the largest improvement in adjusted R^2.

 Example 8.14 Results corresponding to the *full model* for the `mario_kart` data are shown in Table 8.6. How should we proceed under the backward elimination strategy?

Our baseline adjusted R^2 from the full model is $R^2_{adj} = 0.7108$, and we need to determine whether dropping a predictor will improve the adjusted R^2. To check, we fit four models that each drop a different predictor, and we record the adjusted R^2 from each:

Exclude ...	cond_new	stock_photo	duration	wheels
	$R^2_{adj} = 0.6626$	$R^2_{adj} = 0.7107$	$R^2_{adj} = 0.7128$	$R^2_{adj} = 0.3487$

The third model without `duration` has the highest adjusted R^2 of 0.7128, so we compare it to the adjusted R^2 for the full model. Because eliminating `duration` leads to a model with a higher adjusted R^2, we drop `duration` from the model.

Since we eliminated a predictor from the model in the first step, we see whether we should eliminate any additional predictors. Our baseline adjusted R^2 is now $R^2_{adj} = 0.7128$. We now fit three new models, which consider eliminating each of the three remaining predictors:

Exclude duration and ...	cond_new	stock_photo	wheels
	$R^2_{adj} = 0.6587$	$R^2_{adj} = 0.7124$	$R^2_{adj} = 0.3414$

None of these models lead to an improvement in adjusted R^2, so we do not eliminate any of the remaining predictors. That is, after backward elimination, we are left with the model that keeps `cond_new`, `stock_photos`, and `wheels`, which we can summarize using the coefficients from Table 8.7:

$$\hat{y} = b_0 + b_1 x_1 + b_2 x_2 + b_4 x_4$$
$$\widehat{price} = 36.05 + 5.18 \times \texttt{cond\_new} + 1.12 \times \texttt{stock\_photo} + 7.30 \times \texttt{wheels}$$

The **forward selection** strategy is the reverse of the backward elimination technique. Instead of eliminating variables one-at-a-time, we add variables one-at-a-time until we cannot find any variables that improve the model (as measured by adjusted R^2).

● **Example 8.15**   Construct a model for the `mario_kart` data set using the forward selection strategy.

We start with the model that includes no variables. Then we fit each of the possible models with just one variable. That is, we fit the model including just `cond_new`, then the model including just `stock_photo`, then a model with just `duration`, and a model with just `wheels`. Each of the four models provides an adjusted R^2 value:

$$\text{Add ...} \quad \begin{array}{cccc} \texttt{cond\_new} & \texttt{stock\_photo} & \texttt{duration} & \texttt{wheels} \\ R^2_{adj} = 0.3459 & R^2_{adj} = 0.0332 & R^2_{adj} = 0.1338 & R^2_{adj} = 0.6390 \end{array}$$

In this first step, we compare the adjusted R^2 against a baseline model that has no predictors. The no-predictors model always has $R^2_{adj} = 0$. The model with one predictor that has the largest adjusted R^2 is the model with the `wheels` predictor, and because this adjusted R^2 is larger than the adjusted R^2 from the model with no predictors ($R^2_{adj} = 0$), we will add this variable to our model.

We repeat the process again, this time considering 2-predictor models where one of the predictors is `wheels` and with a new baseline of $R^2_{adj} = 0.6390$:

$$\text{Add \texttt{wheels} and ...} \quad \begin{array}{ccc} \texttt{cond\_new} & \texttt{stock\_photo} & \texttt{duration} \\ R^2_{adj} = 0.7124 & R^2_{adj} = 0.6587 & R^2_{adj} = 0.6528 \end{array}$$

The best predictor in this stage, `cond_new`, has a higher adjusted R^2 (0.7124) than the baseline (0.6390), so we also add `cond_new` to the model.

Since we have again added a variable to the model, we continue and see whether it would be beneficial to add a third variable:

$$\text{Add \texttt{wheels}, \texttt{cond\_new}, and ...} \quad \begin{array}{cc} \texttt{stock\_photo} & \texttt{duration} \\ R^2_{adj} = 0.7128 & R^2_{adj} = 0.7107 \end{array}$$

The model adding `stock_photo` improved adjusted R^2 (0.7124 to 0.7128), so we add `stock_photo` to the model.

Because we have again added a predictor, we check whether adding the last variable, `duration`, will improve adjusted R^2. We compare the adjusted R^2 for the model with `duration` and the other three predictors (0.7108) to the model that only considers `wheels`, `cond_new`, and `stock_photo` (0.7128). Adding `duration` does not improve the adjusted R^2, so we do not add it to the model, and we have arrived at the same model that we identified from backward elimination.

Model selection strategies

Backward elimination begins with the largest model and eliminates variables one-by-one until we are satisfied that all remaining variables are important to the model. Forward selection starts with no variables included in the model, then it adds in variables according to their importance until no other important variables are found.

There is no guarantee that backward elimination and forward selection will arrive at the same final model. If both techniques are tried and they arrive at different models, we choose the model with the larger R^2_{adj}; other tie-break options exist but are beyond the scope of this book.

8.2.3 The p-value approach, an alternative to adjusted R^2

The p-value may be used as an alternative to adjusted R^2 for model selection.

In backward elimination, we would identify the predictor corresponding to the largest p-value. If the p-value is above the significance level, usually $\alpha = 0.05$, then we would drop that variable, refit the model, and repeat the process. If the largest p-value is less than $\alpha = 0.05$, then we would not eliminate any predictors and the current model would be our best-fitting model.

In forward selection with p-values, we reverse the process. We begin with a model that has no predictors, then we fit a model for each possible predictor, identifying the model where the corresponding predictor's p-value is smallest. If that p-value is smaller than $\alpha = 0.05$, we add it to the model and repeat the process, considering whether to add more variables one-at-a-time. When none of the remaining predictors can be added to the model and have a p-value less than 0.05, then we stop adding variables and the current model would be our best-fitting model.

⊙ **Guided Practice 8.16** Examine Table 8.7 on page 378, which considers the model including the cond_new, stock_photo, and wheels predictors. If we were using the p-value approach with backward elimination and we were considering this model, which of these three variables would be up for elimination? Would we drop that variable, or would we keep it in the model?[11]

While the adjusted R^2 and p-value approaches are similar, they sometimes lead to different models, with the adjusted R^2 approach tending to include more predictors in the final model. For example, if we had used the p-value approach with the auction data, we would not have included the stock_photo predictor in the final model.

When to use the adjusted R^2 and when to use the p-value approach

When the sole goal is to improve prediction accuracy, use adjusted R^2. This is commonly the case in machine learning applications.

When we care about understanding which variables are statistically significant predictors of the response, or if there is interest in producing a simpler model at the potential cost of a little prediction accuracy, then the p-value approach is preferred.

Regardless of whether you use adjusted R^2 or the p-value approach, or if you use the backward elimination of forward selection strategy, our job is not done after variable selection. We must still verify the model conditions are reasonable.

---

[11]The stock_photo predictor is up for elimination since it has the largest p-value. Additionally, since that p-value is larger than 0.05, we would in fact eliminate stock_photo from the model.

8.3   Checking model assumptions using graphs

Multiple regression methods using the model

$$\hat{y} = \beta_0 + \beta_1 x_1 + \beta_2 x_2 + \cdots + \beta_k x_k$$

generally depend on the following four assumptions:

1. the residuals of the model are nearly normal,
2. the variability of the residuals is nearly constant,
3. the residuals are independent, and
4. each variable is linearly related to the outcome.

Diagnostic plots can be used to check each of these assumptions. We will consider the model from the Mario Kart auction data, and check whether there are any notable concerns:

$$\widehat{price} = 36.05 + 5.18 \times \texttt{cond\_new} + 1.12 \times \texttt{stock\_photo} + 7.30 \times \texttt{wheels}$$

Normal probability plot. A normal probability plot of the residuals is shown in Figure 8.8. While the plot exhibits some minor irregularities, there are no outliers that might be cause for concern. In a normal probability plot for residuals, we tend to be most worried about residuals that appear to be outliers, since these indicate long tails in the distribution of residuals.

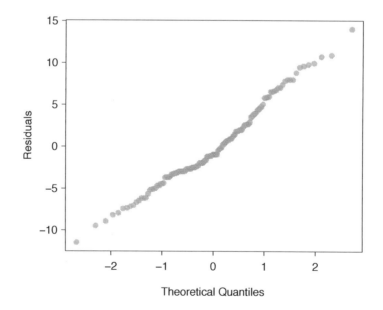

Figure 8.8: A normal probability plot of the residuals is helpful in identifying observations that might be outliers.

Absolute values of residuals against fitted values. A plot of the absolute value of the residuals against their corresponding fitted values $(\hat{y}_i)$ is shown in Figure 8.9. This plot is helpful to check the condition that the variance of the residuals is approximately constant. We don't see any obvious deviations from constant variance in this example.

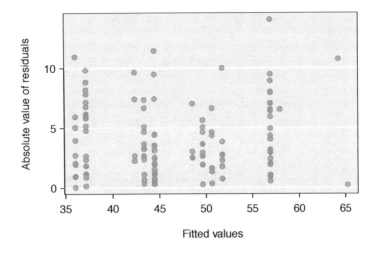

Figure 8.9: Comparing the absolute value of the residuals against the fitted values ($\hat{y}_i$) is helpful in identifying deviations from the constant variance assumption.

Residuals in order of their data collection. A plot of the residuals in the order their corresponding auctions were observed is shown in Figure 8.10. Such a plot is helpful in identifying any connection between cases that are close to one another, e.g. we could look for declining prices over time or if there was a time of the day when auctions tended to fetch a higher price. Here we see no structure that indicates a problem.[12]

Residuals against each predictor variable. We consider a plot of the residuals against the cond_new variable, the residuals against the stock_photo variable, and the residuals against the wheels variable. These plots are shown in Figure 8.11. For the two-level condition variable, we are guaranteed not to see any remaining trend, and instead we are checking that the variability doesn't fluctuate across groups, which it does not. However, looking at the stock photo variable, we find that there is some difference in the variability of the residuals in the two groups. Additionally, when we consider the residuals against the wheels variable, we see some possible structure. There appears to be curvature in the residuals, indicating the relationship is probably not linear.

It is necessary to summarize diagnostics for any model fit. If the diagnostics support the model assumptions, this would improve credibility in the findings. If the diagnostic assessment shows remaining underlying structure in the residuals, we should try to adjust the model to account for that structure. If we are unable to do so, we may still report the model but must also note its shortcomings. In the case of the auction data, we report that there appears to be non-constant variance in the stock photo variable and that there may be a nonlinear relationship between the total price and the number of wheels included for an auction. This information would be important to buyers and sellers who may review the analysis, and omitting this information could be a setback to the very people who the model might assist.

---

[12]An especially rigorous check would use **time series** methods. For instance, we could check whether consecutive residuals are correlated. Doing so with these residuals yields no statistically significant correlations.

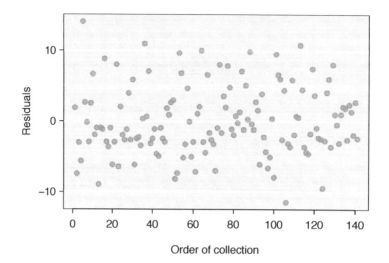

Figure 8.10: Plotting residuals in the order that their corresponding observations were collected helps identify connections between successive observations. If it seems that consecutive observations tend to be close to each other, this indicates the independence assumption of the observations would fail.

"All models are wrong, but some are useful" -George E.P. Box

The truth is that no model is perfect. However, even imperfect models can be useful. Reporting a flawed model can be reasonable so long as we are clear and report the model's shortcomings.

Caution: Don't report results when assumptions are grossly violated

While there is a little leeway in model assumptions, don't go too far. If model assumptions are very clearly violated, consider a new model, even if it means learning more statistical methods or hiring someone who can help.

TIP: Confidence intervals in multiple regression

Confidence intervals for coefficients in multiple regression can be computed using the same formula as in the single predictor model:

$$b_i \; \pm \; t_{df}^{\star} SE_{b_i}$$

where $t_{df}^{\star}$ is the appropriate t-value corresponding to the confidence level and model degrees of freedom, $df = n - k - 1$.

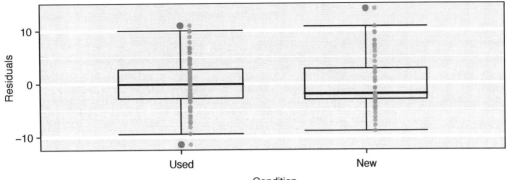

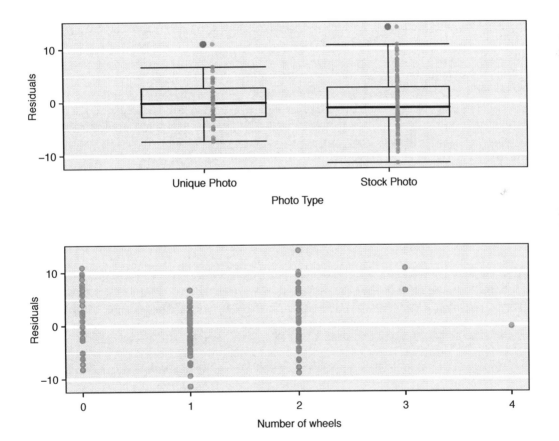

Figure 8.11: For the condition and stock photo variables, we check for differences in the distribution shape or variability of the residuals. In the case of the stock photos variable, we see a little less variability in the unique photo group than the stock photo group. For numerical predictors, we also check for trends or other structure. We see some slight bowing in the residuals against the `wheels` variable in the bottom plot.

8.4   Introduction to logistic regression 📹

In this section we introduce **logistic regression** as a tool for building models when there is a categorical response variable with two levels. Logistic regression is a type of **generalized linear model** (GLM) for response variables where regular multiple regression does not work very well. In particular, the response variable in these settings often takes a form where residuals look completely different from the normal distribution.

GLMs can be thought of as a two-stage modeling approach. We first model the response variable using a probability distribution, such as the binomial or Poisson distribution. Second, we model the parameter of the distribution using a collection of predictors and a special form of multiple regression.

In Section 8.4 we will revisit the `email` data set from Chapter 1. These emails were collected from a single email account, and we will work on developing a basic spam filter using these data. The response variable, `spam`, has been encoded to take value 0 when a message is not spam and 1 when it is spam. Our task will be to build an appropriate model that classifies messages as spam or not spam using email characteristics coded as predictor variables. While this model will not be the same as those used in large-scale spam filters, it shares many of the same features.

8.4.1   Email data

The `email` data set was first presented in Chapter 1 with a relatively small number of variables. In fact, there are many more variables available that might be useful for classifying spam. Descriptions of these variables are presented in Table 8.12. The `spam` variable will be the outcome, and the other 10 variables will be the model predictors. While we have limited the predictors used in this section to be categorical variables (where many are represented as indicator variables), numerical predictors may also be used in logistic regression. See the footnote for an additional discussion on this topic.[13]

8.4.2   Modeling the probability of an event

TIP: Notation for a logistic regression model

The outcome variable for a GLM is denoted by Y_i, where the index i is used to represent observation i. In the email application, Y_i will be used to represent whether email i is spam ($Y_i = 1$) or not ($Y_i = 0$).

The predictor variables are represented as follows: $x_{1,i}$ is the value of variable 1 for observation i, $x_{2,i}$ is the value of variable 2 for observation i, and so on.

Logistic regression is a generalized linear model where the outcome is a two-level categorical variable. The outcome, Y_i, takes the value 1 (in our application, this represents a spam message) with probability p_i and the value 0 with probability $1 - p_i$. It is the probability p_i that we model in relation to the predictor variables.

---

[13]Recall from Chapter 7 that if outliers are present in predictor variables, the corresponding observations may be especially influential on the resulting model. This is the motivation for omitting the numerical variables, such as the number of characters and line breaks in emails, that we saw in Chapter 1. These variables exhibited extreme skew. We could resolve this issue by transforming these variables (e.g. using a log-transformation), but we will omit this further investigation for brevity.

variable	description
spam	Specifies whether the message was spam.
to_multiple	An indicator variable for if more than one person was listed in the *To* field of the email.
cc	An indicator for if someone was CCed on the email.
attach	An indicator for if there was an attachment, such as a document or image.
dollar	An indicator for if the word "dollar" or dollar symbol ($) appeared in the email.
winner	An indicator for if the word "winner" appeared in the email message.
inherit	An indicator for if the word "inherit" (or a variation, like "inheritance") appeared in the email.
password	An indicator for if the word "password" was present in the email.
format	Indicates if the email contained special formatting, such as bolding, tables, or links
re_subj	Indicates whether "Re:" was included at the start of the email subject.
exclaim_subj	Indicates whether any exclamation point was included in the email subject.

Table 8.12: Descriptions for 11 variables in the email data set. Notice that all of the variables are indicator variables, which take the value 1 if the specified characteristic is present and 0 otherwise.

The logistic regression model relates the probability an email is spam (p_i) to the predictors $x_{1,i}, x_{2,i}, ..., x_{k,i}$ through a framework much like that of multiple regression:

$$transformation(p_i) = \beta_0 + \beta_1 x_{1,i} + \beta_2 x_{2,i} + \cdots \beta_k x_{k,i} \tag{8.17}$$

We want to choose a transformation in Equation (8.17) that makes practical and mathematical sense. For example, we want a transformation that makes the range of possibilities on the left hand side of Equation (8.17) equal to the range of possibilities for the right hand side; if there was no transformation for this equation, the left hand side could only take values between 0 and 1, but the right hand side could take values outside of this range. A common transformation for p_i is the **logit transformation**, which may be written as

$$logit(p_i) = \log_e \left(\frac{p_i}{1 - p_i} \right)$$

The logit transformation is shown in Figure 8.13. Below, we rewrite Equation (8.17) using the logit transformation of p_i:

$$\log_e \left(\frac{p_i}{1 - p_i} \right) = \beta_0 + \beta_1 x_{1,i} + \beta_2 x_{2,i} + \cdots + \beta_k x_{k,i}$$

In our spam example, there are 10 predictor variables, so $k = 10$. This model isn't very intuitive, but it still has some resemblance to multiple regression, and we can fit this model using software. In fact, once we look at results from software, it will start to feel like we're back in multiple regression, even if the interpretation of the coefficients is more complex.

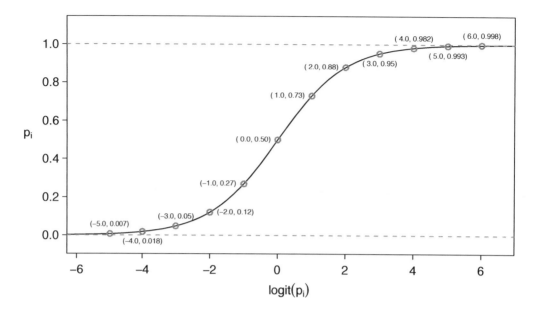

Figure 8.13: Values of p_i against values of $logit(p_i)$.

● **Example 8.18**   Here we create a spam filter with a single predictor: `to_multiple`. This variable indicates whether more than one email address was listed in the *To* field of the email. The following logistic regression model was fit using statistical software:

$$\log\left(\frac{p_i}{1-p_i}\right) = -2.12 - 1.81 \times \texttt{to\_multiple}$$

If an email is randomly selected and it has just one address in the *To* field, what is the probability it is spam? What if more than one address is listed in the *To* field?

If there is only one email in the *To* field, then `to_multiple` takes value 0 and the right side of the model equation equals -2.12. Solving for p_i: $\frac{e^{-2.12}}{1+e^{-2.12}} = 0.11$. Just as we labeled a fitted value of y_i with a "hat" in single-variable and multiple regression, we will do the same for this probability: $\hat{p}_i = 0.11$.

If there is more than one address listed in the *To* field, then the right side of the model equation is $-2.12 - 1.81 \times 1 = -3.93$, which corresponds to a probability $\hat{p}_i = 0.02$.

Notice that we could examine -2.12 and -3.93 in Figure 8.13 to estimate the probability before formally calculating the value.

To convert from values on the regression-scale (e.g. -2.12 and -3.93 in Example 8.18), use the following formula, which is the result of solving for p_i in the regression model:

$$p_i = \frac{e^{\beta_0 + \beta_1 x_{1,i} + \cdots + \beta_k x_{k,i}}}{1 + e^{\beta_0 + \beta_1 x_{1,i} + \cdots + \beta_k x_{k,i}}}$$

As with most applied data problems, we substitute the point estimates for the parameters (the β_i) so that we may make use of this formula. In Example 8.18, the probabilities were calculated as

$$\frac{e^{-2.12}}{1 + e^{-2.12}} = 0.11 \qquad\qquad \frac{e^{-2.12-1.81}}{1 + e^{-2.12-1.81}} = 0.02$$

While the information about whether the email is addressed to multiple people is a helpful start in classifying email as spam or not, the probabilities of 11% and 2% are not dramatically different, and neither provides very strong evidence about which particular email messages are spam. To get more precise estimates, we'll need to include many more variables in the model.

We used statistical software to fit the logistic regression model with all ten predictors described in Table 8.12. Like multiple regression, the result may be presented in a summary table, which is shown in Table 8.14. The structure of this table is almost identical to that of multiple regression; the only notable difference is that the p-values are calculated using the normal distribution rather than the t-distribution.

	Estimate	Std. Error	z value	Pr($>$|z|)
(Intercept)	-0.8362	0.0962	-8.69	0.0000
to_multiple	-2.8836	0.3121	-9.24	0.0000
winner	1.7038	0.3254	5.24	0.0000
format	-1.5902	0.1239	-12.84	0.0000
re_subj	-2.9082	0.3708	-7.84	0.0000
exclaim_subj	0.1355	0.2268	0.60	0.5503
cc	-0.4863	0.3054	-1.59	0.1113
attach	0.9790	0.2170	4.51	0.0000
dollar	-0.0582	0.1589	-0.37	0.7144
inherit	0.2093	0.3197	0.65	0.5127
password	-1.4929	0.5295	-2.82	0.0048

Table 8.14: Summary table for the full logistic regression model for the spam filter example.

Just like multiple regression, we could trim some variables from the model using the p-value. Using backward elimination with a p-value cutoff of 0.05 (start with the full model and trim the predictors with p-values greater than 0.05), we ultimately eliminate the exclaim_subj, dollar, inherit, and cc predictors. The remainder of this section will rely on this smaller model, which is summarized in Table 8.15.

	Estimate	Std. Error	z value	Pr($>$|z|)
(Intercept)	-0.8595	0.0910	-9.44	0.0000
to_multiple	-2.8372	0.3092	-9.18	0.0000
winner	1.7370	0.3218	5.40	0.0000
format	-1.5569	0.1207	-12.90	0.0000
re_subj	-3.0482	0.3630	-8.40	0.0000
attach	0.8643	0.2042	4.23	0.0000
password	-1.4871	0.5290	-2.81	0.0049

Table 8.15: Summary table for the logistic regression model for the spam filter, where variable selection has been performed.

⊙ **Guided Practice 8.19**   Examine the summary of the reduced model in Table 8.15, and in particular, examine the `to_multiple` row. Is the point estimate the same as we found before, -1.81, or is it different? Explain why this might be.[14]

Point estimates will generally change a little – and sometimes a lot – depending on which other variables are included in the model. This is usually due to colinearity in the predictor variables. We previously saw this in the Ebay auction example when we compared the coefficient of `cond_new` in a single-variable model and the corresponding coefficient in the multiple regression model that used three additional variables (see Sections 8.1.1 and 8.1.2).

● **Example 8.20**   Spam filters are built to be automated, meaning a piece of software is written to collect information about emails as they arrive, and this information is put in the form of variables. These variables are then put into an algorithm that uses a statistical model, like the one we've fit, to classify the email. Suppose we write software for a spam filter using the reduced model shown in Table 8.15. If an incoming email has the word "winner" in it, will this raise or lower the model's calculated probability that the incoming email is spam?

The estimated coefficient of `winner` is positive (1.7370). A positive coefficient estimate in logistic regression, just like in multiple regression, corresponds to a positive association between the predictor and response variables when accounting for the other variables in the model. Since the response variable takes value 1 if an email is spam and 0 otherwise, the positive coefficient indicates that the presence of "winner" in an email raises the model probability that the message is spam.

● **Example 8.21**   Suppose the same email from Example 8.20 was in HTML format, meaning the `format` variable took value 1. Does this characteristic increase or decrease the probability that the email is spam according to the model?

Since HTML corresponds to a value of 1 in the `format` variable and the coefficient of this variable is negative (-1.5569), this would lower the probability estimate returned from the model.

8.4.3   Practical decisions in the email application

Examples 8.20 and 8.21 highlight a key feature of logistic and multiple regression. In the spam filter example, some email characteristics will push an email's classification in the direction of spam while other characteristics will push it in the opposite direction.

If we were to implement a spam filter using the model we have fit, then each future email we analyze would fall into one of three categories based on the email's characteristics:

1. The email characteristics generally indicate the email is not spam, and so the resulting probability that the email is spam is quite low, say, under 0.05.

2. The characteristics generally indicate the email is spam, and so the resulting probability that the email is spam is quite large, say, over 0.95.

3. The characteristics roughly balance each other out in terms of evidence for and against the message being classified as spam. Its probability falls in the remaining range, meaning the email cannot be adequately classified as spam or not spam.

---

[14]The new estimate is different: -2.87. This new value represents the estimated coefficient when we are also accounting for other variables in the logistic regression model.

If we were managing an email service, we would have to think about what should be done in each of these three instances. In an email application, there are usually just two possibilities: filter the email out from the regular inbox and put it in a "spambox", or let the email go to the regular inbox.

⊙ **Guided Practice 8.22** The first and second scenarios are intuitive. If the evidence strongly suggests a message is not spam, send it to the inbox. If the evidence strongly suggests the message is spam, send it to the spambox. How should we handle emails in the third category?[15]

⊙ **Guided Practice 8.23** Suppose we apply the logistic model we have built as a spam filter and that 100 messages are placed in the spambox over 3 months. If we used the guidelines above for putting messages into the spambox, about how many legitimate (non-spam) messages would you expect to find among the 100 messages?[16]

Almost any classifier will have some error. In the spam filter guidelines above, we have decided that it is okay to allow up to 5% of the messages in the spambox to be real messages. If we wanted to make it a little harder to classify messages as spam, we could use a cutoff of 0.99. This would have two effects. Because it raises the standard for what can be classified as spam, it reduces the number of good emails that are classified as spam. However, it will also fail to correctly classify an increased fraction of spam messages. No matter the complexity and the confidence we might have in our model, these practical considerations are absolutely crucial to making a helpful spam filter. Without them, we could actually do more harm than good by using our statistical model.

8.4.4 Diagnostics for the email classifier

Logistic regression conditions

There are two key conditions for fitting a logistic regression model:

1. Each predictor x_i is linearly related to $\text{logit}(p_i)$ if all other predictors are held constant.

2. Each outcome Y_i is independent of the other outcomes.

The first condition of the logistic regression model is not easily checked without a fairly sizable amount of data. Luckily, we have 3,921 emails in our data set! Let's first visualize these data by plotting the true classification of the emails against the model's fitted probabilities, as shown in Figure 8.16. The vast majority of emails (spam or not) still have fitted probabilities below 0.5.

This may at first seem very discouraging: we have fit a logistic model to create a spam filter, but no emails have a fitted probability of being spam above 0.75. Don't despair; we will discuss ways to improve the model through the use of better variables in Section 8.4.5.

---

[15] In this particular application, we should err on the side of sending more mail to the inbox rather than mistakenly putting good messages in the spambox. So, in summary: emails in the first and last categories go to the regular inbox, and those in the second scenario go to the spambox.

[16] First, note that we proposed a cutoff for the predicted probability of 0.95 for spam. In a worst case scenario, all the messages in the spambox had the minimum probability equal to about 0.95. Thus, we should expect to find about 5 or fewer legitimate messages among the 100 messages placed in the spambox.

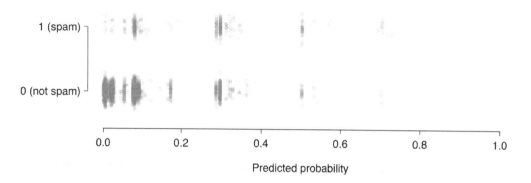

Figure 8.16: The predicted probability that each of the 3,912 emails is spam is classified by their grouping, spam or not. Noise (small, random vertical shifts) have been added to each point so that points with nearly identical values aren't plotted exactly on top of one another. This makes it possible to see more observations.

We'd like to assess the quality of our model. For example, we might ask: if we look at emails that we modeled as having a 10% chance of being spam, do we find about 10% of them actually are spam? To help us out, we'll borrow an advanced statistical method called **natural splines** that estimates the local probability over the region 0.00 to 0.75 (the largest predicted probability was 0.73, so we avoid extrapolating). All you need to know about natural splines to understand what we are doing is that they are used to fit flexible lines rather than straight lines.

The curve fit using natural splines is shown in Figure 8.17 as a solid black line. If the logistic model fits well, the curve should closely follow the dashed $y = x$ line. We have added shading to represent the confidence bound for the curved line to clarify what fluctuations might plausibly be due to chance. Even with this confidence bound, there are weaknesses in the first model assumption. The solid curve and its confidence bound dips below the dashed line from about 0.1 to 0.3, and then it drifts above the dashed line from about 0.35 to 0.55. These deviations indicate the model relating the parameter to the predictors does not closely resemble the true relationship.

We could evaluate the second logistic regression model assumption – independence of the outcomes – using the model residuals. The residuals for a logistic regression model are calculated the same way as with multiple regression: the observed outcome minus the expected outcome. For logistic regression, the expected value of the outcome is the fitted probability for the observation, and the residual may be written as

$$e_i = Y_i - \hat{p}_i$$

We could plot these residuals against a variety of variables or in their order of collection, as we did with the residuals in multiple regression. However, since the model will need to be revised to effectively classify spam and you have already seen similar residual plots in Section 8.3, we won't investigate the residuals here.

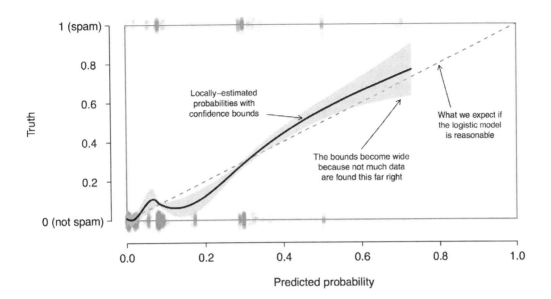

Figure 8.17: The solid black line provides the empirical estimate of the probability for observations based on their predicted probabilities (confidence bounds are also shown for this line), which is fit using natural splines. A small amount of noise was added to the observations in the plot to allow more observations to be seen.

8.4.5   Improving the set of variables for a spam filter

If we were building a spam filter for an email service that managed many accounts (e.g. Gmail or Hotmail), we would spend much more time thinking about additional variables that could be useful in classifying emails as spam or not. We also would use transformations or other techniques that would help us include strongly skewed numerical variables as predictors.

Take a few minutes to think about additional variables that might be useful in identifying spam. Below is a list of variables we think might be useful:

(1) An indicator variable could be used to represent whether there was prior two-way correspondence with a message's sender. For instance, if you sent a message to john@example.com and then John sent you an email, this variable would take value 1 for the email that John sent. If you had never sent John an email, then the variable would be set to 0.

(2) A second indicator variable could utilize an account's past spam flagging information. The variable could take value 1 if the sender of the message has previously sent messages flagged as spam.

(3) A third indicator variable could flag emails that contain links included in previous spam messages. If such a link is found, then set the variable to 1 for the email. Otherwise, set it to 0.

The variables described above take one of two approaches. Variable (1) is specially designed to capitalize on the fact that spam is rarely sent between individuals that have two-way communication. Variables (2) and (3) are specially designed to flag common spammers or spam messages. While we would have to verify using the data that each of the variables is effective, these seem like promising ideas.

Table 8.18 shows a contingency table for spam and also for the new variable described in (1) above. If we look at the 1,090 emails where there was correspondence with the sender in the preceding 30 days, not one of these message was spam. This suggests variable (1) would be very effective at accurately classifying some messages as not spam. With this single variable, we would be able to send about 28% of messages through to the inbox with confidence that almost none are spam.

| | prior correspondence | | |
	no	yes	Total
spam	367	0	367
not spam	2464	1090	3554
Total	2831	1090	3921

Table 8.18: A contingency table for spam and a new variable that represents whether there had been correspondence with the sender in the preceding 30 days.

The variables described in (2) and (3) would provide an excellent foundation for distinguishing messages coming from known spammers or messages that take a known form of spam. To utilize these variables, we would need to build databases: one holding email addresses of known spammers, and one holding URLs found in known spam messages. Our access to such information is limited, so we cannot implement these two variables in this textbook. However, if we were hired by an email service to build a spam filter, these would be important next steps.

In addition to finding more and better predictors, we would need to create a customized logistic regression model for each email account. This may sound like an intimidating task, but its complexity is not as daunting as it may at first seem. We'll save the details for a statistics course where computer programming plays a more central role.

For what is the extremely challenging task of classifying spam messages, we have made a lot of progress. We have seen that simple email variables, such as the format, inclusion of certain words, and other circumstantial characteristics, provide helpful information for spam classification. Many challenges remain, from better understanding logistic regression to carrying out the necessary computer programming, but completing such a task is very nearly within your reach.

8.5 Exercises

8.5.1 Introduction to multiple regression

8.1 Baby weights, Part I. The Child Health and Development Studies investigate a range of topics. One study considered all pregnancies between 1960 and 1967 among women in the Kaiser Foundation Health Plan in the San Francisco East Bay area. Here, we study the relationship between smoking and weight of the baby. The variable smoke is coded 1 if the mother is a smoker, and 0 if not. The summary table below shows the results of a linear regression model for predicting the average birth weight of babies, measured in ounces, based on the smoking status of the mother.[17]

	Estimate	Std. Error	t value	Pr($>$|t|)
(Intercept)	123.05	0.65	189.60	0.0000
smoke	-8.94	1.03	-8.65	0.0000

The variability within the smokers and non-smokers are about equal and the distributions are symmetric. With these conditions satisfied, it is reasonable to apply the model. (Note that we don't need to check linearity since the predictor has only two levels.)

(a) Write the equation of the regression line.

(b) Interpret the slope in this context, and calculate the predicted birth weight of babies born to smoker and non-smoker mothers.

(c) Is there a statistically significant relationship between the average birth weight and smoking?

8.2 Baby weights, Part II. Exercise 8.1 introduces a data set on birth weight of babies. Another variable we consider is parity, which is 0 if the child is the first born, and 1 otherwise. The summary table below shows the results of a linear regression model for predicting the average birth weight of babies, measured in ounces, from parity.

	Estimate	Std. Error	t value	Pr($>$|t|)
(Intercept)	120.07	0.60	199.94	0.0000
parity	-1.93	1.19	-1.62	0.1052

(a) Write the equation of the regression line.

(b) Interpret the slope in this context, and calculate the predicted birth weight of first borns and others.

(c) Is there a statistically significant relationship between the average birth weight and parity?

---

[17]Child Health and Development Studies, Baby weights data set.

8.3 Baby weights, Part III. We considered the variables smoke and parity, one at a time, in modeling birth weights of babies in Exercises 8.1 and 8.2. A more realistic approach to modeling infant weights is to consider all possibly related variables at once. Other variables of interest include length of pregnancy in days (gestation), mother's age in years (age), mother's height in inches (height), and mother's pregnancy weight in pounds (weight). Below are three observations from this data set.

	bwt	gestation	parity	age	height	weight	smoke
1	120	284	0	27	62	100	0
2	113	282	0	33	64	135	0
⋮	⋮	⋮	⋮	⋮	⋮	⋮	⋮
1236	117	297	0	38	65	129	0

The summary table below shows the results of a regression model for predicting the average birth weight of babies based on all of the variables included in the data set.

	Estimate	Std. Error	t value	$Pr(>|t|)$
(Intercept)	-80.41	14.35	-5.60	0.0000
gestation	0.44	0.03	15.26	0.0000
parity	-3.33	1.13	-2.95	0.0033
age	-0.01	0.09	-0.10	0.9170
height	1.15	0.21	5.63	0.0000
weight	0.05	0.03	1.99	0.0471
smoke	-8.40	0.95	-8.81	0.0000

(a) Write the equation of the regression line that includes all of the variables.

(b) Interpret the slopes of gestation and age in this context.

(c) The coefficient for parity is different than in the linear model shown in Exercise 8.2. Why might there be a difference?

(d) Calculate the residual for the first observation in the data set.

(e) The variance of the residuals is 249.28, and the variance of the birth weights of all babies in the data set is 332.57. Calculate the R^2 and the adjusted R^2. Note that there are 1,236 observations in the data set.

8.4 Absenteeism, Part I. Researchers interested in the relationship between absenteeism from school and certain demographic characteristics of children collected data from 146 randomly sampled students in rural New South Wales, Australia, in a particular school year. Below are three observations from this data set.

	eth	sex	lrn	days
1	0	1	1	2
2	0	1	1	11
⋮	⋮	⋮	⋮	⋮
146	1	0	0	37

The summary table below shows the results of a linear regression model for predicting the average number of days absent based on ethnic background (`eth`: 0 - aboriginal, 1 - not aboriginal), sex (`sex`: 0 - female, 1 - male), and learner status (`lrn`: 0 - average learner, 1 - slow learner).[18]

	Estimate	Std. Error	t value	$Pr(>|t|)$
(Intercept)	18.93	2.57	7.37	0.0000
eth	-9.11	2.60	-3.51	0.0000
sex	3.10	2.64	1.18	0.2411
lrn	2.15	2.65	0.81	0.4177

(a) Write the equation of the regression line.

(b) Interpret each one of the slopes in this context.

(c) Calculate the residual for the first observation in the data set: a student who is aboriginal, male, a slow learner, and missed 2 days of school.

(d) The variance of the residuals is 240.57, and the variance of the number of absent days for all students in the data set is 264.17. Calculate the R^2 and the adjusted R^2. Note that there are 146 observations in the data set.

8.5 GPA. A survey of 55 Duke University students asked about their GPA, number of hours they study at night, number of nights they go out, and their gender. Summary output of the regression model is shown below. Note that male is coded as 1.

	Estimate	Std. Error	t value	$Pr(>|t|)$
(Intercept)	3.45	0.35	9.85	0.00
studyweek	0.00	0.00	0.27	0.79
sleepnight	0.01	0.05	0.11	0.91
outnight	0.05	0.05	1.01	0.32
gender	-0.08	0.12	-0.68	0.50

(a) Calculate a 95% confidence interval for the coefficient of gender in the model, and interpret it in the context of the data.

(b) Would you expect a 95% confidence interval for the slope of the remaining variables to include 0? Explain

---

[18]W. N. Venables and B. D. Ripley. *Modern Applied Statistics with S*. Fourth Edition. Data can also be found in the R MASS package. New York: Springer, 2002.

8.6 Cherry trees. Timber yield is approximately equal to the volume of a tree, however, this value is difficult to measure without first cutting the tree down. Instead, other variables, such as height and diameter, may be used to predict a tree's volume and yield. Researchers wanting to understand the relationship between these variables for black cherry trees collected data from 31 such trees in the Allegheny National Forest, Pennsylvania. Height is measured in feet, diameter in inches (at 54 inches above ground), and volume in cubic feet.[19]

	Estimate	Std. Error	t value	Pr($>$|t|)
(Intercept)	-57.99	8.64	-6.71	0.00
height	0.34	0.13	2.61	0.01
diameter	4.71	0.26	17.82	0.00

(a) Calculate a 95% confidence interval for the coefficient of height, and interpret it in the context of the data.

(b) One tree in this sample is 79 feet tall, has a diameter of 11.3 inches, and is 24.2 cubic feet in volume. Determine if the model overestimates or underestimates the volume of this tree, and by how much.

8.5.2 Model selection

8.7 Baby weights, Part IV. Exercise 8.3 considers a model that predicts a newborn's weight using several predictors (gestation length, parity, age of mother, height of mother, weight of mother, smoking status of mother). The table below shows the adjusted R-squared for the full model as well as adjusted R-squared values for all models we evaluate in the first step of the backwards elimination process.

	Model	Adjusted R^2
1	Full model	0.2541
2	No gestation	0.1031
3	No parity	0.2492
4	No age	0.2547
5	No height	0.2311
6	No weight	0.2536
7	No smoking status	0.2072

Which, if any, variable should be removed from the model first?

---

[19]D.J. Hand. *A handbook of small data sets.* Chapman & Hall/CRC, 1994.

8.8 Absenteeism, Part II. Exercise 8.4 considers a model that predicts the number of days absent using three predictors: ethnic background (`eth`), gender (`sex`), and learner status (`lrn`). The table below shows the adjusted R-squared for the model as well as adjusted R-squared values for all models we evaluate in the first step of the backwards elimination process.

	Model	Adjusted R^2
1	Full model	0.0701
2	No ethnicity	-0.0033
3	No sex	0.0676
4	No learner status	0.0723

Which, if any, variable should be removed from the model first?

8.9 Baby weights, Part V. Exercise 8.3 provides regression output for the full model (including all explanatory variables available in the data set) for predicting birth weight of babies. In this exercise we consider a forward-selection algorithm and add variables to the model one-at-a-time. The table below shows the p-value and adjusted R^2 of each model where we include only the corresponding predictor. Based on this table, which variable should be added to the model first?

variable	gestation	parity	age	height	weight	smoke
p-value	2.2×10^{-16}	0.1052	0.2375	2.97×10^{-12}	8.2×10^{-8}	2.2×10^{-16}
R^2_{adj}	0.1657	0.0013	0.0003	0.0386	0.0229	0.0569

8.10 Absenteeism, Part III. Exercise 8.4 provides regression output for the full model, including all explanatory variables available in the data set, for predicting the number of days absent from school. In this exercise we consider a forward-selection algorithm and add variables to the model one-at-a-time. The table below shows the p-value and adjusted R^2 of each model where we include only the corresponding predictor. Based on this table, which variable should be added to the model first?

variable	ethnicity	sex	learner status
p-value	0.0007	0.3142	0.5870
R^2_{adj}	0.0714	0.0001	0

8.11 Movie lovers, Part I. Suppose a social scientist is interested in studying what makes audiences love or hate a movie. She collects a random sample of movies (genre, length, cast, director, budget, etc.) as well as a measure of the success of the movie (score on a film review aggregator website). If as part of her research she is interested in finding out which variables are significant predictors of movie success, what type of model selection method should she use?

8.12 Movie lovers, Part II. Suppose an online media streaming company is interested in building a movie recommendation system. The website maintains data on the movies in their database (genre, length, cast, director, budget, etc.) and additionally collects data from their subscribers (demographic information, previously watched movies, how they rated previously watched movies, etc.). The recommendation system will be deemed successful if subscribers actually watch, and rate highly, the movies recommended to them. Should the company use the adjusted R^2 or the p-value approach in selecting variables for their recommendation system?

8.5.3   Checking model assumptions using graphs

8.13 Baby weights, Part V. Exercise 8.3 presents a regression model for predicting the average birth weight of babies based on length of gestation, parity, height, weight, and smoking status of the mother. Determine if the model assumptions are met using the plots below. If not, describe how to proceed with the analysis.

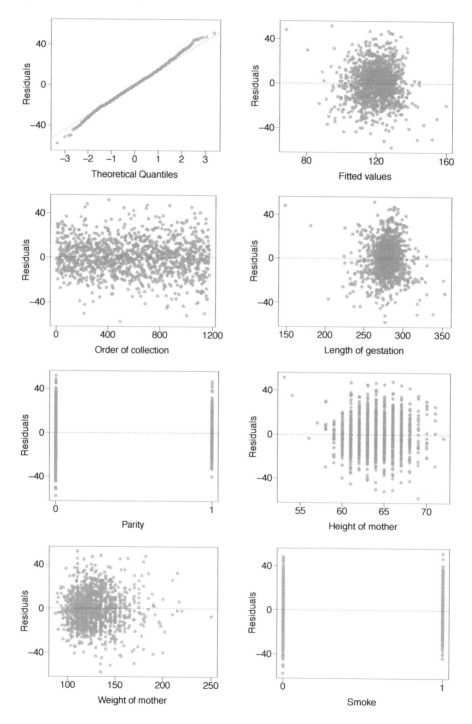

8.14   GPA and IQ. A regression model for predicting GPA from gender and IQ was fit, and both predictors were found to be statistically significant. Using the plots given below, determine if this regression model is appropriate for these data.

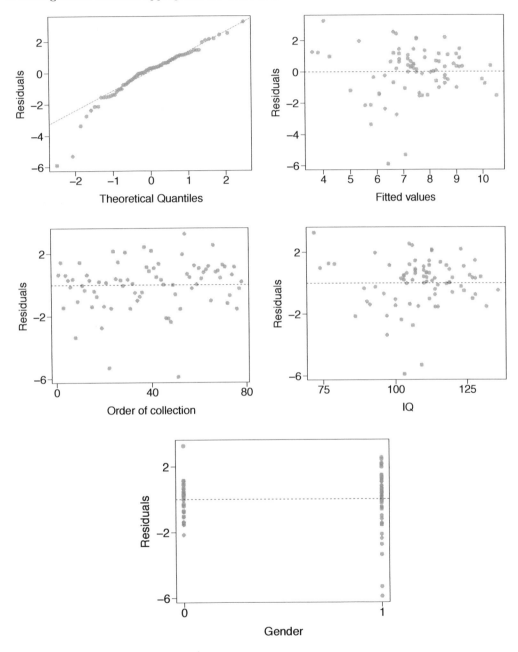

8.5.4 Introduction to logistic regression

8.15 Possum classification, Part I. The common brushtail possum of the Australia region is a bit cuter than its distant cousin, the American opossum (see Figure 7.5 on page 334). We consider 104 brushtail possums from two regions in Australia, where the possums may be considered a random sample from the population. The first region is Victoria, which is in the eastern half of Australia and traverses the southern coast. The second region consists of New South Wales and Queensland, which make up eastern and northeastern Australia.

We use logistic regression to differentiate between possums in these two regions. The outcome variable, called population, takes value 1 when a possum is from Victoria and 0 when it is from New South Wales or Queensland. We consider five predictors: sex_male (an indicator for a possum being male), head_length, skull_width, total_length, and tail_length. Each variable is summarized in a histogram. The full logistic regression model and a reduced model after variable selection are summarized in the table.

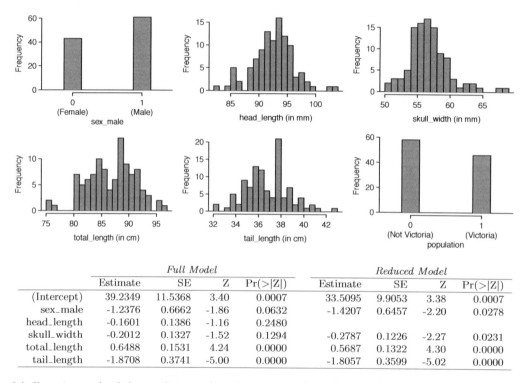

	Full Model				*Reduced Model*			
	Estimate	SE	Z	Pr(>|Z|)	Estimate	SE	Z	Pr(>|Z|)
(Intercept)	39.2349	11.5368	3.40	0.0007	33.5095	9.9053	3.38	0.0007
sex_male	-1.2376	0.6662	-1.86	0.0632	-1.4207	0.6457	-2.20	0.0278
head_length	-0.1601	0.1386	-1.16	0.2480				
skull_width	-0.2012	0.1327	-1.52	0.1294	-0.2787	0.1226	-2.27	0.0231
total_length	0.6488	0.1531	4.24	0.0000	0.5687	0.1322	4.30	0.0000
tail_length	-1.8708	0.3741	-5.00	0.0000	-1.8057	0.3599	-5.02	0.0000

(a) Examine each of the predictors. Are there any outliers that are likely to have a very large influence on the logistic regression model?

(b) The summary table for the full model indicates that at least one variable should be eliminated when using the p-value approach for variable selection: head_length. The second component of the table summarizes the reduced model following variable selection. Explain why the remaining estimates change between the two models.

8.16 Challenger disaster, Part I. On January 28, 1986, a routine launch was anticipated for the Challenger space shuttle. Seventy-three seconds into the flight, disaster happened: the shuttle broke apart, killing all seven crew members on board. An investigation into the cause of the disaster focused on a critical seal called an O-ring, and it is believed that damage to these O-rings during a shuttle launch may be related to the ambient temperature during the launch. The table below summarizes observational data on O-rings for 23 shuttle missions, where the mission order is based on the temperature at the time of the launch. *Temp* gives the temperature in Fahrenheit, *Damaged* represents the number of damaged O-rings, and *Undamaged* represents the number of O-rings that were not damaged.

Shuttle Mission	1	2	3	4	5	6	7	8	9	10	11	12
Temperature	53	57	58	63	66	67	67	67	68	69	70	70
Damaged	5	1	1	1	0	0	0	0	0	0	1	0
Undamaged	1	5	5	5	6	6	6	6	6	6	5	6

Shuttle Mission	13	14	15	16	17	18	19	20	21	22	23
Temperature	70	70	72	73	75	75	76	76	78	79	81
Damaged	1	0	0	0	0	1	0	0	0	0	0
Undamaged	5	6	6	6	6	5	6	6	6	6	6

(a) Each column of the table above represents a different shuttle mission. Examine these data and describe what you observe with respect to the relationship between temperatures and damaged O-rings.

(b) Failures have been coded as 1 for a damaged O-ring and 0 for an undamaged O-ring, and a logistic regression model was fit to these data. A summary of this model is given below. Describe the key components of this summary table in words.

	Estimate	Std. Error	z value	Pr(>|z|)
(Intercept)	11.6630	3.2963	3.54	0.0004
Temperature	-0.2162	0.0532	-4.07	0.0000

(c) Write out the logistic model using the point estimates of the model parameters.

(d) Based on the model, do you think concerns regarding O-rings are justified? Explain.

8.17 Possum classification, Part II. A logistic regression model was proposed for classifying common brushtail possums into their two regions in Exercise 8.15. The outcome variable took value 1 if the possum was from Victoria and 0 otherwise.

	Estimate	SE	Z	Pr(>|Z|)
(Intercept)	33.5095	9.9053	3.38	0.0007
sex_male	-1.4207	0.6457	-2.20	0.0278
skull_width	-0.2787	0.1226	-2.27	0.0231
total_length	0.5687	0.1322	4.30	0.0000
tail_length	-1.8057	0.3599	-5.02	0.0000

(a) Write out the form of the model. Also identify which of the variables are positively associated when controlling for other variables.

(b) Suppose we see a brushtail possum at a zoo in the US, and a sign says the possum had been captured in the wild in Australia, but it doesn't say which part of Australia. However, the sign does indicate that the possum is male, its skull is about 63 mm wide, its tail is 37 cm long, and its total length is 83 cm. What is the reduced model's computed probability that this possum is from Victoria? How confident are you in the model's accuracy of this probability calculation?

8.18　Challenger disaster, Part II. Exercise 8.16 introduced us to O-rings that were identified as a plausible explanation for the breakup of the Challenger space shuttle 73 seconds into takeoff in 1986. The investigation found that the ambient temperature at the time of the shuttle launch was closely related to the damage of O-rings, which are a critical component of the shuttle. See this earlier exercise if you would like to browse the original data.

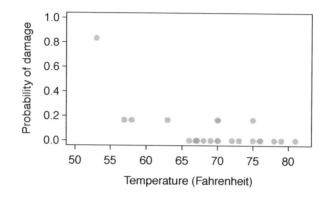

(a) The data provided in the previous exercise are shown in the plot. The logistic model fit to these data may be written as

$$\log\left(\frac{\hat{p}}{1-\hat{p}}\right) = 11.6630 - 0.2162 \times Temperature$$

where $\hat{p}$ is the model-estimated probability that an O-ring will become damaged. Use the model to calculate the probability that an O-ring will become damaged at each of the following ambient temperatures: 51, 53, and 55 degrees Fahrenheit. The model-estimated probabilities for several additional ambient temperatures are provided below, where subscripts indicate the temperature:

$$\hat{p}_{57} = 0.341 \qquad \hat{p}_{59} = 0.251 \qquad \hat{p}_{61} = 0.179 \qquad \hat{p}_{63} = 0.124$$
$$\hat{p}_{65} = 0.084 \qquad \hat{p}_{67} = 0.056 \qquad \hat{p}_{69} = 0.037 \qquad \hat{p}_{71} = 0.024$$

(b) Add the model-estimated probabilities from part (a) on the plot, then connect these dots using a smooth curve to represent the model-estimated probabilities.

(c) Describe any concerns you may have regarding applying logistic regression in this application, and note any assumptions that are required to accept the model's validity.

Appendix A

End of chapter exercise solutions

1 Introduction to data

1.1 (a) Treatment: $10/43 = 0.23 \rightarrow 23\%$. Control: $2/46 = 0.04 \rightarrow 4\%$. (b) There is a 19% difference between the pain reduction rates in the two groups. At first glance, it appears patients in the treatment group are more likely to experience pain reduction from the acupuncture treatment. (c) Answers may vary but should be sensible. Two possible answers: [1]Though the groups' difference is big, I'm skeptical the results show a real difference and think this might be due to chance. [2]The difference in these rates looks pretty big, so I suspect acupuncture is having a positive impact on pain.

1.3 (a) 143,196 eligible study subjects born in Southern California between 1989 and 1993. (b) Measurements of carbon monoxide, nitrogen dioxide, ozone, and particulate matter less than $10\mu g/m^3$ (PM_{10}) collected at air-quality-monitoring stations as well as length of gestation. Continuous numerical variables. (c) "Is there an association between air pollution exposure and preterm births?"

1.5 (a) 160 children. (b) Age (numerical, continuous), sex (categorical), whether they were an only child or not (categorical), and whether they cheated or not (categorical). (c) Research question: "Does explicitly telling children not to cheat affect their likelihood to cheat?"

1.7 (a) $50 \times 3 = 150$. (b) Four continuous numerical variables: sepal length, sepal width, petal length, and petal width. (c) One categorical variable, species, with three levels: *setosa*, *versicolor*, and *virginica*.

1.9 (a) Population: all births, sample: 143,196 births between 1989 and 1993 in Southern California. (b) If births in this time span at the geography can be considered to be representative of all births, then the results are generalizable to the population of Southern California. However, since the study is observational the findings cannot be used to establish causal relationships.

1.11 (a) Population: all asthma patients aged 18-69 who rely on medication for asthma treatment. Sample: 600 such patients. (b) If the patients in this sample, who are likely not randomly sampled, can be considered to be representative of all asthma patients aged 18-69 who rely on medication for asthma treatment, then the results are generalizable to the population defined above. Additionally, since the study is experimental, the findings can be used to establish causal relationships.

1.13 (a) Observation. (b) Variable. (c) Sample statistic (mean). (d) Population parameter (mean).

1.15 (a) Explanatory: number of study hours per week. Response: GPA. (b) Somewhat weak positive relationship with data becoming more sparse as the number of study hours increases. One responded reported a GPA above 4.0, which is clearly a data error. There are a few respondents who reported unusually high study hours (60 and 70 hours/week). Variability in GPA is much higher for students who study less than those who study more, which might be due to the fact that there aren't many respondents who reported studying higher hours. (c) Observational. (d) Since observational, cannot infer causation.

1.17 (a) Observational. (b) Use stratified sampling to randomly sample a fixed number of students, say 10, from each section for a total sample size of 40 students.

1.19 (a) Positive, non-linear, somewhat strong. Countries in which a higher percentage of the population have access to the internet also tend to have higher average life expectancies, however rise in life expectancy trails off before around 80 years old. (b) Observational. (c) Wealth: countries with individuals who can widely afford the internet can probably also afford basic medical care. (Note: Answers may vary.)

1.21 (a) Simple random sampling is okay. In fact, it's rare for simple random sampling to not be a reasonable sampling method! (b) The student opinions may vary by field of study, so the stratifying by this variable makes sense and would be reasonable. (c) Students of similar ages are probably going to have more similar opinions, and we want clusters to be diverse with respect to the outcome of interest, so this would **not** be a good approach. (Additional thought: the clusters in this case may also have very different numbers of people, which can also create unexpected sample sizes.)

1.23 (a) The cases are 200 randomly sampled men and women. (b) The response variable is attitude towards a fictional microwave oven. (c) The explanatory variable is dispositional attitude. (d) Yes, the cases are sampled randomly. (e) This is an observational study since there is no random assignment to treatments. (f) No, we cannot establish a causal link between the explanatory and response variables since the study is observational. (g) Yes, the results of the study can be generalized to the population at large since the sample is random.

1.25 (a) Non-responders may have a different response to this question, e.g. parents who returned the surveys likely don't have difficulty spending time with their children. (b) It is unlikely that the women who were reached at the same address 3 years later are a random sample. These missing responders are probably renters (as opposed to homeowners) which means that they might be in a lower socio-economic status than the respondents. (c) There is no control group in this study, this is an observational study, and there may be confounding variables, e.g. these people may go running because they are generally healthier and/or do other exercises.

1.27 (a) Simple random sample. Non-response bias, if only those people who have strong opinions about the survey responds his sample may not be representative of the population. (b) Convenience sample. Under coverage bias, his sample may not be representative of the population since it consists only of his friends. It is also possible that the study will have non-response bias if some choose to not bring back the survey. (c) Convenience sample. This will have a similar issues to handing out surveys to friends. (d) Multi-stage sampling. If the classes are similar to each other with respect to student composition this approach should not introduce bias, other than potential non-response bias.

1.29 No, students were not randomly sampled (voluntary sample) and the sample only contains college students at a university in Ontario.

1.31 (a) Exam performance. (b) Light level: fluorescent overhead lighting, yellow overhead lighting, no overhead lighting (only desk lamps). (c) Sex: man, woman.

1.33 (a) Exam performance. (b) Light level (overhead lighting, yellow overhead lighting, no overhead lighting) and noise level (no noise, construction noise, and human chatter noise). (c) Since the researchers want to ensure equal gender representation, sex will be a blocking variable.

1.35 Need randomization and blinding. One possible outline: (1) Prepare two cups for each participant, one containing regular Coke and the other containing Diet Coke. Make sure the cups are identical and contain equal amounts of soda. Label the cups A (regular) and B (diet). (Be sure to randomize A and B for each trial!) (2) Give each participant the two cups, one cup at a time, in random order, and ask the participant to record a value that indicates how much she liked the beverage. Be sure that neither the participant nor the person handing out the cups knows the identity of the beverage to make this a double-blind experiment. (Answers may vary.)

1.37 (a) Experiment. (b) Treatment: 25 grams of chia seeds twice a day, control: placebo. (c) Yes, gender. (d) Yes, single blind since the patients were blinded to the treatment they received. (e) Since this is an experiment, we can make a causal statement. However, since the sample is not random, the causal statement cannot be generalized to the population at large.

1.39 (a) 1: linear. 3: nonlinear. (b) 4: linear. (c) 2.

1.41

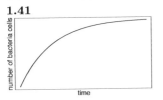

1.43 (a) Population mean, $\mu_{2007} = 52$; sample mean, $\bar{x}_{2008} = 58$. (b) Population mean, $\mu_{2001} = 3.37$; sample mean, $\bar{x}_{2012} = 3.59$.

1.45 Any 10 employees whose average number of days off is between the minimum and the mean number of days off for the entire workforce at this plant.

1.47 (a) Dist 2 has a higher mean since $20 > 13$, and a higher standard deviation since 20 is further from the rest of the data than 13. (b) Dist 1 has a higher mean since $-20 > -40$, and Dist 2 has a higher standard deviation since -40 is farther away from the rest of the data than -20. (c) Dist 2 has a higher mean since all values in this distribution are higher than those in Dist 1, but both distribution have the same standard deviation since they are equally variable around their respective means. (d) Both

distributions have the same mean since they're both centered at 300, but Dist 2 has a higher standard deviation since the observations are farther from the mean than in Dist 1.

1.49 (a) Q1 $\approx$ 5, median $\approx$ 15, Q3 $\approx$ 35 (b) Since the distribution is right skewed, we would expect the mean to be higher than the median.

1.51 (a) About 30. (b) Since the distribution is right skewed the mean is higher than the median. (c) Q1: between 15 and 20, Q3: between 35 and 40, IQR: about 20. (d) Values that are considered to be unusually low or high lie more than 1.5×IQR away from the quartiles. Upper fence: Q3 + 1.5 × IQR = $37.5 + 1.5 \times 20 = 67.5$; Lower fence: Q1 - 1.5 × IQR = $17.5 + 1.5 \times 20 = -12.5$; The lowest AQI recorded is not lower than 5 and the highest AQI recorded is not higher than 65, which are both within the fences. Therefore none of the days in this sample would be considered to have an unusually low or high AQI.

1.53 The histogram shows that the distribution is bimodal, which is not apparent in the box plot. The box plot makes it easy to identify more precise values of observations outside of the whiskers.

1.55 (a) The distribution of number of pets per household is likely right skewed as there is a natural boundary at 0 and only a few people have many pets. Therefore the center would be best described by the median, and variability would be best described by the IQR. (b) The distribution of number of distance to work is likely right skewed as there is a natural boundary at 0 and only a few people live a very long distance from work. Therefore the center would be best described by the median, and variability would be best described by the IQR. (c) The distribution of heights of males is likely symmetric. Therefore the center would be best described by the mean, and variability would be best described by the standard deviation.

1.57 No, we would expect this distribution to be right skewed. There are two reasons for this: (1) there is a natural boundary at 0 (it is not possible to watch less than 0 hours of TV), (2) the standard deviation of the distribution is very large compared to the mean.

1.59 The statement "50% of Facebook users have over 100 friends" means that the median number of friends is 100, which is lower than the mean number of friends (190), which suggests a right skewed distribution for the number of friends of Facebook users.

1.61 (a) The median is a much better measure of the typical amount earned by these 42 people. The mean is much higher than the income of 40 of the 42 people. This is because the mean is an arithmetic average and gets affected by the two extreme observations. The median does not get effected as much since it is robust to outliers. (b) The IQR is a much better measure of variability in the amounts earned by nearly all of the 42 people. The standard deviation gets affected greatly by the two high salaries, but the IQR is robust to these extreme observations.

1.63 (a) The distribution is unimodal and symmetric with a mean of about 25 minutes and a standard deviation of about 5 minutes. There does not appear to be any counties with unusually high or low mean travel times. Since the distribution is already unimodal and symmetric, a log transformation is not necessary. (b) Answers will vary. There are pockets of longer travel time around DC, Southeastern NY, Chicago, Minneapolis, Los Angeles, and many other big cities. There is also a large section of shorter average commute times that overlap with farmland in the Midwest. Many farmers' homes are adjacent to their farmland, so their commute would be brief, which may explain why the average commute time for these counties is relatively low.

1.65 (a) We see the order of the categories and the relative frequencies in the bar plot. (b) There are no features that are apparent in the pie chart but not in the bar plot. (c) We usually prefer to use a bar plot as we can also see the relative frequencies of the categories in this graph.

1.67 The vertical locations at which the ideological groups break into the Yes, No, and Not Sure categories differ, which indicates that like- lihood of supporting the DREAM act varies by political ideology. This suggests that the two variables may be dependent.

1.69 (a) (i) False. Instead of comparing counts, we should compare percentages of people in each group who suffered cardiovascular problems. (ii) True. (iii) False. Association does not imply causation. We cannot infer a causal relationship based on an observational study. The difference from part (ii) is subtle. (iv) True.
(b) Proportion of all patients who had cardiovascular problems: $\frac{7,979}{227,571} \approx 0.035$
(c) The expected number of heart attacks in the rosiglitazone group, if having cardiovascular problems and treatment were independent, can be calculated as the number of patients in that group multiplied by the overall cardiovascular problem rate in the study: $67,593 * \frac{7,979}{227,571} \approx 2370$.
(d) (i) H_0: The treatment and cardiovascular problems are independent. They have no relationship, and the difference in incidence rates between the rosiglitazone and pioglitazone groups is due to chance. H_A: The treatment and cardiovascular problems are not independent. The difference in the incidence rates between the rosiglitazone and pioglitazone groups is not due to chance and rosiglitazone is associated with an increased risk of serious cardiovascular problems. (ii) A higher number of patients with cardiovascular problems than expected under the assumption of independence would provide support for the alternative hypothesis as this would suggest that rosiglitazone increases the risk of such problems. (iii) In the actual study, we observed 2,593 cardiovascular events in the rosiglitazone group. In the 1,000 simulations under the independence model, we observed somewhat less than 2,593 in every single simulation, which suggests that the actual results did not come from the independence model. That is, the variables do not appear to be independent, and we reject the independence model in favor of the alternative. The study's results provide convincing evidence that rosiglitazone is associated with an increased risk of cardiovascular problems.

2 Probability

2.1 (a) False. These are independent trials. (b) False. There are red face cards. (c) True. A card cannot be both a face card and an ace.

2.3 (a) 10 tosses. Fewer tosses mean more variability in the sample fraction of heads, meaning there's a better chance of getting at least 60% heads. (b) 100 tosses. More flips means the observed proportion of heads would often be closer to the average, 0.50, and therefore also above 0.40. (c) 100 tosses. With more flips, the observed proportion of heads would often be closer to the average, 0.50. (d) 10 tosses. Fewer flips would increase variability in the fraction of tosses that are heads.

2.5 (a) $0.5^{10} = 0.00098$. (b) $0.5^{10} = 0.00098$. (c) $P(\text{at least one tails}) = 1 - P(\text{no tails}) = 1 - (0.5^{10}) \approx 1 - 0.001 = 0.999$.

2.7 (a) No, there are voters who are both independent and swing voters.
(b)

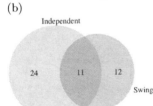

(c) Each Independent voter is either a swing voter or not. Since 35% of voters are Independents and 11% are both Independent and swing voters, the other 24% must not be swing voters. (d) 0.47. (e) 0.53. (f) P(Independent) × P(swing) $= 0.35 \times 0.23 = 0.08$, which does not equal P(Independent and swing) $= 0.11$, so the events are dependent.

2.9 (a) If the class is not graded on a curve, they are independent. If graded on a curve, then neither independent nor disjoint – unless the instructor will only give one A, which is a situation we will ignore in parts (b) and (c). (b) They are probably not independent: if you study together, your study habits would be related, which suggests your course performances are also related. (c) No. See the answer to part (a) when the course is not graded on a curve. More generally: if two things are un-related (independent), then one occurring does not preclude the other from occurring.

2.11 (a) $0.16 + 0.09 = 0.25$. (b) $0.17 + 0.09 = 0.26$. (c) Assuming that the education level of the husband and wife are independent: $0.25 \times 0.26 = 0.065$. You might also notice we actually made a second assumption: that the decision to get married is unrelated to education level. (d) The husband/wife independence assumption is probably not reasonable, because people often marry another person with a comparable level of education. We will leave it to you to think about whether the second assumption noted in part (c) is reasonable.

2.13 (a) Invalid. Sum is greater than 1. (b) Valid. Probabilities are between 0 and 1, and they sum to 1. In this class, every student gets a C. (c) Invalid. Sum is less than 1. (d) Invalid. There is a negative probability. (e) Valid. Probabilities are between 0 and 1, and they sum to 1. (f) Invalid. There is a negative probability.

2.15 (a) No, but we could if A and B are independent. (b-i) 0.21. (b-ii) 0.79. (b-iii) 0.3. (c) No, because $0.1 \neq 0.21$, where 0.21 was the value computed under independence from part (a). (d) 0.143.

2.17 (a) No, 0.18 of respondents fall into this combination. (b) $0.60 + 0.20 - 0.18 = 0.62$. (c) $0.18/0.20 = 0.9$. (d) $0.11/0.33 \approx 0.33$. (e) No, otherwise the answers to (c) and (d) would be the same. (f) $0.06/0.34 \approx 0.18$.

2.19 (a) No. There are 6 females who like Five Guys Burgers. (b) $162/248 = 0.65$. (c) $181/252 = 0.72$. (d) Under the assumption of a dating choices being independent of hamburger preference, which on the surface seems reasonable: $0.65 \times 0.72 = 0.468$. (e) $(252 + 6 - 1)/500 = 0.514$.

2.21 (a)

(b) 0.84

2.23 0.8247.

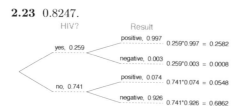

2.25 0.0714. Even when a patient tests positive for lupus, there is only a 7.14% chance that he actually has lupus. House may be right.

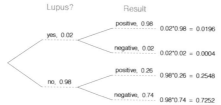

2.27 (a) 0.3. (b) 0.3. (c) 0.3. (d) $0.3 \times 0.3 = 0.09$. (e) Yes, the population that is being sampled from is identical in each draw.

2.29 (a) $2/9 \approx 0.22$. (b) $3/9 \approx 0.33$. (c) $\frac{3}{10} \times \frac{2}{9} \approx 0.067$. (d) No, e.g. in this exercise, removing one chip meaningfully changes the probability of what might be drawn next.

2.31 $P(^1\text{leggings}, {}^2\text{jeans}, {}^3\text{jeans}) = \frac{5}{24} \times \frac{7}{23} \times \frac{6}{22} = 0.0173$. However, the person with leggings could have come 2nd or 3rd, and these each have this same probability, so $3 \times 0.0173 = 0.0519$.

2.33 (a) 13. (b) No, these 27 students are not a random sample from the university's student population. For example, it might be argued that the proportion of smokers among students who go to the gym at 9 am on a Saturday morning would be lower than the proportion of smokers in the university as a whole.

2.35 (a) E(X) = 3.59. SD(X) = 9.64. (b) E(X) = -1.41. SD(X) = 9.64. (c) No, the expected net profit is negative, so on average you expect to lose money.

2.37 5% increase in value.

2.39 E = -0.0526. SD = 0.9986.

2.41 (a) E = \$3.90. SD = \$0.34. (b) E = \$27.30. SD = \$0.89.

2.43 Approximate answers are OK. (a) $(29 + 32)/144 = 0.42$. (b) $21/144 = 0.15$. (c) $(26 + 12 + 15)/144 = 0.37$.

3 Distributions of random variables

3.1 (a) 8.85%. (b) 6.94%. (c) 58.86%. (d) 4.56%.

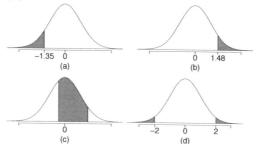

3.3 (a) Verbal: $N(\mu = 151, \sigma = 7)$, Quant: $N(\mu = 153, \sigma = 7.67)$. (b) $Z_{VR} = 1.29$, $Z_{QR} = 0.52$.

(c) She scored 1.29 standard deviations above the mean on the Verbal Reasoning section and 0.52 standard deviations above the mean on the Quantitative Reasoning section. (d) She did better on the Verbal Reasoning section since her Z-score on that section was higher. (e) $Perc_{VR} = 0.9007 \approx 90\%$, $Perc_{QR} = 0.6990 \approx 70\%$. (f) $100\% - 90\% = 10\%$ did better than her on VR, and $100\% - 70\% = 30\%$ did better than her on QR. (g) We cannot compare the raw scores since they are on different scales. Comparing her percentile scores is more appropriate when comparing her performance to others. (h) Answer to part (b) would not change as Z-scores can be calculated for distributions that are not normal. However, we could not answer parts (d)-(f) since we cannot use the normal probability table to calculate probabilities and percentiles without a normal model.

3.5 (a) $Z = 0.84$, which corresponds to approximately 160 on QR. (b) $Z = -0.52$, which corresponds to approximately 147 on VR.

3.7 (a) $Z = 1.2 \rightarrow 0.1151$. (b) $Z = -1.28 \rightarrow 70.6°\text{F}$ or colder.

3.9 (a) $N(25, 2.78)$. (b) $Z = 1.08 \to 0.1401$. (c) The answers are very close because only the units were changed. (The only reason why they differ at all is because 28°C is 82.4°F, not precisely 83°F.) (d) Since $IQR = Q3 - Q1$, we first need to find $Q3$ and $Q1$ and take the difference between the two. Remember that $Q3$ is the 75th and $Q1$ is the 25th percentile of a distribution. $Q1 = 23.13$, $Q3 = 26.86$, $IQR = 26.86 - 23.13 = 3.73$.

3.11 (a) $Z = 0.67$. (b) $\mu = \$1650$, $x = \$1800$. (c) $0.67 = \frac{1800 - 1650}{\sigma} \to \sigma = \$223.88$.

3.13 $Z = 1.56 \to 0.0594$, i.e. 6%.

3.15 (a) $Z = 0.73 \to 0.2327$. (b) If you are bidding on only one auction and set a low maximum bid price, someone will probably outbid you. If you set a high maximum bid price, you may win the auction but pay more than is necessary. If bidding on more than one auction, and you set your maximum bid price very low, you probably won't win any of the auctions. However, if the maximum bid price is even modestly high, you are likely to win multiple auctions. (c) An answer roughly equal to the 10th percentile would be reasonable. Regrettably, no percentile cut-off point guarantees beyond any possible event that you win at least one auction. However, you may pick a higher percentile if you want to be more sure of winning an auction. (d) Answers will vary a little but should correspond to the answer in part (c). We use the 10th percentile: $Z = -1.28 \to \$69.80$.

3.17 (a) 70% of the data are within 1 standard deviation of the mean, 95% are within 2 and 100% are within 3 standard deviations of the mean. Therefore, we can say that the data approximately follow the 68-95-99.7% Rule. (b) The distribution is unimodal and symmetric. The superimposed normal curve seems to approximate the distribution pretty well. The points on the normal probability plot also seem to follow a straight line. There is one possible outlier on the lower end that is apparent in both graphs, but it is not too extreme. We can say that the distribution is nearly normal.

3.19 (a) No. The cards are not independent. For example, if the first card is an ace of clubs, that implies the second card cannot be an ace of clubs. Additionally, there are many possible categories, which would need to be simplified. (b) No. There are six events under consideration. The Bernoulli distribution allows for only two events or categories. Note that rolling a die could be a Bernoulli trial if we simply to two events, e.g. rolling a 6 and not rolling a 6, though specifying such details would be necessary.

3.21 (a) $(1 - 0.471)^2 \times 0.471 = 0.1318$. (b) $0.471^3 = 0.1045$. (c) $\mu = 1/0.471 = 2.12$, $\sigma = \sqrt{2.38} = 1.54$. (d) $\mu = 1/0.30 = 3.33$, $\sigma = 2.79$. (e) When p is smaller, the event is rarer, meaning the expected number of trials before a success and the standard deviation of the waiting time are higher.

3.23 (a) $0.875^2 \times 0.125 = 0.096$. (b) $\mu = 8$, $\sigma = 7.48$.

3.25 (a) Binomial conditions are met: (1) Independent trials: In a random sample, whether or not one 18-20 year old has consumed alcohol does not depend on whether or not another one has. (2) Fixed number of trials: $n = 10$. (3) Only two outcomes at each trial: Consumed or did not consume alcohol. (4) Probability of a success is the same for each trial: $p = 0.697$. (b) 0.203. (c) 0.203. (d) 0.167. (e) 0.997.

3.27 (a) $\mu = 34.85$, $\sigma = 3.25$ (b) $Z = \frac{45 - 34.85}{3.25} = 3.12$. 45 is more than 3 standard deviations away from the mean, we can assume that it is an unusual observation. Therefore yes, we would be surprised. (c) Using the normal approximation, 0.0009. With 0.5 correction, 0.0015.

3.29 Want to find the probability that there will be 1,786 or more enrollees. Using the normal approximation: 0.0582. With a 0.5 correction: 0.0559.

3.31 (a) $1 - 0.75^3 = 0.5781$. (b) 0.1406. (c) 0.4219. (d) $1 - 0.25^3 = 0.9844$.

3.33 (a) Geometric distribution: 0.109. (b) Binomial: 0.219. (c) Binomial: 0.137. (d) $1 - 0.875^6 = 0.551$. (e) Geometric: 0.084. (f) Using a binomial distribution with $n = 6$ and $p = 0.75$, we see that $\mu = 4.5$, $\sigma = 1.06$, and $Z = 2.36$. Since this is not within 2 SD, it may be considered unusual.

3.35 0 wins (-\$3): 0.1458. 1 win (-\$1): 0.3936. 2 wins (+\$1): 0.3543. 3 wins (+\$3): 0.1063.

3.37 (a) $\overset{Anna}{1/5} \times \overset{Ben}{1/4} \times \overset{Carl}{1/3} \times \overset{Damian}{1/2} \times \overset{Eddy}{1/1} = 1/5! = 1/120$. (b) Since the probabilities must add to 1, there must be $5! = 120$ possible orderings. (c) $8! = 40,320$.

3.39 (a) 0.0804. (b) 0.0322. (c) 0.0193.

3.41 (a) Negative binomial with $n = 4$ and $p = 0.55$, where a success is defined here as a female student. The negative binomial setting is appropriate since the last trial is fixed but the order of the first 3 trials is unknown. (b) 0.1838. (c) $\binom{3}{1} = 3$. (d) In the binomial model there are no restrictions on the outcome of the last trial. In the negative binomial model the last trial is fixed. Therefore we are interested in the number of ways of orderings of the other $k - 1$ successes in the first $n - 1$ trials.

3.43 (a) Poisson with $\lambda = 75$. (b) $\mu = \lambda = 75$, $\sigma = \sqrt{\lambda} = 8.66$. (c) $Z = -1.73$. Since 60 is within 2 standard deviations of the mean, it would not generally be considered unusual. Note that we often use this rule of thumb even when the normal model does not apply. (d) Using Poisson with $\lambda = 75$: 0.0402.

4 Foundations for inference

4.1 (a) Mean. Each student reports a numerical value: a number of hours. (b) Mean. Each student reports a number, which is a percentage, and we can average over these percentages. (c) Proportion. Each student reports Yes or No, so this is a categorical variable and we use a proportion. (d) Mean. Each student reports a number, which is a percentage like in part (b). (e) Proportion. Each student reports whether or not s/he expects to get a job, so this is a categorical variable and we use a proportion.

4.3 (a) Mean: 13.65. Median: 14. (b) SD: 1.91. IQR: $15 - 13 = 2$. (c) $Z_{16} = 1.23$, which is not unusual since it is within 2 SD of the mean. $Z_{18} = 2.28$, which is generally considered unusual. (d) No. Point estimates that are based on samples only approximate the population parameter, and they vary from one sample to another. (e) We use the SE, which is $1.91/\sqrt{100} = 0.191$ for this sample's mean.

4.5 (a) We are building a distribution of sample statistics, in this case the sample mean. Such a distribution is called a sampling distribution. (b) Because we are dealing with the distribution of sample means, we need to check to see if the Central Limit Theorem applies. Our sample size is greater than 30, and we are told that random sampling is employed. With these conditions met, we expect that the distribution of the sample mean will be nearly normal and therefore symmetric. (c) Because we are dealing with a sampling distribution, we measure its variability with the standard error. $SE = 18.2/\sqrt{45} = 2.713$. (d) The sample means will be more variable with the smaller sample size.

4.7 Recall that the general formula is

$$\text{point estimate} \pm Z^\star \times SE$$

First, identify the three different values. The point estimate is 45%, $Z^\star = 1.96$ for a 95% confidence level, and $SE = 1.2\%$. Then, plug the values into the formula:

$$45\% \pm 1.96 \times 1.2\% \quad \rightarrow \quad (42.6\%, 47.4\%)$$

We are 95% confident that the proportion of US adults who live with one or more chronic conditions is between 42.6% and 47.4%.

4.9 (a) False. Confidence intervals provide a range of plausible values, and sometimes the truth is missed. A 95% confidence interval "misses" about 5% of the time. (b) True. Notice that the description focuses on the true population value. (c) True. If we examine the 95% confidence interval computed in Exercise 4.9, we can see that 50% is not included in this interval. This means that in a hypothesis test, we would reject the null hypothesis that the proportion is 0.5. (d) False. The standard error describes the uncertainty in the overall estimate from natural fluctuations due to randomness, not the uncertainty corresponding to individuals' responses.

4.11 (a) We are 95% confident that Americans spend an average of 1.38 to 1.92 hours per day relaxing or pursuing activities they enjoy. (b) Their confidence level must be higher as the width of the confidence interval increases as the confidence level increases. (c) The new margin of error will be smaller since as the sample size increases the standard error decreases, which will decrease the margin of error.

4.13 (a) False. Provided the data distribution is not very strongly skewed ($n = 64$ in this sample, so we can be slightly lenient with the skew), the sample mean will be nearly normal, allowing for the method normal approximation described. (b) False. Inference is made on the population parameter, not the point estimate. The point estimate is always in the confidence interval. (c) True. (d) False. The confidence interval is not about a sample mean. (e) False. To be more confident that we capture the parameter, we need a wider interval. Think about needing a bigger net to be more sure of catching a fish in a murky lake. (f) True. Optional explanation: This is true since the normal model was used to model the sample mean. The margin of error is half the width of the interval, and the sample mean is the midpoint of the interval. (g) False. In the calculation of the standard error, we divide the standard deviation by the square root of the sample size. To cut the SE (or margin of error) in half, we would need to sample $2^2 = 4$ times the number of people in the initial sample.

4.15 Independence: sample from $< 10\%$ of population, and it is a random sample. We can assume that the students in this sample are independent of each other with respect to number of exclusive relationships they have been in. Notice that there are no students who have had no exclusive relationships in the sample, which suggests some student responses are likely missing (perhaps only positive values were reported). The sample size is at least 30. The skew is strong, but the sample is very large so this is not a concern. 90% CI: (2.97, 3.43). We are 90% confident that undergraduate students have been in 2.97 to 3.43 exclusive relationships, on average.

4.17 (a) $H_0 : \mu = 8$ (On average, New Yorkers sleep 8 hours a night.)
$H_A : \mu < 8$ (On average, New Yorkers sleep less than 8 hours a night.)
(b) $H_0 : \mu = 15$ (The average amount of company time each employee spends not working is 15 minutes for March Madness.)
$H_A : \mu > 15$ (The average amount of company time each employee spends not working is greater than 15 minutes for March Madness.)

4.19 The hypotheses should be about the population mean (μ), not the sample mean. The null hypothesis should have an equal sign and the alternative hypothesis should be about the null hypothesized value, not the observed sample mean. Correction:

$$H_0 : \mu = 10 \ hours$$
$$H_A : \mu > 10 \ hours$$

The one-sided test indicates that we are only interested in showing that 10 is an underestimate. Here the interest is in only one direction, so a one-sided test seems most appropriate. If we would also be interested if the data showed strong evidence that 10 was an overestimate, then the test should be two-sided.

4.21 (a) This claim does is not supported since 3 hours (180 minutes) is not in the interval. (b) 2.2 hours (132 minutes) is in the 95% confidence interval, so we do not have evidence to say she is wrong. However, it would be more appropriate to use the point estimate of the sample. (c) A 99% confidence interval will be wider than a 95% confidence interval, meaning it would enclose this smaller interval. This means 132 minutes would be in the wider interval, and we would not reject her claim based on a 99% confidence level.

4.23 $H_0 : \mu = 130$. $H_A : \mu \neq 130$. $Z = 1.39 \rightarrow$ p-value $= 0.1646$, which is larger than $\alpha = 0.05$. The data do not provide convincing evidence that the true average calorie content in bags of potato chips is different than 130 calories.

4.25 (a) Independence: The sample is random and 64 patients would almost certainly make up less than 10% of the ER residents. The sample size is at least 30. No information is provided about the skew. In practice, we would ask to see the data to check this condition, but here we will make the assumption that the skew is not very strong. (b) $H_0 : \mu = 127$. $H_A : \mu \neq 127$. $Z = 2.15 \rightarrow$ p-value $= 0.0316$. Since the p-value is less than $\alpha = 0.05$, we reject H_0. The data provide convincing evidence that the average ER wait time has increased over the last year. (c) Yes, it would change. The p-value is greater than 0.01, meaning we would fail to reject H_0 at $\alpha = 0.01$.

4.27 $Z = 1.65 = \frac{\bar{x}-30}{10/\sqrt{70}} \rightarrow \bar{x} = 31.97$.

4.29 (a) H_0: Anti-depressants do not help symptoms of Fibromyalgia. H_A: Anti- depressants do treat symptoms of Fibromyalgia. Remark: Diana might also have taken special note if her symptoms got much worse, so a more scientific approach would have been to use a two-sided test. If you proposed a two-sided approach, your answers in (b) and (c) will be different. (b) Concluding that anti-depressants work for the treatment of Fibromyalgia symptoms when they actually do not. (c) Concluding that anti-depressants do not work for the treatment of Fibromyalgia symptoms when they actually do.

4.31 (a) Scenario I is higher. Recall that a sample mean based on less data tends to be less accurate and have larger standard errors. (b) Scenario I is higher. The higher the confidence level, the higher the corresponding margin of error. (c) They are equal. The sample size does not affect the calculation of the p- value for a given Z-score. (d) Scenario I is higher. If the null hypothesis is harder to reject (lower α), then we are more likely to make a Type 2 Error when the alternative hypothesis is true.

4.33 (a) The distribution is unimodal and strongly right skewed with a median between 5 and 10 years old. Ages range from 0 to slightly over 50 years old, and the middle 50% of the distribution is roughly between 5 and 15 years old. There are potential outliers on the higher end. (b) When the sample size is small, the sampling distribution is right skewed, just like the population distribution. As the sample size increases, the sampling distribution gets more unimodal, symmetric, and approaches normality. The variability also decreases. This is consistent with the Central Limit Theorem. (c) n = 5: $\mu_{\bar{x}} = 10.44$, $\sigma_{\bar{x}} = 4.11$; n = 30: $\mu_{\bar{x}} = 10.44$, $\sigma_{\bar{x}} = 1.68$; n = 100: $\mu_{\bar{x}} = 10.44$, $\sigma_{\bar{x}} = 0.92$. The centers of the sampling distributions shown in part (b) appear to be around 10. It is difficult to estimate the standard deviation for the sampling distribution when $n = 5$ from the histogram (since the distribution is somewhat skewed). If 1.68 is a plausible estimate for the standard deviation of the sampling distribution when $n = 30$, then using the 68-95-99.7% Rule, we would expect the values to range roughly between $10.44 \pm 3*1.68 = (5.4, 15.48)$, which seems to be the case. Similarly, when $n = 100$, we would expect the values to range roughly be-

tween $10.44 \pm 3 * 0.92 = (7.68, 13.2)$, which also seems to be the case.

4.35 (a) Right skewed. There is a long tail on the higher end of the distribution but a much shorter tail on the lower end. (b) Less than, as the median would be less than the mean in a right skewed distribution. (c) We should not. (d) Even though the population distribution is not normal, the conditions for inference are reasonably satisfied, with the possible exception of skew. If the skew isn't very strong (we should ask to see the data), then we can use the Central Limit Theorem to estimate this probability. For now, we'll assume the skew isn't very strong, though the description suggests it is at least moderate to strong. Use $N(1.3, SD_{\bar{x}} = 0.3/\sqrt{60})$: $Z = 2.58 \rightarrow 0.0049$. (e) It would decrease it by a factor of $1/\sqrt{2}$.

4.37 The centers are the same in each plot, and each data set is from a nearly normal distribution, though the histograms may not look very normal since each represents only 100 data points. The only way to tell which plot corresponds to which scenario is to examine the variability of each distribution. Plot B is the most variable, followed by Plot A, then Plot C. This means Plot B will correspond to the original data, Plot A to the sample means with size 5, and Plot C to the sample means with size 25.

4.39 (a) $Z = -3.33 \rightarrow 0.0004$. (b) The population SD is known and the data are nearly normal, so the sample mean will be nearly normal with distribution $N(\mu, \sigma/\sqrt{n})$, i.e. $N(2.5, 0.0095)$. (c) $Z = -10.54 \rightarrow \approx 0$. (d) See below:

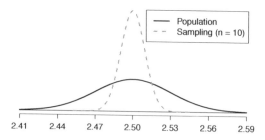

(e) We could not estimate (a) without a nearly normal population distribution. We also could not estimate (c) since the sample size is not sufficient to yield a nearly normal sampling distribution if the population distribution is not nearly normal.

4.41 (a) We cannot use the normal model for this calculation, but we can use the histogram. About 500 songs are shown to be longer than 5 minutes, so the probability is about $500/3000 = 0.167$. (b) Two different answers are reasonable. $^{Option\ 1}$Since the population distribution is only slightly skewed to the right, even a small sample size will yield a nearly normal sampling distribution. We also know that the songs are sampled randomly and the sample size is less than 10% of the population, so the length of one song in the sample is independent of another. We are looking for the probability that the total length of 15 songs is more than 60 minutes, which means that the average song should last at least $60/15 = 4$ minutes. Using $SD_{\bar{x}} = 1.63/\sqrt{15}$, $Z = 1.31 \rightarrow 0.0951$. $^{Option\ 2}$Since the population distribution is not normal, a small sample size may not be sufficient to yield a nearly normal sampling distribution. Therefore, we cannot estimate the probability using the tools we have learned so far. (c) We can now be confident that the conditions are satisfied. $Z = 0.92 \rightarrow 0.1788$.

4.43 (a) $H_0 : \mu_{2009} = \mu_{2004}$. $H_A : \mu_{2009} \neq \mu_{2004}$. (b) $\bar{x}_{2009} - \bar{x}_{2004} = -3.6$ spam emails per day. (c) The null hypothesis was not rejected, and the data do not provide convincing evidence that the true average number of spam emails per day in years 2004 and 2009 are different. The observed difference is about what we might expect from sampling variability alone. (d) Yes, since the hypothesis of no difference was not rejected in part (c).

4.45 (a) $H_0 : p_{2009} = p_{2004}$. $H_A : p_{2009} \neq p_{2004}$. (b) -7%. (c) The null hypothesis was rejected. The data provide strong evidence that the true proportion of those who once a month or less frequently delete their spam email was higher in 2004 than in 2009. The difference is so large that it cannot easily be explained as being due to chance. (d) No, since the null difference, 0, was rejected in part (c).

4.47 True. If the sample size is large, then the standard error will be small, meaning even relatively small differences between the null value and point estimate can be statistically significant.

5 Inference for numerical data

5.1 (a) $df = 6 - 1 = 5$, $t_5^\star = 2.02$ (column with two tails of 0.10, row with $df = 5$). (b) $df = 21 - 1 = 20$, $t_{20}^\star = 2.53$ (column with two tails of 0.02, row with $df = 20$). (c) $df = 28$, $t_{28}^\star = 2.05$. (d) $df = 11$, $t_{11}^\star = 3.11$.

5.3 (a) between 0.025 and 0.05 (b) less than 0.005 (c) greater than 0.2 (d) between 0.01 and 0.025

5.5 The mean is the midpoint: $\bar{x} = 20$. Identify the margin of error: $ME = 1.015$, then use $t_{35}^\star = 2.03$ and $SE = s/\sqrt{n}$ in the formula for margin of error to identify $s = 3$.

5.7 (a) H_0: $\mu = 8$ (New Yorkers sleep 8 hrs per night on average.) H_A: $\mu < 8$ (New Yorkers sleep less than 8 hrs per night on average.) (b) Independence: The sample is random and from less than 10% of New Yorkers. The sample is small, so we will use a t-distribution. For this size sample, slight skew is acceptable, and the min/max suggest there is not much skew in the data. $T = -1.75$. $df = 25 - 1 = 24$. (c) $0.025 <$ p-value < 0.05. If in fact the true population mean of the amount New Yorkers sleep per night was 8 hours, the probability of getting a random sample of 25 New Yorkers where the average amount of sleep is 7.73 hrs per night or less is between 0.025 and 0.05. (d) Since p-value < 0.05, reject H_0. The data provide strong evidence that New Yorkers sleep less than 8 hours per night on average. (e) No, as we rejected H_0.

5.9 $t_{19}^\star$ is 1.73 for a one-tail. We want the lower tail, so set -1.73 equal to the T-score, then solve for $\bar{x}$: 56.91.

5.11 (a) We will conduct a 1-sample t-test. H_0: $\mu = 5$. H_A: $\mu < 5$. We'll use $\alpha = 0.05$. This is a random sample, so the observations are independent. To proceed, we assume the distribution of years of piano lessons is approximately normal. $SE = 2.2/\sqrt{20} = 0.4919$. The test statistic is $T = (4.6 - 5)/SE = -0.81$. $df = 20 - 1 = 19$. The one-tail p-value is about 0.21, which is bigger than $\alpha = 0.05$, so we do not reject H_0. That is, we do not have sufficiently strong evidence to reject Georgianna's claim.
(b) Using $SE = 0.4919$ and $t^{\star}_{df=19} = 2.093$, the confidence interval is (3.57, 5.63). We are 95% confident that the average number of years a child takes piano lessons in this city is 3.57 to 5.63 years.
(c) They agree, since we did not reject the null hypothesis and the null value of 5 was in the t-interval.

5.13 If the sample is large, then the margin of error will be about $1.96 \times 100/\sqrt{n}$. We want this value to be less than 10, which leads to $n \geq 384.16$, meaning we need a sample size of at least 385 (round up for sample size calculations!).

5.15 (a) Two-sided, we are evaluating a difference, not in a particular direction. (b) Paired, data are recorded in the same cities at two different time points. The temperature in a city at one point is not independent of the temperature in the same city at another time point. (c) t-test, sample is small and population standard deviation is unknown.

5.17 (a) Since it's the same students at the beginning and the end of the semester, there is a pairing between the datasets, for a given student their beginning and end of semester grades are dependent. (b) Since the subjects were sampled randomly, each observation in the men's group does not have a special correspondence with exactly one observation in the other (women's) group. (c) Since it's the same subjects at the beginning and the end of the study, there is a pairing between the datasets, for a subject student their beginning and end of semester artery thickness are dependent. (d) Since it's the same subjects at the beginning and the end of the study, there is a pairing between the datasets, for a subject student their beginning and end of semester weights are dependent.

5.19 (a) For each observation in one data set, there is exactly one specially-corresponding observation in the other data set for the same geographic location. The data are paired. (b) H_0 : $\mu_{diff} = 0$ (There is no difference in average daily high temperature between January 1, 1968 and January 1, 2008 in the continental US.) H_A : $\mu_{diff} > 0$ (Average daily high temperature in January 1, 1968 was lower than average daily high temperature in January, 2008 in the continental US.) If you chose a two-sided test, that would also be acceptable. If this is the case, note that your p-value will be a little bigger than what is reported here in part (d). (c) Locations are random and represent less than 10% of all possible locations in the US. The sample size is at least 30. We are not given the distribution to check the skew. In practice, we would ask to see the data to check this condition, but here we will move forward under the assumption that it is not strongly skewed. (d) $T_{50} \approx 1.60 \to 0.05 <$ p-value < 0.10. (e) Since the p-value $> \alpha$ (since not given use 0.05), fail to reject H_0. The data do not provide strong evidence of temperature warming in the continental US. However it should be noted that the p-value is very close to 0.05. (f) Type 2 Error, since we may have incorrectly failed to reject H_0. There may be an increase, but we were unable to detect it. (g) Yes, since we failed to reject H_0, which had a null value of 0.

5.21 (a) (-0.05, 2.25). (b) We are 90% confident that the average daily high on January 1, 2008 in the continental US was 0.05 degrees lower to 2.25 degrees higher than the average daily high on January 1, 1968. (c) No, since 0 is included in the interval.

5.23 (a) Each of the 36 mothers is related to exactly one of the 36 fathers (and vice-versa), so there is a special correspondence between the mothers and fathers. (b) $H_0 : \mu_{diff} = 0$. $H_A : \mu_{diff} \neq 0$. Independence: random sample from less than 10% of population. Sample size of at least 30. The skew of the differences is, at worst, slight. $T_{35} = 2.72 \rightarrow$ p-value $= 0.01$. Since p-value < 0.05, reject H_0. The data provide strong evidence that the average IQ scores of mothers and fathers of gifted children are different, and the data indicate that mothers' scores are higher than fathers' scores for the parents of gifted children.

5.25 No, he should not move forward with the test since the distributions of total personal income are very strongly skewed. When sample sizes are large, we can be a bit lenient with skew. However, such strong skew observed in this exercise would require somewhat large sample sizes, somewhat higher than 30.

5.27 (a) These data are paired. For example, the Friday the 13th in say, September 1991, would probably be more similar to the Friday the 6th in September 1991 than to Friday the 6th in another month or year. (b) Let $\mu_{diff} = \mu_{sixth} - \mu_{thirteenth}$. $H_0 : \mu_{diff} = 0$. $H_A : \mu_{diff} \neq 0$. (c) Independence: The months selected are not random. However, if we think these dates are roughly equivalent to a simple random sample of all such Friday 6th/13th date pairs, then independence is reasonable. To proceed, we must make this strong assumption, though we should note this assumption in any reported results. Normality: With fewer than 10 observations, we would need to use the t-distribution to model the sample mean. The normal probability plot of the differences shows an approximately straight line. There isn't a clear reason why this distribution would be skewed, and since the normal quantile plot looks reasonable, we can mark this condition as reasonably satisfied. (d) $T = 4.94$ for $df = 10 - 1 = 9 \rightarrow$ p-value < 0.01. (e) Since p-value < 0.05, reject H_0. The data provide strong evidence that the average number of cars at the intersection is higher on Friday the 6th than on Friday the 13th. (We might believe this intersection is representative of all roads, i.e. there is higher traffic on Friday the 6th relative to Friday the 13th.

However, we should be cautious of the required assumption for such a generalization.) (f) If the average number of cars passing the intersection actually was the same on Friday the 6th and 13th, then the probability that we would observe a test statistic so far from zero is less than 0.01. (g) We might have made a Type 1 Error, i.e. incorrectly rejected the null hypothesis.

5.29 (a) $H_0 : \mu_{diff} = 0$. $H_A : \mu_{diff} \neq 0$. $T = -2.71$. $df = 5$. $0.02 <$ p-value < 0.05. Since p-value < 0.05, reject H_0. The data provide strong evidence that the average number of traffic accident related emergency room admissions are different between Friday the 6th and Friday the 13th. Furthermore, the data indicate that the direction of that difference is that accidents are lower on Friday the 6th relative to Friday the 13th. (b) (-6.49, -0.17). (c) This is an observational study, not an experiment, so we cannot so easily infer a causal intervention implied by this statement. It is true that there is a difference. However, for example, this does not mean that a responsible adult going out on Friday the 13th has a higher chance of harm than on any other night.

5.31 (a) Chicken fed linseed weighed an average of 218.75 grams while those fed horsebean weighed an average of 160.20 grams. Both distributions are relatively symmetric with no apparent outliers. There is more variability in the weights of chicken fed linseed. (b) $H_0 : \mu_{ls} = \mu_{hb}$. $H_A : \mu_{ls} \neq \mu_{hb}$. We leave the conditions to you to consider. $T = 3.02$, $df = min(11, 9) = 9 \rightarrow 0.01 <$ p-value < 0.02. Since p-value < 0.05, reject H_0. The data provide strong evidence that there is a significant difference between the average weights of chickens that were fed linseed and horsebean. (c) Type 1 Error, since we rejected H_0. (d) Yes, since p-value > 0.01, we would have failed to reject H_0.

5.33 $H_0 : \mu_C = \mu_S$. $H_A : \mu_C \neq \mu_S$. $T = 3.27$, $df = 11 \rightarrow$ p-value < 0.01. Since p-value < 0.05, reject H_0. The data provide strong evidence that the average weight of chickens that were fed casein is different than the average weight of chickens that were fed soybean (with weights from casein being higher). Since this is a randomized experiment, the observed difference can be attributed to the diet.

5.35 $H_0 : \mu_T = \mu_C$. $H_A : \mu_T \neq \mu_C$. $T = 2.24$, $df = 21 \rightarrow 0.02 <$ p-value < 0.05. Since p-value < 0.05, reject H_0. The data provide strong evidence that the average food consumption by the patients in the treatment and control groups are different. Furthermore, the data indicate patients in the distracted eating (treatment) group consume more food than patients in the control group.

5.37 Let $\mu_{diff} = \mu_{pre} - \mu_{post}$. $H_0 : \mu_{diff} = 0$: Treatment has no effect. $H_A : \mu_{diff} > 0$: Treatment is effective in reducing P.D.T. scores, the average pre-treatment score is higher than the average post-treatment score. Note that the reported values are pre minus post, so we are looking for a positive difference, which would correspond to a reduction in the P.D.T. score. Conditions are checked as follows. Independence: The subjects are randomly assigned to treatments, so the patients in each group are independent. All three sample sizes are smaller than 30, so we use t-tests. Distributions of differences are somewhat skewed. The sample sizes are small, so we cannot reliably relax this assumption. (We will proceed, but we would not report the results of this specific analysis, at least for treatment group 1.) For all three groups: $df = 13$. $T_1 = 1.89$ ($0.025 <$ p-value < 0.05), $T_2 = 1.35$ (p-value $= 0.10$), $T_3 = -1.40$ (p-value > 0.10). The only significant test reduction is found in Treatment 1, however, we had earlier noted that this result might not be reliable due to the skew in the distribution. Note that the calculation of the p-value for Treatment 3 was unnecessary: the sample mean indicated a increase in P.D.T. scores under this treatment (as opposed to a decrease, which was the result of interest). That is, we could tell without formally completing the hypothesis test that the p-value would be large for this treatment group.

5.39 Difference we care about: 40. Single tail of 90%: $1.28 \times SE$. Rejection region bounds: $\pm 1.96 \times SE$ (if 5% significance level). Setting $3.24 \times SE = 40$, subbing in $SE = \sqrt{\frac{94^2}{n} + \frac{94^2}{n}}$, and solving for the sample size n gives 116 plots of land for each fertilizer.

5.41 Alternative.

5.43 $H_0 : \mu_1 = \mu_2 = \cdots = \mu_6$. H_A: The average weight varies across some (or all) groups. Independence: Chicks are randomly assigned to feed types (presumably kept separate from one another), therefore independence of observations is reasonable. Approx. normal: the distributions of weights within each feed type appear to be fairly symmetric. Constant variance: Based on the side-by-side box plots, the constant variance assumption appears to be reasonable. There are differences in the actual computed standard deviations, but these might be due to chance as these are quite small samples. $F_{5,65} = 15.36$ and the p-value is approximately 0. With such a small p-value, we reject H_0. The data provide convincing evidence that the average weight of chicks varies across some (or all) feed supplement groups.

5.45 (a) H_0: The population mean of MET for each group is equal to the others. H_A: At least one pair of means is different. (b) Independence: We don't have any information on how the data were collected, so we cannot assess independence. To proceed, we must assume the subjects in each group are independent. In practice, we would inquire for more details. Approx. normal: The data are bound below by zero and the standard deviations are larger than the means, indicating very strong skew. However, since the sample sizes are extremely large, even extreme skew is acceptable. Constant variance: This condition is sufficiently met, as the standard deviations are reasonably consistent across groups. (c) See below, with the last column omitted:

	Df	Sum Sq	Mean Sq	F value
coffee	4	10508	2627	5.2
Residuals	50734	25564819	504	
Total	50738	25575327		

(d) Since p-value is very small, reject H_0. The data provide convincing evidence that the average MET differs between at least one pair of groups.

5.47 (a) H_0: Average GPA is the same for all majors. H_A: At least one pair of means are different. (b) Since p-value > 0.05, fail to reject H_0. The data do not provide convincing evidence of a difference between the average GPAs across three groups of majors. (c) The total degrees of freedom is $195 + 2 = 197$, so the sample size is $197 + 1 = 198$.

5.49 (a) False. As the number of groups increases, so does the number of comparisons and hence the modified significance level decreases. (b) True. (c) True. (d) False. We need observations to be independent regardless of sample size.

5.51 (a) H_0: Average score difference is the same for all treatments. H_A: At least one pair of means are different. (b) We should check conditions. If we look back to the earlier exercise, we will see that the patients were randomized, so independence is satisfied. There are some minor concerns about skew, especially with the third group, though this may be ac-

ceptable. The standard deviations across the groups are reasonably similar. Since the p-value is less than 0.05, reject H_0. The data provide convincing evidence of a difference between the average reduction in score among treatments. (c) We determined that at least two means are different in part (b), so we now conduct $K = 3 \times 2/2 = 3$ pairwise t-tests that each use $\alpha = 0.05/3 = 0.0167$ for a significance level. Use the following hypotheses for each pairwise test. H_0: The two means are equal. H_A: The two means are different. The sample sizes are equal and we use the pooled SD, so we can compute $SE = 3.7$ with the pooled $df = 39$. The p-value only for Trmt 1 vs. Trmt 3 may be statistically significant: $0.01 <$ p-value < 0.02. Since we cannot tell, we should use a computer to get the p-value, 0.015, which is statistically significant for the adjusted significance level. That is, we have identified Treatment 1 and Treatment 3 as having different effects. Checking the other two comparisons, the differences are not statistically significant.

6 Inference for categorical data

6.1 (a) False. Doesn't satisfy success-failure condition. (b) True. The success-failure condition is not satisfied. In most samples we would expect $\hat{p}$ to be close to 0.08, the true population proportion. While $\hat{p}$ can be much above 0.08, it is bound below by 0, suggesting it would take on a right skewed shape. Plotting the sampling distribution would confirm this suspicion. (c) False. $SE_{\hat{p}} = 0.0243$, and $\hat{p} = 0.12$ is only $\frac{0.12 - 0.08}{0.0243} = 1.65$ SEs away from the mean, which would not be considered unusual. (d) True. $\hat{p} = 0.12$ is 2.32 standard errors away from the mean, which is often considered unusual. (e) False. Decreases the SE by a factor of $1/\sqrt{2}$.

6.3 (a) True. See the reasoning of 6.1(b). (b) True. We take the square root of the sample

size in the SE formula. (c) True. The independence and success-failure conditions are satisfied. (d) True. The independence and success-failure conditions are satisfied.

6.5 (a) False. A confidence interval is constructed to estimate the population proportion, not the sample proportion. (b) True. 95% CI: $70\% \pm 8\%$. (c) True. By the definition of the confidence level. (d) True. Quadrupling the sample size decreases the SE and ME by a factor of $1/\sqrt{4}$. (e) True. The 95% CI is entirely above 50%.

6.7 With a random sample from $< 10\%$ of the population, independence is satisfied. The success-failure condition is also satisfied. $ME = z^{\star} \sqrt{\frac{\hat{p}(1-\hat{p})}{n}} = 1.96 \sqrt{\frac{0.56 \times 0.44}{600}} = 0.0397 \approx 4\%$

6.9 (a) Proportion of graduates from this university who found a job within one year of graduating. $\hat{p} = 348/400 = 0.87$. (b) This is a random sample from less than 10% of the population, so the observations are independent. Success-failure condition is satisfied: 348 successes, 52 failures, both well above 10. (c) (0.8371, 0.9029). We are 95% confident that approximately 84% to 90% of graduates from this university found a job within one year of completing their undergraduate degree. (d) 95% of such random samples would produce a 95% confidence interval that includes the true proportion of students at this university who found a job within one year of graduating from college. (e) (0.8267, 0.9133). Similar interpretation as before. (f) 99% CI is wider, as we are more confident that the true proportion is within the interval and so need to cover a wider range.

6.11 (a) No. The sample only represents students who took the SAT, and this was also an online survey. (b) (0.5289, 0.5711). We are 90% confident that 53% to 57% of high school seniors who took the SAT are fairly certain that they will participate in a study abroad program in college. (c) 90% of such random samples would produce a 90% confidence interval that includes the true proportion. (d) Yes. The interval lies entirely above 50%.

6.13 (a) This is an appropriate setting for a hypothesis test. $H_0 : p = 0.50$. $H_A : p > 0.50$. Both independence and the success-failure condition are satisfied. $Z = 1.12 \rightarrow$ p-value = 0.1314. Since the p-value $> \alpha = 0.05$, we fail to reject H_0. The data do not provide strong evidence that more than half of all Independents oppose the public option plan. (b) Yes, since we did not reject H_0 in part (a).

6.15 (a) $H_0 : p = 0.38$. $H_A : p \neq 0.38$. Independence (random sample, < 10% of population) and the success-failure condition are satisfied. $Z = -20.5 \rightarrow$ p-value ≈ 0. Since the p-value is very small, we reject H_0. The data provide strong evidence that the proportion of Americans who only use their cell phones to access the internet is different than the Chinese proportion of 38%, and the data indicate that the proportion is lower in the US. (b) If in fact 38% of Americans used their cell phones as a primary access point to the internet, the probability of obtaining a random sample of 2,254 Americans where 17% or less or 59% or more use their only their cell phones to access the internet would be approximately 0. (c) (0.1545, 0.1855). We are 95% confident that approximately 15.5% to 18.6% of all Americans primarily use their cell phones to browse the internet.

6.17 (a) $H_0 : p = 0.5$. $H_A : p > 0.5$. Independence (random sample, < 10% of population) is satisfied, as is the success-failure conditions (using $p_0 = 0.5$, we expect 40 successes and 40 failures). $Z = 2.91 \rightarrow$ p-value = 0.0018. Since the p-value < 0.05, we reject the null hypothesis. The data provide strong evidence that the rate of correctly identifying a soda for these people is significantly better than just by random guessing. (b) If in fact people cannot tell the difference between diet and regular soda and they randomly guess, the probability of getting a random sample of 80 people where 53 or more identify a soda correctly would be 0.0018.

6.19 (a) Independence is satisfied (random sample from < 10% of the population), as is the success-failure condition (40 smokers, 160 non-smokers). The 95% CI: (0.145, 0.255). We are 95% confident that 14.5% to 25.5% of all students at this university smoke. (b) We want $z^\star SE$ to be no larger than 0.02 for a 95% confidence level. We use $z^\star = 1.96$ and plug in the point estimate $\hat{p} = 0.2$ within the SE formula: $1.96\sqrt{0.2(1 - 0.2)/n} \leq 0.02$. The sample size n should be at least 1,537.

6.21 The margin of error, which is computed as $z^\star SE$, must be smaller than 0.01 for a 90% confidence level. We use $z^\star = 1.65$ for a 90% confidence level, and we can use the point estimate $\hat{p} = 0.52$ in the formula for SE. $1.65\sqrt{0.52(1 - 0.52)/n} \leq 0.01$. Therefore, the sample size n must be at least 6,796.

6.23 This is not a randomized experiment, and it is unclear whether people would be affected by the behavior of their peers. That is, independence may not hold. Additionally, there are only 5 interventions under the provocative scenario, so the success-failure condition does not hold. Even if we consider a hypothesis test where we pool the proportions, the success-failure condition will not be satisfied. Since one condition is questionable and the other is not satisfied, the difference in sample proportions will not follow a nearly normal distribution.

6.25 (a) False. The entire confidence interval is above 0. (b) True. (c) True. (d) True. (e) False. It is simply the negated and reordered values: (-0.06,-0.02).

6.27 (a) (0.23, 0.33). We are 95% confident that the proportion of Democrats who support the plan is 23% to 33% higher than the proportion of Independents who do. (b) True.

6.29 (a) College grads: 23.7%. Non-college grads: 33.7%. (b) Let p_{CG} and p_{NCG} represent the proportion of college graduates and non-college graduates who responded "do not know". $H_0 : p_{CG} = p_{NCG}$. $H_A : p_{CG} \neq p_{NCG}$. Independence is satisfied (random sample, $< 10\%$ of the population), and the success-failure condition, which we would check using the pooled proportion ($\hat{p} = 235/827 = 0.284$), is also satisfied. $Z = -3.18 \rightarrow$ p-value = 0.0014. Since the p-value is very small, we reject H_0. The data provide strong evidence that the proportion of college graduates who do not have an opinion on this issue is different than that of non-college graduates. The data also indicate that fewer college grads say they "do not know" than non-college grads (i.e. the data indicate the direction after we reject H_0).

6.31 (a) College grads: 35.2%. Non-college grads: 33.9%. (b) Let p_{CG} and p_{NCG} represent the proportion of college graduates and non-college grads who support offshore drilling. $H_0 : p_{CG} = p_{NCG}$. $H_A : p_{CG} \neq p_{NCG}$. Independence is satisfied (random sample, $< 10\%$ of the population), and the success-failure condition, which we would check using the pooled proportion ($\hat{p} = 286/827 = 0.346$), is also satisfied. $Z = 0.39 \rightarrow$ p-value = 0.6966. Since the p-value $> \alpha$ (0.05), we fail to reject H_0. The data do not provide strong evidence of a difference between the proportions of college graduates and non-college graduates who support offshore drilling in California.

6.33 Subscript C means control group. Subscript T means truck drivers. $H_0 : p_C = p_T$. $H_A : p_C \neq p_T$. Independence is satisfied (random samples, $< 10\%$ of the population), as is the success-failure condition, which we would check using the pooled proportion ($\hat{p} = 70/495 = 0.141$). $Z = -1.65 \rightarrow$ p-value = 0.0989. Since the p-value is high (default to $\alpha = 0.05$), we fail to reject H_0. The data do not provide strong evidence that the rates of sleep deprivation are different for non-transportation workers and truck drivers.

6.35 (a) Summary of the study:

		Virol. failure		
		Yes	No	Total
Treatment	Nevaripine	26	94	120
	Lopinavir	10	110	120
	Total	36	204	240

(b) $H_0 : p_N = p_L$. There is no difference in virologic failure rates between the Nevaripine and Lopinavir groups. $H_A : p_N \neq p_L$. There is some difference in virologic failure rates between the Nevaripine and Lopinavir groups. (c) Random assignment was used, so the observations in each group are independent. If the patients in the study are representative of those in the general population (something impossible to check with the given information), then we can also confidently generalize the findings to the population. The success-failure condition, which we would check using the pooled proportion ($\hat{p} = 36/240 = 0.15$), is satisfied. $Z = 2.89 \rightarrow$ p-value = 0.0039. Since the p-value is low, we reject H_0. There is strong evidence of a difference in virologic failure rates between the Nevaripine and Lopinavir groups do not appear to be independent.

6.37 No. The samples at the beginning and at the end of the semester are not independent since the survey is conducted on the same students.

6.39 (a) False. The chi-square distribution has one parameter called degrees of freedom. (b) True. (c) True. (d) False. As the degrees of freedom increases, the shape of the chi-square distribution becomes more symmetric.

6.41 (a) H_0: The distribution of the format of the book used by the students follows the professor's predictions. H_A: The distribution of the format of the book used by the students does not follow the professor's predictions. (b) $E_{hard\ copy} = 126 \times 0.60 = 75.6$. $E_{print} = 126 \times 0.25 = 31.5$. $E_{online} = 126 \times 0.15 = 18.9$. (c) Independence: The sample is not random. However, if the professor has reason to believe that the proportions are stable from one term to the next and students are not affecting each other's study habits, independence is probably reasonable. Sample size: All expected counts are at least 5. (d) $\chi^2 = 2.32$, $df = 2$, p-value > 0.3. (e) Since the p-value is large, we fail to reject H_0. The data do not provide strong evidence indicating the professor's predictions were statistically inaccurate.

6.43 Use a chi-squared goodness of fit test. H_0: Each option is equally likely. H_A: Some options are preferred over others. Total sample size: 99. Expected counts: $(1/3) * 99 = 33$ for each option. These are all above 5, so conditions are satisfied. $df = 3 - 1 = 2$ and $\chi^2 = \frac{(43-33)^2}{33} + \frac{(21-33)^2}{33} + \frac{(35-33)^2}{33} = 7.52 \to 0.02 <$ p-value < 0.05. Since the p-value is less than 5%, we reject H_0. The data provide convincing evidence that some options are preferred over others.

6.45 (a) Two-way table:

Treatment	Quit Yes	No	Total
Patch + support group	40	110	150
Only patch	30	120	150
Total	70	230	300

(b-i) $E_{row_1, col_1} = \frac{(row\ 1\ total) \times (col\ 1\ total)}{table\ total} = 35$. This is lower than the observed value.
(b-ii) $E_{row_2, col_2} = \frac{(row\ 2\ total) \times (col\ 2\ total)}{table\ total} = 115$. This is lower than the observed value.

6.47 H_0: The opinion of college grads and non-grads is not different on the topic of drilling for oil and natural gas off the coast of California. H_A: Opinions regarding the drilling for oil and natural gas off the coast of California has an association with earning a college degree.

$$E_{row\ 1, col\ 1} = 151.5 \quad E_{row\ 1, col\ 2} = 134.5$$
$$E_{row\ 2, col\ 1} = 162.1 \quad E_{row\ 2, col\ 2} = 143.9$$
$$E_{row\ 3, col\ 1} = 124.5 \quad E_{row\ 3, col\ 2} = 110.5$$

Independence: The samples are both random, unrelated, and from less than 10% of the population, so independence between observations is reasonable. Sample size: All expected counts are at least 5. $\chi^2 = 11.47$, $df = 2 \to 0.001 <$ p-value < 0.005. Since the p-value $< \alpha$, we reject H_0. There is strong evidence that there is an association between support for off-shore drilling and having a college degree.

6.49 (a) H_0: The age of Los Angeles residents is independent of shipping carrier preference variable. H_A: The age of Los Angeles residents is associated with the shipping carrier preference variable. (b) The conditions are not satisfied since some expected counts are below 5.

6.51 No. For a confidence interval, we check the success-failure condition using the data, and there are only 9 respondents who said bullying is no problem at all.

6.53 (a) $H_0 : p = 0.69$. $H_A : p \neq 0.69$. (b) $\hat{p} = \frac{17}{30} = 0.57$. (c) The success-failure condition is not satisfied; note that it is appropriate to use the null value ($p_0 = 0.69$) to compute the expected number of successes and failures. (d) Answers may vary. Each student can be represented with a card. Take 100 cards, 69 black cards representing those who follow the news about Egypt and 31 red cards representing those who do not. Shuffle the cards and draw with replacement (shuffling each time in between draws) 30 cards representing the 30 high school students. Calculate the proportion of black cards in this sample, $\hat{p}_{sim}$, i.e. the proportion of those who follow the news in the simulation. Repeat this many times (e.g. 10,000 times) and plot the resulting sample proportions. The p-value will be two times the proportion of simulations where $\hat{p}_{sim} \leq 0.57$. (Note: we would generally use a computer to perform these simulations.) (e) The p-value is about $0.001 + 0.005 + 0.020 + 0.035 + 0.075 = 0.136$, meaning the two-sided p-value is about 0.272. Your p-value may vary slightly since it is based on a visual estimate. Since the p-value is greater than 0.05, we fail to reject H_0. The data do not provide strong evidence that the proportion of high school students who followed the news about Egypt is different than the proportion of American adults who did.

6.55 The subscript $_{pr}$ corresponds to provocative and $_{con}$ to conservative. (a) $H_0 : p_{pr} = p_{con}$. $H_A : p_{pr} \neq p_{con}$. (b) -0.35. (c) The left tail for the p-value is calculated by adding up the two left bins: $0.005 + 0.015 = 0.02$. Doubling the one tail, the p-value is 0.04. (Students may have approximate results, and a small number of students may have a p-value of about 0.05.) Since the p-value is low, we reject H_0. The data provide strong evidence that people react differently under the two scenarios.

7 Introduction to linear regression

7.1 (a) The residual plot will show randomly distributed residuals around 0. The variance is also approximately constant. (b) The residuals will show a fan shape, with higher variability for smaller x. There will also be many points on the right above the line. There is trouble with the model being fit here.

7.3 (a) Strong relationship, but a straight line would not fit the data. (b) Strong relationship, and a linear fit would be reasonable. (c) Weak relationship, and trying a linear fit would be reasonable. (d) Moderate relationship, but a straight line would not fit the data. (e) Strong relationship, and a linear fit would be reasonable. (f) Weak relationship, and trying a linear fit would be reasonable.

7.5 (a) Exam 2 since there is less of a scatter in the plot of final exam grade versus exam 2. Notice that the relationship between Exam 1 and the Final Exam appears to be slightly nonlinear. (b) Exam 2 and the final are relatively close to each other chronologically, or Exam 2 may be cumulative so has greater similarities in material to the final exam. Answers may vary for part (b).

7.7 (a) $r = -0.7 \to$ (4). (b) $r = 0.45 \to$ (3). (c) $r = 0.06 \to$ (1). (d) $r = 0.92 \to$ (2).

7.9 (a) True. (b) False, correlation is a measure of the linear association between any two numerical variables.

7.11 (a) The relationship is positive, weak, and possibly linear. However, there do appear to be some anomalous observations along the left where several students have the same height that is notably far from the cloud of the other points. Additionally, there are many students who appear not to have driven a car, and they are represented by a set of points along the bottom of the scatterplot. (b) There is no obvious explanation why simply being tall should lead a person to drive faster. However, one confounding factor is gender. Males tend to be taller than females on average, and personal experiences (anecdotal) may suggest they drive faster. If we were to follow-up on this suspicion, we would find that sociological studies confirm this suspicion. (c) Males are taller on average and they drive faster. The gender variable is indeed an important confounding variable.

7.13 (a) There is a somewhat weak, positive, possibly linear relationship between the distance traveled and travel time. There is clustering near the lower left corner that we should take special note of. (b) Changing the units will not change the form, direction or strength of the relationship between the two variables. If longer distances measured in miles are associated with longer travel time measured in minutes, longer distances measured in kilometers will be associated with longer travel time measured in hours. (c) Changing units doesn't affect correlation: $r = 0.636$.

7.15 (a) There is a moderate, positive, and linear relationship between shoulder girth and height. (b) Changing the units, even if just for one of the variables, will not change the form, direction or strength of the relationship between the two variables.

7.17 In each part, we can write the husband ages as a linear function of the wife ages.
(a) $age_H = age_W + 3$.
(b) $age_H = age_W - 2$.
(c) $age_H = 2 \times age_W$.
Since the slopes are positive and these are perfect linear relationships, the correlation will be exactly 1 in all three parts. An alternative way to gain insight into this solution is to create a mock data set, e.g. 5 women aged 26, 27, 28, 29, and 30, then find the husband ages for each wife in each part and create a scatterplot.

7.19 Correlation: no units. Intercept: kg. Slope: kg/cm.

7.21 Over-estimate. Since the residual is calculated as *observed − predicted*, a negative residual means that the predicted value is higher than the observed value.

7.23 (a) There is a positive, very strong, linear association between the number of tourists and spending. (b) Explanatory: number of tourists (in thousands). Response: spending (in millions of US dollars). (c) We can predict spending for a given number of tourists using a regression line. This may be useful information for determining how much the country may want to spend in advertising abroad, or to forecast expected revenues from tourism. (d) Even though the relationship appears linear in the scatterplot, the residual plot actually shows a nonlinear relationship. This is not a contradiction: residual plots can show divergences from linearity that can be difficult to see in a scatterplot. A simple linear model is inadequate for modeling these data. It is also important to consider that these data are observed sequentially, which means there may be a hidden structure not evident in the current plots but that is important to consider.

7.25 (a) First calculate the slope: $b_1 = R \times s_y/s_x = 0.636 \times 113/99 = 0.726$. Next, make use of the fact that the regression line passes through the point $(\bar{x}, \bar{y})$: $\bar{y} = b_0 + b_1 \times \bar{x}$. Plug in $\bar{x}$, $\bar{y}$, and b_1, and solve for b_0: 51. Solution: *travel time* $= 51 + 0.726 \times distance$. (b) b_1: For each additional mile in distance, the model predicts an additional 0.726 minutes in travel time. b_0: When the distance traveled is 0 miles, the travel time is expected to be 51 minutes. It does not make sense to have a travel distance of 0 miles in this context. Here, the y-intercept serves only to adjust the height of the line and is meaningless by itself. (c) $R^2 = 0.636^2 = 0.40$. About 40% of the variability in travel time is accounted for by the model, i.e. explained by the distance traveled. (d) *travel time* $= 51 + 0.726 \times distance = 51 + 0.726 \times 103 \approx 126$ minutes. (Note: we should be cautious in our predictions with this model since we have not yet evaluated whether it is a well-fit model.) (e) $e_i = y_i - \hat{y}_i = 168 - 126 = 42$ minutes. A positive residual means that the model underestimates the travel time. (f) No, this calculation would require extrapolation.

7.27 There is an upwards trend. However, the variability is higher for higher calorie counts, and it looks like there might be two clusters of observations above and below the line on the right, so we should be cautious about fitting a linear model to these data.

7.29 (a) $\widehat{murder} = -29.901 + 2.559 \times poverty\%$ (b) Expected murder rate in metropolitan areas with no poverty is -29.901 per million. This is obviously not a meaningful value, it just serves to adjust the height of the regression line. (c) For each additional percentage increase in poverty, we expect murders per million to be higher on average by 2.559. (d) Poverty level explains 70.52% of the variability in murder rates in metropolitan areas. (e) $\sqrt{0.7052} = 0.8398$

7.31 (a) There is an outlier in the bottom right. Since it is far from the center of the data, it is a point with high leverage. It is also an influential point since, without that observation, the regression line would have a very different slope. (b) There is an outlier in the bottom right. Since it is far from the center of the data, it is a point with high leverage. However, it does not appear to be affecting the line much, so it is not an influential point. (c) The observation is in the center of the data (in the x-axis direction), so this point does *not* have high leverage. This means the point won't have much effect on the slope of the line and so is not an influential point.

7.33 (a) There is a negative, moderate-to-strong, somewhat linear relationship between percent of families who own their home and the percent of the population living in urban areas in 2010. There is one outlier: a state where 100% of the population is urban. The variability in the percent of homeownership also increases as we move from left to right in the plot. (b) The outlier is located in the bottom right corner, horizontally far from the center of the other points, so it is a point with high leverage. It is an influential point since excluding this point from the analysis would greatly affect the slope of the regression line.

7.35 (a) The relationship is positive, moderate-to-strong, and linear. There are a few outliers but no points that appear to be influential. (b) $\widehat{weight} = -105.0113 + 1.0176 \times height$. Slope: For each additional centimeter in height, the model predicts the average weight to be 1.0176 additional kilograms (about 2.2 pounds). Intercept: People who are 0 centimeters tall are expected to weigh -105.0113 kilograms. This is obviously not possible. Here, the y-intercept serves only to adjust the height of the line and is meaningless by itself. (c) H_0: The true slope coefficient of height is zero ($\beta_1 = 0$). H_A: The true slope coefficient of height is greater than zero ($\beta_1 > 0$). A two-sided test would also be acceptable for this application. The p-value for the two-sided alternative hypothesis ($\beta_1 \neq 0$) is incredibly small, so the p-value for the one-sided hypothesis will be even smaller. That is, we reject H_0. The data provide convincing evidence that height and weight are positively correlated. The true slope parameter is indeed greater than 0. (d) $R^2 = 0.72^2 = 0.52$. Approximately 52% of the variability in weight can be explained by the height of individuals.

7.37 (a) $H_0: \beta_1 = 0$. $H_A: \beta_1 > 0$. A two-sided test would also be acceptable for this application. The p-value, as reported in the table, is incredibly small. Thus, for a one-sided test, the p-value will also be incredibly small, and we reject H_0. The data provide convincing evidence that wives' and husbands' heights are positively correlated. (b) $\widehat{height}_W = 43.5755 + 0.2863 \times height_H$. (c) Slope: For each additional inch

in husband's height, the average wife's height is expected to be an additional 0.2863 inches on average. Intercept: Men who are 0 inches tall are expected to have wives who are, on average, 43.5755 inches tall. The intercept here is meaningless, and it serves only to adjust the height of the line. (d) The slope is positive, so r must also be positive. $r = \sqrt{0.09} = 0.30$. (e) 63.2612. Since R^2 is low, the prediction based on this regression model is not very reliable. (f) No, we should avoid extrapolating.

7.39 (a) $r = \sqrt{0.28} \approx -0.53$. We know the correlation is negative due to the negative association shown in the scatterplot. (b) The residuals appear to be fan shaped, indicating non-constant variance. Therefore a simple least squares fit is not appropriate for these data.

7.41 (a) $H_0 : \beta_1 = 0; H_A : \beta_1 \neq 0$ (b) The p-value for this test is approximately 0, therefore we reject H_0. The data provide convincing evidence that poverty percentage is a significant predictor of murder rate. (c) $n = 20, df = 18, T_{18}^* = 2.10; 2.559 \pm 2.10 \times 0.390 = (1.74, 3.378)$; For each percentage point poverty is higher, murder rate is expected to be higher on average by 1.74 to 3.378 per million. (d) Yes, we rejected H_0 and the confidence interval does not include 0.

7.43 This is a one-sided test, so the p-value should be half of the p-value given in the regression table, which will be approximately 0. Therefore the data provide convincing evidence that poverty percentage is positively associated with murder rate.

8 Multiple and logistic regression

8.1 (a) $\widehat{baby\_weight} = 123.05 - 8.94 \times smoke$ (b) The estimated body weight of babies born to smoking mothers is 8.94 ounces lower than babies born to non-smoking mothers. Smoker: $123.05 - 8.94 \times 1 = 114.11$ ounces. Non-smoker: $123.05 - 8.94 \times 0 = 123.05$ ounces. (c) $H_0: \beta_1 = 0$. $H_A: \beta_1 \neq 0$. $T = -8.65$, and the p-value is approximately 0. Since the p-value is very small, we reject H_0. The data provide strong evidence that the true slope parameter is different than 0 and that there is an association between birth weight and smoking. Furthermore, having rejected H_0, we can conclude that smoking is associated with lower birth weights.

8.3 (a) $\widehat{baby\_weight} = -80.41 + 0.44 \times gestation - 3.33 \times parity - 0.01 \times age + 1.15 \times height + 0.05 \times weight - 8.40 \times smoke$. (b) $\beta_{gestation}$: The model predicts a 0.44 ounce increase in the birth weight of the baby for each additional day of pregnancy, all else held constant. β_{age}: The model predicts a 0.01 ounce decrease in the birth weight of the baby for each additional year in mother's age, all else held constant. (c) Parity might be correlated with one of the other variables in the model, which complicates model estimation. (d) $\widehat{baby\_weight} = 120.58$. $e = 120 - 120.58 = -0.58$. The model over-predicts this baby's birth weight. (e) $R^2 = 0.2504$. $R_{adj}^2 = 0.2468$.

8.5 (a) (-0.32, 0.16). We are 95% confident that male students on average have GPAs 0.32 points lower to 0.16 points higher than females when controlling for the other variables in the model. (b) Yes, since the p-value is larger than 0.05 in all cases (not including the intercept).

8.7 Remove age.

8.9 Based on the p-value alone, either gestation or smoke should be added to the model first. However, since the adjusted R^2 for the model with gestation is higher, it would be preferable to add gestation in the first step of the forward-selection algorithm. (Other explanations are possible. For instance, it would be reasonable to only use the adjusted R^2.)

8.11 She should use p-value selection since she is interested in finding out about significant predictors, not just optimizing predictions.

8.13 Nearly normal residuals: The normal probability plot shows a nearly normal distribution of the residuals, however, there are some minor irregularities at the tails. With a data set so large, these would not be a concern.
Constant variability of residuals: The scatterplot of the residuals versus the fitted values does not show any overall structure. However, values that have very low or very high fitted values appear to also have somewhat larger outliers. In addition, the residuals do appear to have constant variability between the two parity and smoking status groups, though these items are relatively minor.
Independent residuals: The scatterplot of residuals versus the order of data collection shows a random scatter, suggesting that there is no apparent structures related to the order the data were collected.
Linear relationships between the response variable and numerical explanatory variables: The residuals vs. height and weight of mother are randomly distributed around 0. The residuals

vs. length of gestation plot also does not show any clear or strong remaining structures, with the possible exception of very short or long gestations. The rest of the residuals do appear to be randomly distributed around 0.
All concerns raised here are relatively mild. There are some outliers, but there is so much data that the influence of such observations will be minor.

8.15 (a) There are a few potential outliers, e.g. on the left in the `total_length` variable, but nothing that will be of serious concern in a data set this large. (b) When coefficient estimates are sensitive to which variables are included in the model, this typically indicates that some variables are collinear. For example, a possum's gender may be related to its head length, which would explain why the coefficient (and p-value) for `sex_male` changed when we removed the `head_length` variable. Likewise, a possum's skull width is likely to be related to its head length, probably even much more closely related than the head length was to gender.

8.17 (a) The logistic model relating $\hat{p}_i$ to the predictors may be written as $\log\left(\frac{\hat{p}_i}{1-\hat{p}_i}\right) = 33.5095 - 1.4207 \times sex\_male_i - 0.2787 \times skull\_width_i + 0.5687 \times total\_length_i - 1.8057 \times tail\_length_i$. Only `total_length` has a positive association with a possum being from Victoria. (b) $\hat{p} = 0.0062$. While the probability is very near zero, we have not run diagnostics on the model. We might also be a little skeptical that the model will remain accurate for a possum found in a US zoo. For example, perhaps the zoo selected a possum with specific characteristics but only looked in one region. On the other hand, it is encouraging that the possum was caught in the wild. (Answers regarding the reliability of the model probability will vary.)

Appendix B

Distribution tables

B.1 Normal Probability Table

The area to the left of Z represents the percentile of the observation. The normal probability table always lists percentiles.

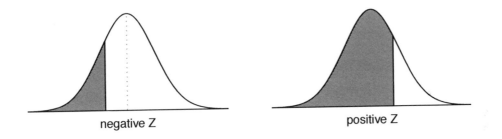

negative Z   positive Z

To find the area to the right, calculate 1 minus the area to the left.

1.0000   −   0.6664   =   0.3336

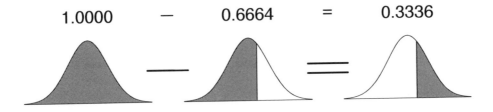

For additional details about working with the normal distribution and the normal probability table, see Section 3.1, which starts on page 127.

negative Z

Second decimal place of Z										
0.09	0.08	0.07	0.06	0.05	0.04	0.03	0.02	0.01	0.00	Z
0.0002	0.0003	0.0003	0.0003	0.0003	0.0003	0.0003	0.0003	0.0003	0.0003	−3.4
0.0003	0.0004	0.0004	0.0004	0.0004	0.0004	0.0004	0.0005	0.0005	0.0005	−3.3
0.0005	0.0005	0.0005	0.0006	0.0006	0.0006	0.0006	0.0006	0.0007	0.0007	−3.2
0.0007	0.0007	0.0008	0.0008	0.0008	0.0008	0.0009	0.0009	0.0009	0.0010	−3.1
0.0010	0.0010	0.0011	0.0011	0.0011	0.0012	0.0012	0.0013	0.0013	0.0013	−3.0
0.0014	0.0014	0.0015	0.0015	0.0016	0.0016	0.0017	0.0018	0.0018	0.0019	−2.9
0.0019	0.0020	0.0021	0.0021	0.0022	0.0023	0.0023	0.0024	0.0025	0.0026	−2.8
0.0026	0.0027	0.0028	0.0029	0.0030	0.0031	0.0032	0.0033	0.0034	0.0035	−2.7
0.0036	0.0037	0.0038	0.0039	0.0040	0.0041	0.0043	0.0044	0.0045	0.0047	−2.6
0.0048	0.0049	0.0051	0.0052	0.0054	0.0055	0.0057	0.0059	0.0060	0.0062	−2.5
0.0064	0.0066	0.0068	0.0069	0.0071	0.0073	0.0075	0.0078	0.0080	0.0082	−2.4
0.0084	0.0087	0.0089	0.0091	0.0094	0.0096	0.0099	0.0102	0.0104	0.0107	−2.3
0.0110	0.0113	0.0116	0.0119	0.0122	0.0125	0.0129	0.0132	0.0136	0.0139	−2.2
0.0143	0.0146	0.0150	0.0154	0.0158	0.0162	0.0166	0.0170	0.0174	0.0179	−2.1
0.0183	0.0188	0.0192	0.0197	0.0202	0.0207	0.0212	0.0217	0.0222	0.0228	−2.0
0.0233	0.0239	0.0244	0.0250	0.0256	0.0262	0.0268	0.0274	0.0281	0.0287	−1.9
0.0294	0.0301	0.0307	0.0314	0.0322	0.0329	0.0336	0.0344	0.0351	0.0359	−1.8
0.0367	0.0375	0.0384	0.0392	0.0401	0.0409	0.0418	0.0427	0.0436	0.0446	−1.7
0.0455	0.0465	0.0475	0.0485	0.0495	0.0505	0.0516	0.0526	0.0537	0.0548	−1.6
0.0559	0.0571	0.0582	0.0594	0.0606	0.0618	0.0630	0.0643	0.0655	0.0668	−1.5
0.0681	0.0694	0.0708	0.0721	0.0735	0.0749	0.0764	0.0778	0.0793	0.0808	−1.4
0.0823	0.0838	0.0853	0.0869	0.0885	0.0901	0.0918	0.0934	0.0951	0.0968	−1.3
0.0985	0.1003	0.1020	0.1038	0.1056	0.1075	0.1093	0.1112	0.1131	0.1151	−1.2
0.1170	0.1190	0.1210	0.1230	0.1251	0.1271	0.1292	0.1314	0.1335	0.1357	−1.1
0.1379	0.1401	0.1423	0.1446	0.1469	0.1492	0.1515	0.1539	0.1562	0.1587	−1.0
0.1611	0.1635	0.1660	0.1685	0.1711	0.1736	0.1762	0.1788	0.1814	0.1841	−0.9
0.1867	0.1894	0.1922	0.1949	0.1977	0.2005	0.2033	0.2061	0.2090	0.2119	−0.8
0.2148	0.2177	0.2206	0.2236	0.2266	0.2296	0.2327	0.2358	0.2389	0.2420	−0.7
0.2451	0.2483	0.2514	0.2546	0.2578	0.2611	0.2643	0.2676	0.2709	0.2743	−0.6
0.2776	0.2810	0.2843	0.2877	0.2912	0.2946	0.2981	0.3015	0.3050	0.3085	−0.5
0.3121	0.3156	0.3192	0.3228	0.3264	0.3300	0.3336	0.3372	0.3409	0.3446	−0.4
0.3483	0.3520	0.3557	0.3594	0.3632	0.3669	0.3707	0.3745	0.3783	0.3821	−0.3
0.3859	0.3897	0.3936	0.3974	0.4013	0.4052	0.4090	0.4129	0.4168	0.4207	−0.2
0.4247	0.4286	0.4325	0.4364	0.4404	0.4443	0.4483	0.4522	0.4562	0.4602	−0.1
0.4641	0.4681	0.4721	0.4761	0.4801	0.4840	0.4880	0.4920	0.4960	0.5000	−0.0

*For $Z \leq -3.50$, the probability is less than or equal to 0.0002.

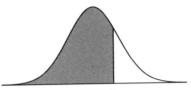

positive Z

	Second decimal place of Z									
Z	0.00	0.01	0.02	0.03	0.04	0.05	0.06	0.07	0.08	0.09
0.0	0.5000	0.5040	0.5080	0.5120	0.5160	0.5199	0.5239	0.5279	0.5319	0.5359
0.1	0.5398	0.5438	0.5478	0.5517	0.5557	0.5596	0.5636	0.5675	0.5714	0.5753
0.2	0.5793	0.5832	0.5871	0.5910	0.5948	0.5987	0.6026	0.6064	0.6103	0.6141
0.3	0.6179	0.6217	0.6255	0.6293	0.6331	0.6368	0.6406	0.6443	0.6480	0.6517
0.4	0.6554	0.6591	0.6628	0.6664	0.6700	0.6736	0.6772	0.6808	0.6844	0.6879
0.5	0.6915	0.6950	0.6985	0.7019	0.7054	0.7088	0.7123	0.7157	0.7190	0.7224
0.6	0.7257	0.7291	0.7324	0.7357	0.7389	0.7422	0.7454	0.7486	0.7517	0.7549
0.7	0.7580	0.7611	0.7642	0.7673	0.7704	0.7734	0.7764	0.7794	0.7823	0.7852
0.8	0.7881	0.7910	0.7939	0.7967	0.7995	0.8023	0.8051	0.8078	0.8106	0.8133
0.9	0.8159	0.8186	0.8212	0.8238	0.8264	0.8289	0.8315	0.8340	0.8365	0.8389
1.0	0.8413	0.8438	0.8461	0.8485	0.8508	0.8531	0.8554	0.8577	0.8599	0.8621
1.1	0.8643	0.8665	0.8686	0.8708	0.8729	0.8749	0.8770	0.8790	0.8810	0.8830
1.2	0.8849	0.8869	0.8888	0.8907	0.8925	0.8944	0.8962	0.8980	0.8997	0.9015
1.3	0.9032	0.9049	0.9066	0.9082	0.9099	0.9115	0.9131	0.9147	0.9162	0.9177
1.4	0.9192	0.9207	0.9222	0.9236	0.9251	0.9265	0.9279	0.9292	0.9306	0.9319
1.5	0.9332	0.9345	0.9357	0.9370	0.9382	0.9394	0.9406	0.9418	0.9429	0.9441
1.6	0.9452	0.9463	0.9474	0.9484	0.9495	0.9505	0.9515	0.9525	0.9535	0.9545
1.7	0.9554	0.9564	0.9573	0.9582	0.9591	0.9599	0.9608	0.9616	0.9625	0.9633
1.8	0.9641	0.9649	0.9656	0.9664	0.9671	0.9678	0.9686	0.9693	0.9699	0.9706
1.9	0.9713	0.9719	0.9726	0.9732	0.9738	0.9744	0.9750	0.9756	0.9761	0.9767
2.0	0.9772	0.9778	0.9783	0.9788	0.9793	0.9798	0.9803	0.9808	0.9812	0.9817
2.1	0.9821	0.9826	0.9830	0.9834	0.9838	0.9842	0.9846	0.9850	0.9854	0.9857
2.2	0.9861	0.9864	0.9868	0.9871	0.9875	0.9878	0.9881	0.9884	0.9887	0.9890
2.3	0.9893	0.9896	0.9898	0.9901	0.9904	0.9906	0.9909	0.9911	0.9913	0.9916
2.4	0.9918	0.9920	0.9922	0.9925	0.9927	0.9929	0.9931	0.9932	0.9934	0.9936
2.5	0.9938	0.9940	0.9941	0.9943	0.9945	0.9946	0.9948	0.9949	0.9951	0.9952
2.6	0.9953	0.9955	0.9956	0.9957	0.9959	0.9960	0.9961	0.9962	0.9963	0.9964
2.7	0.9965	0.9966	0.9967	0.9968	0.9969	0.9970	0.9971	0.9972	0.9973	0.9974
2.8	0.9974	0.9975	0.9976	0.9977	0.9977	0.9978	0.9979	0.9979	0.9980	0.9981
2.9	0.9981	0.9982	0.9982	0.9983	0.9984	0.9984	0.9985	0.9985	0.9986	0.9986
3.0	0.9987	0.9987	0.9987	0.9988	0.9988	0.9989	0.9989	0.9989	0.9990	0.9990
3.1	0.9990	0.9991	0.9991	0.9991	0.9992	0.9992	0.9992	0.9992	0.9993	0.9993
3.2	0.9993	0.9993	0.9994	0.9994	0.9994	0.9994	0.9994	0.9995	0.9995	0.9995
3.3	0.9995	0.9995	0.9995	0.9996	0.9996	0.9996	0.9996	0.9996	0.9996	0.9997
3.4	0.9997	0.9997	0.9997	0.9997	0.9997	0.9997	0.9997	0.9997	0.9997	0.9998

*For $Z \geq 3.50$, the probability is greater than or equal to 0.9998.

B.2   t-Probability Table

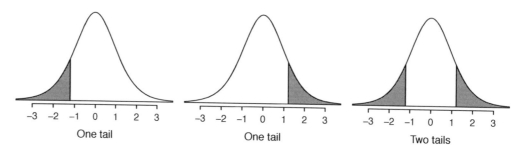

Figure B.1: Tails for the t-distribution.

one tail	0.100	0.050	0.025	0.010	0.005
two tails	0.200	0.100	0.050	0.020	0.010
df 1	3.08	6.31	12.71	31.82	63.66
2	1.89	2.92	4.30	6.96	9.92
3	1.64	2.35	3.18	4.54	5.84
4	1.53	2.13	2.78	3.75	4.60
5	1.48	2.02	2.57	3.36	4.03
6	1.44	1.94	2.45	3.14	3.71
7	1.41	1.89	2.36	3.00	3.50
8	1.40	1.86	2.31	2.90	3.36
9	1.38	1.83	2.26	2.82	3.25
10	1.37	1.81	2.23	2.76	3.17
11	1.36	1.80	2.20	2.72	3.11
12	1.36	1.78	2.18	2.68	3.05
13	1.35	1.77	2.16	2.65	3.01
14	1.35	1.76	2.14	2.62	2.98
15	1.34	1.75	2.13	2.60	2.95
16	1.34	1.75	2.12	2.58	2.92
17	1.33	1.74	2.11	2.57	2.90
18	1.33	1.73	2.10	2.55	2.88
19	1.33	1.73	2.09	2.54	2.86
20	1.33	1.72	2.09	2.53	2.85
21	1.32	1.72	2.08	2.52	2.83
22	1.32	1.72	2.07	2.51	2.82
23	1.32	1.71	2.07	2.50	2.81
24	1.32	1.71	2.06	2.49	2.80
25	1.32	1.71	2.06	2.49	2.79
26	1.31	1.71	2.06	2.48	2.78
27	1.31	1.70	2.05	2.47	2.77
28	1.31	1.70	2.05	2.47	2.76
29	1.31	1.70	2.05	2.46	2.76
30	1.31	1.70	2.04	2.46	2.75

one tail	0.100	0.050	0.025	0.010	0.005
two tails	0.200	0.100	0.050	0.020	0.010
df 31	1.31	1.70	2.04	2.45	2.74
32	1.31	1.69	2.04	2.45	2.74
33	1.31	1.69	2.03	2.44	2.73
34	1.31	1.69	2.03	2.44	2.73
35	1.31	1.69	2.03	2.44	2.72
36	1.31	1.69	2.03	2.43	2.72
37	1.30	1.69	2.03	2.43	2.72
38	1.30	1.69	2.02	2.43	2.71
39	1.30	1.68	2.02	2.43	2.71
40	1.30	1.68	2.02	2.42	2.70
41	1.30	1.68	2.02	2.42	2.70
42	1.30	1.68	2.02	2.42	2.70
43	1.30	1.68	2.02	2.42	2.70
44	1.30	1.68	2.02	2.41	2.69
45	1.30	1.68	2.01	2.41	2.69
46	1.30	1.68	2.01	2.41	2.69
47	1.30	1.68	2.01	2.41	2.68
48	1.30	1.68	2.01	2.41	2.68
49	1.30	1.68	2.01	2.40	2.68
50	1.30	1.68	2.01	2.40	2.68
60	1.30	1.67	2.00	2.39	2.66
70	1.29	1.67	1.99	2.38	2.65
80	1.29	1.66	1.99	2.37	2.64
90	1.29	1.66	1.99	2.37	2.63
100	1.29	1.66	1.98	2.36	2.63
150	1.29	1.66	1.98	2.35	2.61
200	1.29	1.65	1.97	2.35	2.60
300	1.28	1.65	1.97	2.34	2.59
400	1.28	1.65	1.97	2.34	2.59
500	1.28	1.65	1.96	2.33	2.59
∞	1.28	1.65	1.96	2.33	2.58

B.3   Chi-Square Probability Table

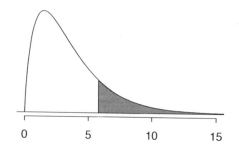

Figure B.2: Areas in the chi-square table always refer to the right tail.

Upper tail	0.3	0.2	0.1	0.05	0.02	0.01	0.005	0.001
df 1	1.07	1.64	2.71	3.84	5.41	6.63	7.88	10.83
2	2.41	3.22	4.61	5.99	7.82	9.21	10.60	13.82
3	3.66	4.64	6.25	7.81	9.84	11.34	12.84	16.27
4	4.88	5.99	7.78	9.49	11.67	13.28	14.86	18.47
5	6.06	7.29	9.24	11.07	13.39	15.09	16.75	20.52
6	7.23	8.56	10.64	12.59	15.03	16.81	18.55	22.46
7	8.38	9.80	12.02	14.07	16.62	18.48	20.28	24.32
8	9.52	11.03	13.36	15.51	18.17	20.09	21.95	26.12
9	10.66	12.24	14.68	16.92	19.68	21.67	23.59	27.88
10	11.78	13.44	15.99	18.31	21.16	23.21	25.19	29.59
11	12.90	14.63	17.28	19.68	22.62	24.72	26.76	31.26
12	14.01	15.81	18.55	21.03	24.05	26.22	28.30	32.91
13	15.12	16.98	19.81	22.36	25.47	27.69	29.82	34.53
14	16.22	18.15	21.06	23.68	26.87	29.14	31.32	36.12
15	17.32	19.31	22.31	25.00	28.26	30.58	32.80	37.70
16	18.42	20.47	23.54	26.30	29.63	32.00	34.27	39.25
17	19.51	21.61	24.77	27.59	31.00	33.41	35.72	40.79
18	20.60	22.76	25.99	28.87	32.35	34.81	37.16	42.31
19	21.69	23.90	27.20	30.14	33.69	36.19	38.58	43.82
20	22.77	25.04	28.41	31.41	35.02	37.57	40.00	45.31
25	28.17	30.68	34.38	37.65	41.57	44.31	46.93	52.62
30	33.53	36.25	40.26	43.77	47.96	50.89	53.67	59.70
40	44.16	47.27	51.81	55.76	60.44	63.69	66.77	73.40
50	54.72	58.16	63.17	67.50	72.61	76.15	79.49	86.66

Index

52575673R00242